Transcriptional Regulation of Cardiac Development and Disease

Transcriptional Regulation of Cardiac Development and Disease

Editors

Nicole Wagner
Kay-Dietrich Wagner

MDPI • Basel • Beijing • Wuhan • Barcelona • Belgrade • Manchester • Tokyo • Cluj • Tianjin

Editors
Nicole Wagner
Institute of Biology Valrose
Université Côte d'Azur
Nice
France

Kay-Dietrich Wagner
Institute of Biology Valrose
Université Côte d'Azur
Nice
France

Editorial Office
MDPI
St. Alban-Anlage 66
4052 Basel, Switzerland

This is a reprint of articles from the Special Issue published online in the open access journal *International Journal of Molecular Sciences* (ISSN 1422-0067) (available at: www.mdpi.com/journal/ijms/special_issues/Cardiac_Development).

For citation purposes, cite each article independently as indicated on the article page online and as indicated below:

LastName, A.A.; LastName, B.B.; LastName, C.C. Article Title. *Journal Name* **Year**, *Volume Number*, Page Range.

ISBN 978-3-0365-3998-0 (Hbk)
ISBN 978-3-0365-3997-3 (PDF)

Contents

International Journal of
Molecular Sciences

Editorial

Transcriptional Regulation of Cardiac Development and Disease

Nicole Wagner * and Kay-Dietrich Wagner

Institute of Biology Valrose, University of Nice Sophia Antipolis, 06107 Nice, France; kwagner@unice.fr
* Correspondence: nwagner@unice.fr; Tel.: +33-493-377665

Citation: Wagner, N.; Wagner, K.-D. Transcriptional Regulation of Cardiac Development and Disease. *Int. J. Mol. Sci.* **2022**, 23, 2945. https://doi.org/10.3390/ijms23062945

Received: 25 February 2022
Accepted: 7 March 2022
Published: 9 March 2022

Publisher's Note: MDPI stays neutral with regard to jurisdictional claims in published maps and institutional affiliations.

The heart, which is the first organ to develop in the embryo, is indispensable for vital functions throughout life. Cardiovascular diseases are the leading cause of mortality worldwide, indicating the importance of the proper development and function of this very specialized muscle. The complex biological events in cardiac development, including very different cell types which contribute to heart formation and the action of transcriptional regulators, are often recapitulated in the cardiac repair mechanisms upon cardiac disease. A profound understanding of cardiac development is therefore necessary to develop efficient therapeutic strategies for cardiac diseases. The present Special Issue of *International Journal of Molecular Science* analyzes transcriptional regulation processes, cardiac-tissue complexity regarding advances in cardiac tissue regeneration and novel aspects in human cardiac diseases, thereby providing new insights in features of cardiac development and diseases.

In this Special Issue, Yun Wang et al. describe that the bone morphogenetic protein (Bmp) signaling pathway directly regulates the basic helix–loop–helix (bHLH) transcription factor Hand1 in the development of the cardiac outflow tract (OFT). OFT defects are the most common congenital heart defects. OFT formation is mainly orchestrated by two different progenitor cell populations, namely the second heart field (SHF) progenitors and cardiac neural crest cells (NCCs). Using an inducible conditional Bmp2 and Bmp4 double-knockout SHF cell-type specific approach, the authors elegantly demonstrated that Bmp signaling regulates Hand1 expression. Although the knockout of Bmp signaling concerns only the SHF progenitor cells, a reduced Hand1 expression was also found in NCC progenitor cells, implying that Bmp signaling regulates Hand1 in OFT formation through cell-autonomous and non-cell-autonomous mechanisms. In contrast to Bmp loss-of-function, Bmp gain-of-function increased Hand1 expression, which further confirmed that Hand1 expression responds to the Bmp signaling dosage. Smad transcription factors are the major signal transducers for receptors of the Bmp signaling pathway and can interact with specific DNA motifs to regulate gene expression. The authors further confirmed the transcriptional activation of Hand1 by Bmp/Smad using transfection assays and by the direct binding of Bmp/Smad to Hand1 with chromatin immunoprecipitation (ChIP) experiments. This study demonstrates well that the canonical Bmp/Smad signaling pathway in the SHF directly activates Hand1 expression in a dose-dependent manner during OFT development, providing novel insights into the molecular regulation of OFT development [1].

Wilms' tumor 1 (Wt1) gene, which encodes a zinc finger protein, is an important regulator during embryogenesis but is also involved in pathological processes, such as carcinogenesis [2]. It is important to note that Wt1 has a crucial role in heart formation, as the homozygous deletion of Wt1 in mouse embryos has been proven to be lethal due to disturbed cardiac development [3]. Furthermore, Wt1 expression in the heart has been described in various cell types, including epicardial, endothelial and smooth muscle cells, and fibroblasts. In this Special Issue, we present a review that provides an overview of general cardiac development and summarize the current knowledge regarding the expression and function of Wt1 in heart development and disease in detail. We focus on the expression of Wt1 in different cardiac cell types and its regulatory mechanisms. We further detail the implication of Wt1 in human cardiac pathologies. Given the importance of Wt1 for cardiac development, it seems obvious that Wt1 is strongly involved in cardiac repair after injury, as the re-activation of developmental programs can be considered to be a paradigm for regeneration. The understanding of the role of Wt1 in these processes and the molecules involved therein is essential for the development of therapeutic strategies [4]. As Wt1 expression in cardiomyocytes remains a controversial issue in the developing and/or diseased heart and has not been reported in adult healthy hearts, we focused our research work for this Special Issue on the potential expression and function of Wt1, specifically in cardiomyocytes. We first investigated cardiac Wt1 expression levels during development, in the adult, and under pathological conditions after myocardial infarction. We found that Wt1 expression was elevated during heart development and declined after birth. Interestingly, we were able to show that Wt1 was highly expressed in cardiomyocytes during development and persisted in some adult cardiomyocytes, probably suggesting that low levels of Wt1 expression are sufficient to maintain a cardiac progenitor subset from terminal differentiation. Following myocardial infarction, the number of Wt1-expressing cardiomyocytes, as well as individual nuclear Wt1 expression in cardiomyocytes, was strongly upregulated. We further used mouse embryonic stem cell (mESC) differentiation in vitro to obtain additional insights into the Wt1 function in the process of cardiomyocyte development. It was found that Wt1 expression levels increased along with the cardiac differentiation of mESCs. We showed that Wt1 overexpression reduces the phenotypic cardiomyocyte differentiation of ES cell clones, keeping the cells in a more progenitor-like stage which is associated with modified expression levels of stem cell and cardiomyocyte marker genes [5].

Additionally, the Hippo-Yap pathway is strongly implicated in cardiac development. For this Special Issue, Zhiquiang Lin et al. investigated the regulation of cardiac Toll-like receptor genes by the YAP/TEAD1 complex, the terminal effector of Hippo-Yap signaling. Toll-like receptors (TLRs) are involved in the pathogenesis of heart failure as they modulate innate immune responses. The authors determined that the expression of TLRs postnatally increases with age and is strongly induced in pressure overload (PO) and ischemia/reperfusion (IR) stressed mouse hearts. They demonstrate that the YAP/TEAD1 complex is a repressor of cardiac TLR genes, as TEAD1 directly bound genomic regions adjacent to Tlr1, Tlr2, Tlr3, Tlr4, Tlr5, Tlr6, Tlr7 and Tlr9. Furthermore, cardiomyocyte-specific YAP depletion in vivo increased the expression of most TLR genes. This was accompanied by an increase in pro-inflammatory cytokines and an increased susceptibility and worsened outcome in response to LPS-induced stress. In conclusion, Hippo-Yap signaling helps to impede the cardiomyocyte innate immune responses upon cardiac stress [6].

Additionally, Ashraf Yusuf Rangrez et al. sought to elucidate a potential role of the SH3 domain-binding glutamic acid-rich (SH3BGR) gene in cardiomyocyte pathophysiology. They demonstrated an upregulation of SH3BGR in human and mouse cardiac hypertrophy samples. Using in vitro overexpression and knockdown experiments for SH3BGR in neonatal rat cardiomyocytes, they observed that enhanced levels of SH3BGR favor cellular hypertrophy as well as an increase in hypertrophic markers. Knockdown of SH3BGR caused a decrease in hypertrophic marker expression and cell viability, accompanied by an activation of apoptosis. On a molecular level, the authors collected evidence supporting the idea that SH3BGR mediates these effects via the induction of Serum response factor (SRF) signaling [7].

Arrhythmogenic cardiomyopathy (ACM) is caused by mutations in genes predominantly encoding for desmosomal proteins, but Lamin A/C gene (*LMNA*) mutations are also frequently observed in patients with ACM [8]. Veronique Lachaize and coworkers detailed, for this Special Issue, the biophysical and biomechanical impact of the *LMNA* D192G missense mutation on neonatal rat ventricular fibroblasts (NRVF). They evidenced a decreased elasticity, a disturbed cytoskeleton organization and altered cell-to-cell adhesion properties in the cardiac fibroblasts with *LMNA* D192G mutation [9]. As similar observations have been made before in cardiomyocytes with a mutation of *LMNA* D192G [10], the recent findings of Lachaize et al. clearly indicate the importance of *LMNA* mutations for

ACM, as in addition to cardiomyocytes, cardiac fibroblasts are also highly biomechanically impacted by this mutation [9].

Cardiac regeneration studies are widely employed and aim to repair irreversibly damaged heart tissue, and often include stem-cell therapy. Daiva Bironaite et al. investigated the ability of human dilated myocardium-derived human mesenchymal stem/stromal cells (hmMSC) and their healthy, non-dilated myocardium-derived counterparts, to differentiate into a cardiomyogenic cell type. Dilated hmMSCS expressed higher levels of Histone deacetylase (HDAC) compared to hmMSCS from non-dilated myocardium. An inhibition of HDAC resulted in the downregulation of focal adhesion kinase (PTK2), and an increased expression of the cardiomyogenic differentiation-specific genes alpha cardiac actin (ACTC1) and cardiac troponin T (TNNT2), which were more pronounced in dilated hmMSCS. These data indicate that hmMSCS from pathological cardiac material might be useful in cardiac regeneration attempts once pharmacologically re-tuned [11].

Gavin Y. Oudit's group investigated the transcriptomic modifications occurring in human end-stage dilated human cardiomyopathy (DCM), also emphasizing the transcriptome changes that occur with left ventricular assist device support (LVAD). On a histological level, the authors observed an enhanced fibrosis and larger cardiomyocyte cross-sectional areas in DCM tissues without LVAD. On the RNA level, they observed a higher expression of hypertrophic markers brain natriuretic peptide (BNP) and β-myosin heavy chain (β-MHC) in DCM without LVAD. Gene expression analysis using microarrays evidenced oxygen homeostasis, immune response and cellular growth, proliferation, and apoptosis as the most enriched pathways in both ventricles of end-stage DCM hearts. For left ventricles, they further observed a differential expression of genes implicated in the circadian rhythm, muscle contraction, cellular hypertrophy, and extracellular matrix (ECM) remodeling, whereas in right ventricles, genes involved in apelin signaling were differentially expressed. Upon LVAD, the authors observed a normalization of the immune response genes in both ventricles, while the expression of ECM remodeling and oxygen homeostasis genes improved specifically in the left ventricle. Furthermore, the expression level of four miRNAs returned to normal. This study contributes to a completion of transcriptomic analysis in human heart disease and provides valuable information not only regarding the impact of DCM on both ventricles but especially in the context of LVAD [12].

Karine Tadevosyan and coauthors present a highly interesting and detailed review in this Special Issue, entitled "Engineering and Assessing Cardiac Tissue Complexity". They present the general approaches used to develop functional cardiac tissue, which includes a description of cardiac-tissue engineering systems, cell sources, maturation and a functional assessment of the procedures applied. This review will help researchers who seek to conduct or improve cardiac-tissue engineering [13].

Finally, Yevgeniy Kim from Arman Saparov's group reviews the latest findings on gene therapies in cardiovascular diseases. Recent applications, benefits and pitfalls from several preclinical and clinical studies are discussed and ongoing trials are described. The limitations of the current clinical approaches for gene therapy in cardiovascular diseases are presented and valuable ways of improvement are suggested [14].

Conclusions

Taken together, the works included in this Special Issue "Transcriptional Regulation of Cardiac Development and Disease" comprise the most recent studies that elucidate transcriptional mechanisms in cardiac development, repair, and regeneration, as well as recent advances and perspectives in the treatment of cardiovascular diseases. The articles in this Special Issue further improve our understanding of cardiac physiology and pathology, which will assist in the continuous development of efficient cardiovascular therapies in the future.

Author Contributions: Conceptualization, K.-D.W. and N.W., writing—original draft preparation, K.-D.W. and N.W.; writing—review and editing, K.-D.W. and N.W., funding acquisition, K.-D.W. and N.W. All authors have read and agreed to the published version of the manuscript.

Funding: This work was funded by Fondation pour la Recherche Medicale, grant number FRM DPC20170139474 (K.D.W.), Fondation ARC pour la recherche sur le cancer", grant number n°PJA 20161204650 (N.W.), Gemluc (N.W.), and Plan Cancer INSERM (K.D.W.).

Informed Consent Statement: Not applicable.

Data Availability Statement: Not applicable.

Conflicts of Interest: The authors declare no conflict of interests.

References

1. Zheng, M.; Erhardt, S.; Ai, D.; Wang, J. Bmp Signaling Regulates Hand1 in a Dose-Dependent Manner during Heart Development. *Int. J. Mol. Sci.* **2021**, *22*, 9835. [CrossRef] [PubMed]
2. Wagner, K.D.; Cherfils-Vicini, J.; Hosen, N.; Hohenstein, P.; Gilson, E.; Hastie, N.D.; Michiels, J.F.; Wagner, N. The Wilms' tumour suppressor Wt1 is a major regulator of tumour angiogenesis and progression. *Nat. Commun.* **2014**, *5*, 5852. [CrossRef]
3. Kreidberg, J.A.; Sariola, H.; Loring, J.M.; Maeda, M.; Pelletier, J.; Housman, D.; Jaenisch, R. WT-1 is required for early kidney development. *Cell* **1993**, *74*, 679–691. [CrossRef]
4. Wagner, N.; Wagner, K.D. Every Beat You Take-The Wilms' Tumor Suppressor WT1 and the Heart. *Int. J. Mol. Sci.* **2021**, *22*, 7675. [CrossRef]
5. Wagner, N.; Ninkov, M.; Vukolic, A.; Cubukcuoglu Deniz, G.; Rassoulzadegan, M.; Michiels, J.F.; Wagner, K.D. Implications of the Wilms' Tumor Suppressor Wt1 in Cardiomyocyte Differentiation. *Int. J. Mol. Sci.* **2021**, *22*, 4346. [CrossRef]
6. Gao, Y.; Sun, Y.; Ercan-Sencicek, A.G.; King, J.S.; Akerberg, B.N.; Ma, Q.; Kontaridis, M.I.; Pu, W.T.; Lin, Z. YAP/TEAD1 Complex Is a Default Repressor of Cardiac Toll-Like Receptor Genes. *Int. J. Mol. Sci.* **2021**, *22*, 6649. [CrossRef] [PubMed]
7. Deshpande, A.; Borlepawar, A.; Rosskopf, A.; Frank, D.; Frey, N.; Rangrez, A.Y. SH3-Binding Glutamic Acid Rich-Deficiency Augments Apoptosis in Neonatal Rat Cardiomyocytes. *Int. J. Mol. Sci.* **2021**, *22*, 1042. [CrossRef] [PubMed]
8. Quarta, G.; Syrris, P.; Ashworth, M.; Jenkins, S.; Zuborne Alapi, K.; Morgan, J.; Muir, A.; Pantazis, A.; McKenna, W.J.; Elliott, P.M. Mutations in the Lamin A/C gene mimic arrhythmogenic right ventricular cardiomyopathy. *Eur. Heart J.* **2012**, *33*, 1128–1136. [CrossRef] [PubMed]
9. Lachaize, V.; Peña, B.; Ciubotaru, C.; Cojoc, D.; Chen, S.N.; Taylor, M.R.G.; Mestroni, L.; Sbaizero, O. Compromised Biomechanical Properties, Cell-Cell Adhesion and Nanotubes Communication in Cardiac Fibroblasts Carrying the Lamin A/C D192G Mutation. *Int. J. Mol. Sci.* **2021**, *22*, 9193. [CrossRef] [PubMed]
10. Lanzicher, T.; Martinelli, V.; Puzzi, L.; Del Favero, G.; Codan, B.; Long, C.S.; Mestroni, L.; Taylor, M.R.; Sbaizero, O. The Cardiomyopathy Lamin A/C D192G Mutation Disrupts Whole-Cell Biomechanics in Cardiomyocytes as Measured by Atomic Force Microscopy Loading-Unloading Curve Analysis. *Sci. Rep.* **2015**, *5*, 13388. [CrossRef] [PubMed]
11. Miksiunas, R.; Aldonyte, R.; Vailionyte, A.; Jelinskas, T.; Eimont, R.; Stankeviciene, G.; Cepla, V.; Valiokas, R.; Rucinskas, K.; Janusauskas, V.; et al. Cardiomyogenic Differentiation Potential of Human Dilated Myocardium-Derived Mesenchymal Stem/Stromal Cells: The Impact of HDAC Inhibitor SAHA and Biomimetic Matrices. *Int. J. Mol. Sci.* **2021**, *22*, 2702. [CrossRef] [PubMed]
12. Parikh, M.; Shah, S.; Basu, R.; Famulski, K.S.; Kim, D.; Mullen, J.C.; Halloran, P.F.; Oudit, G.Y. Transcriptomic Signatures of End-Stage Human Dilated Cardiomyopathy Hearts with and without Left Ventricular Assist Device Support. *Int. J. Mol. Sci.* **2022**, *23*, 2050. [CrossRef] [PubMed]
13. Tadevosyan, K.; Iglesias-García, O.; Mazo, M.M.; Prósper, F.; Raya, A. Engineering and Assessing Cardiac Tissue Complexity. *Int. J. Mol. Sci.* **2021**, *22*, 1479. [CrossRef] [PubMed]
14. Kim, Y.; Zharkinbekov, Z.; Sarsenova, M.; Yeltay, G.; Saparov, A. Recent Advances in Gene Therapy for Cardiac Tissue Regeneration. *Int. J. Mol. Sci.* **2021**, *22*, 9206. [CrossRef] [PubMed]

International Journal of
Molecular Sciences

Article

Transcriptomic Signatures of End-Stage Human Dilated Cardiomyopathy Hearts with and without Left Ventricular Assist Device Support

Mihir Parikh [1,2], Saumya Shah [1,2], Ratnadeep Basu [1,2], Konrad S. Famulski [3], Daniel Kim [1,2], John C. Mullen [2,4], Philip F. Halloran [3] and Gavin Y. Oudit [1,4,5,*]

1   Division of Cardiology, Department of Medicine, Faculty of Medicine and Dentistry, University of Alberta, Edmonton, AB T6G 2R3, Canada; parikhmihirp@outlook.com (M.P.); saumya@ualberta.ca (S.S.); ratnadee@ualberta.ca (R.B.); dk@ualberta.ca (D.K.)
2   Mazankowski Alberta Heart Institute, Edmonton, AB T6G 2R3, Canada; jmullen@ualberta.ca
3   Division of Nephrology, Department of Medicine, Faculty of Medicine and Dentistry, University of Alberta, Edmonton, AB T6G 2R3, Canada; konrad.famulski@ualberta.ca (K.S.F.); phil.halloran@ualberta.ca (P.F.H.)
4   Division of Cardiac Surgery, Department of Medicine, Faculty of Medicine and Dentistry, University of Alberta, Edmonton, AB T6G 2R3, Canada
5   Department of Physiology, Faculty of Medicine and Dentistry, University of Alberta, Edmonton, AB T6G 2R3, Canada
*   Correspondence: gavin.oudit@ualberta.ca; Tel.: +1-780-407-8569

Citation: Parikh, M.; Shah, S.; Basu, R.; Famulski, K.S.; Kim, D.; Mullen, J.C.; Halloran, P.F.; Oudit, G.Y. Transcriptomic Signatures of End-Stage Human Dilated Cardiomyopathy Hearts with and without Left Ventricular Assist Device Support. *Int. J. Mol. Sci.* **2022**, *23*, 2050. https://doi.org/10.3390/ijms23042050

Academic Editors: Nicole Wagner and Kay-Dietrich Wagner

Received: 8 January 2022
Accepted: 10 February 2022
Published: 12 February 2022

Publisher's Note: MDPI stays neutral with regard to jurisdictional claims in published maps and institutional affiliations.

Abstract: Left ventricular assist device (LVAD) use in patients with dilated cardiomyopathy (DCM) can lead to a differential response in the LV and right ventricle (RV), and RV failure remains the most common complication post-LVAD insertion. We assessed transcriptomic signatures in end-stage DCM, and evaluated changes in gene expression (mRNA) and regulation (microRNA/miRNA) following LVAD. LV and RV free-wall tissues were collected from end-stage DCM hearts with (n = 8) and without LVAD (n = 8). Non-failing control tissues were collected from donated hearts (n = 6). Gene expression (for mRNAs/miRNAs) was determined using microarrays. Our results demonstrate that immune response, oxygen homeostasis, and cellular physiological processes were the most enriched pathways among differentially expressed genes in both ventricles of end-stage DCM hearts. LV genes involved in circadian rhythm, muscle contraction, cellular hypertrophy, and extracellular matrix (ECM) remodelling were differentially expressed. In the RV, genes related to the apelin signalling pathway were affected. Following LVAD use, immune response genes improved in both ventricles; oxygen homeostasis and ECM remodelling genes improved in the LV and, four miRNAs normalized. We conclude that LVAD reduced the expression and induced additional transcriptomic changes of various mRNAs and miRNAs as an integral component of the reverse ventricular remodelling in a chamber-specific manner.

Keywords: translational studies; gene expression and regulation; cardiomyopathy; heart failure; reverse remodelling; left ventricular assist device

1. Introduction

Dilated cardiomyopathy (DCM) is a common manifestation of end-stage heart disease, characterized by left ventricle (LV) dilation, systolic dysfunction, and heart failure (HF) [1,2]. DCM is currently the leading cause of cardiac transplantation in adults. The improved survival of patients with HF, coupled with the overall rise in the prevalence of heart diseases, has led to an increase in the number of patients with advanced HF [3,4]. This creates a supply–demand imbalance for cardiac transplantation, with the number of recipients far exceeding the number of available donor hearts. Left ventricular assist devices (LVAD) represents a critical therapy as bridge to transplant or recovery, and potentially as destination therapy in patients with contraindications for transplantation [5–7]. The

REMATCH [8] and the INTrEPID [9] trial revealed superior survival rates in patients with LVAD over conventional therapy and provided convincing evidence for the use of LVAD as bridge to transplant or recovery and also as potential destination therapy [10]. Treatment with continuous-flow LVAD in patients with advanced HF improves the probability of survival, quality of life and functional capacity compared with a pulsatile device [11,12].

The hemodynamic alteration with LVAD triggers beneficial remodelling at multiple molecular and cellular levels, leading to improved systolic and diastolic function [6,7]. However, the LV and right ventricle (RV) respond differently to the unloading effects of LVAD. In fact, RV dysfunction is the predominant complication post-LVAD implantation and the major cause of morbidity and mortality in these patients, indicating unique differences in RV and LV remodelling [13,14]. The molecular and cellular changes that occur in the heart as a result of LVAD therapy can provide important insight into the differential ventricular response while also clearly identifying the therapeutic benefits of LVAD. Our study aimed to understand chamber-specific transcriptomic changes in explanted human hearts with DCM post-LVAD implantation. LV and RV samples from explanted DCM hearts with and without LVAD implantation, as well as non-failing control hearts, were used for a microarray-based analysis of global mRNA and microRNA (miRNA) expression. In this study, several genes and pathways involved in the pathogenesis of DCM, as well as distinct remodelling and gene regulation patterns in the LV and RV of DCM hearts in response to LVAD use, were identified.

2. Results

2.1. Patient Clinical Characteristics

The clinical characteristics of DCM patients with LVAD (VAD group) and without LVAD implantation (no LVAD, NVAD group) are summarized in Table 1. Both groups had similar demographics, physical exam assessments, past medical histories, medical therapies, laboratory values, and echocardiographic parameters. For VAD group, the median duration of LVAD support was 156 days (IQR: 131–268 days). Limited information was available on donors of non-failing hearts (NFC group). The median age of these six donors was 45 (IQR: 40–51) years old, and all of them had normal LV ejection fraction.

Table 1. Clinical characteristics of DCM patients with and without LVAD implantation.

Criteria	No LVAD (*n* = 8)	LVAD (*n* = 8)	*p* Value
Age at transplant, years	50 (43–54)	57 (45–59)	0.7814
Female sex	1 (12.5)	1 (12.5)	>0.9999
SBP, mmHg	92 (89–97)	91 (80–103)	0.8131
HR, bpm	86 (69–106)	80 (77–96)	0.8935
BMI, kg/m^2	24 (23–27)	25 (24–29)	0.3949
NYHA class			
III	6 (75)	4 (50)	0.5594
IV	1 (12.5)	3 (37.5)	0.5594
Medical history			
HF duration, mo	39 (6–108)	48 (30–120)	0.5493
Hypertension	1 (12.5)	1 (12.5)	>0.9999
Dyslipidemia	0 (0)	2 (25)	0.4667
Kidney disease	2 (25)	2 (25)	>0.9999
Liver disease	4 (50)	2 (25)	0.3147
Diabetes	0 (0)	1 (12.5)	>0.9999
COPD	1 (12.5)	1 (12.5)	>0.9999
Discharge medication			
ACEi/ARB/sacubitril/valsartan	7 (87.5)	6 (75)	0.3147
Beta-blocker	7 (87.5)	6 (75)	>0.9999
MRA	2 (25)	4 (50)	0.6084

Table 1. *Cont.*

Criteria	No LVAD (*n* = 8)	LVAD (*n* = 8)	*p* Value
Laboratory			
Cr, umol/L	108 (95–118)	83 (78–112)	0.9511
eGFR, mL/m in/1.73 m^2	61 (56–69)	89 (65–112)	0.2811
Hb, g/L	118 (107–131)	98 (92–109)	0.1229
Echocardiography			
EF $\leq$ 50%	7 (87.5)	4 (50)	>0.9999
EF, %	18 (15–26)	13 (10–20)	0.6223
LVEDD, mm	64 (59–69)	57 (51–62)	0.3556
LVESD, mm	60 (54–66)	44 (38–59)	0.1523

SBP, systolic blood pressure; HR, heart rate; BMI, body mass index; NYHA, New York Heart Association; HF, heart failure; COPD, chronic obstructive pulmonary disease; ICD, implantable cardioverter defibrillator; CRT-D, cardiac resynchronisation therapy defibrillator; ACEi, angiotensin-converting enzyme inhibitor; ARB, angiotensin receptor blocker; MRA, mineralocorticoid receptor antagonist; Cr, creatinine; eGFR, estimated glomerular filtration rate; Hb, haemoglobin; EF, ejection fraction; LVEDD, left ventricular end diastolic diameter; LVESD, left ventricular end systolic diameter.

2.2. Histological Characteristics of Explanted DCM Hearts with and without LVAD

Increased cardiomyocyte hypertrophy and myocardial fibrosis are prominent features of adverse myocardial remodelling. Histological characteristics of VAD, NVAD and NFC hearts were analysed, with representative images shown in Figure 1. Fibrosis was increased in both the LV and RV of NVAD DCM hearts, at levels approximately 2-fold and 1.5-fold, respectively. Upon LVAD implantation, fibrosis was decreased in the VAD group compared to the NVAD group (Figure 1A,B). Similarly, cardiomyocyte cross-sectional area (CSA) was significantly larger in the LV and RV of NVAD DCM hearts ($p < 0.05$). Upon LVAD implantation, cardiomyocyte CSA was decreased in the VAD group compared to the NVAD group (Figure 1C,D). The cardiac enlargement was also confirmed in the NVAD by significantly up-regulated mRNA levels of the hypertrophic markers brain natriuretic peptide (BNP) and β-myosin heavy chain (β-MHC) compared to NFC. However, the levels of these biomarkers were reduced in VAD compared to NVAD (Figure 1E,F).

2.3. Pathological Genes of DCM Hearts

The gene expression profile of DCM hearts without LVAD implantation (NVAD group) was compared to the gene expression profile of healthy hearts (NFC group) to identify transcripts and genes that were altered as a result of the disease. In the LV, 922 transcripts were differentially expressed (Figure 2A), 392 up-regulated and 530 down-regulated (Figure 2B), in the NVAD compared to the NFC groups. In the RV, 858 transcripts were differentially expressed (Figure 2A), 238 up-regulated and 620 down-regulated (Figure 2B), in the NVAD compared to the NFC groups. Of the differentially expressed transcripts found, 567 were commonly altered in both the LV and RV.

The most significantly enriched KEGG pathways and GO terms related to BP and MF for the up- and down-regulated genes encoding for the observed pathological transcripts are presented in Figure 2C–F. In the LV, 392 up-regulated pathological transcripts were the products of 238 genes (Figure 2C). The most enriched KEGG pathways for this gene set were involved in cellular physiology such as cell growth, proliferation, apoptosis, metabolism, and cell cycle regulation. The other enriched pathways included cardiomyocyte hypertrophy and circadian rhythm, while the majority of enriched BPs controlled myocardial structure and contractility. The enriched MFs reflect findings from pathways and BP analyses, with the majority of MFs involved in transcription, ion channels, cellular growth and muscle contraction. On the other hand, 530 down-regulated pathological transcripts were the products of 330 genes (Figure 2D). The majority of enriched KEGG pathways and BPs were involved in the immune response. Genes for the hypoxia-inducible factor 1 (HIF-1) pathway, which functions in oxygen homeostasis, were also enriched. With regard to MFs, the most enriched terms were related to innate immunity, regulation of the extracellular matrix (ECM), and cellular metabolism.

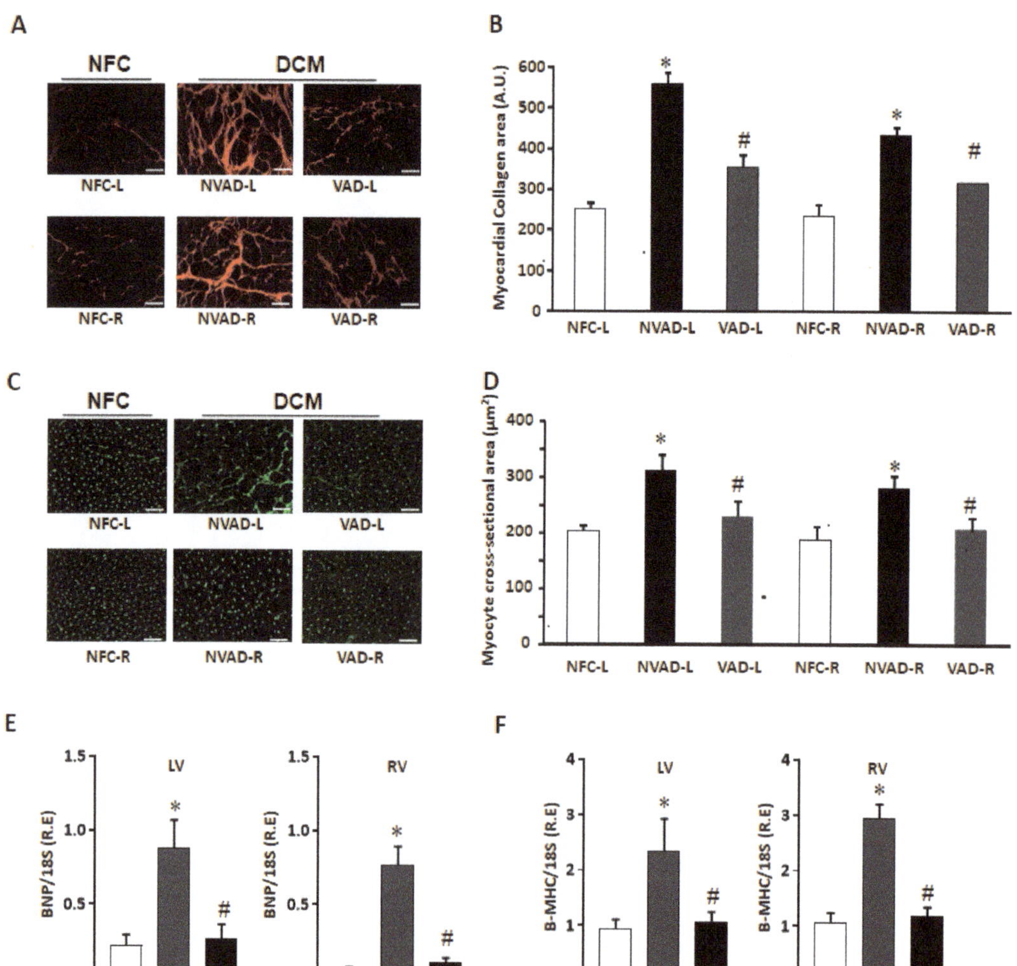

Figure 1. Histological characterization of non-failing control (NFC) and dilated cardiomyopathy (DCM) hearts. (**A**) Representative Picrosirius red staining images for collagen detection in NFC hearts, as well as DCM hearts with left ventricle assist device (VAD) and without LVAD support (NVAD). (**B**) Myocardial collagen content quantified from (**A**). (**C**) Representative Wheat germ agglutinin staining for NFC, DCM-VAD and DCM-NVAD hearts. (**D**) Cardiomyocyte cross-sectional area (CCA) quantified from (**C**). (**E**,**F**) Real-time quantitative PCR results showing the relative mRNA levels of hypertrophic markers brain natriuretic peptide (BNP) and β-myosin heavy chain (β-MHC) in NFC, NVAD, or VAD. $n = 8$, * $p < 0.05$ vs. NFC, # $p < 0.05$ vs. NVAD, unpaired two-tailed t-test. L, left; R, right. Scale bar represents 50 μm.

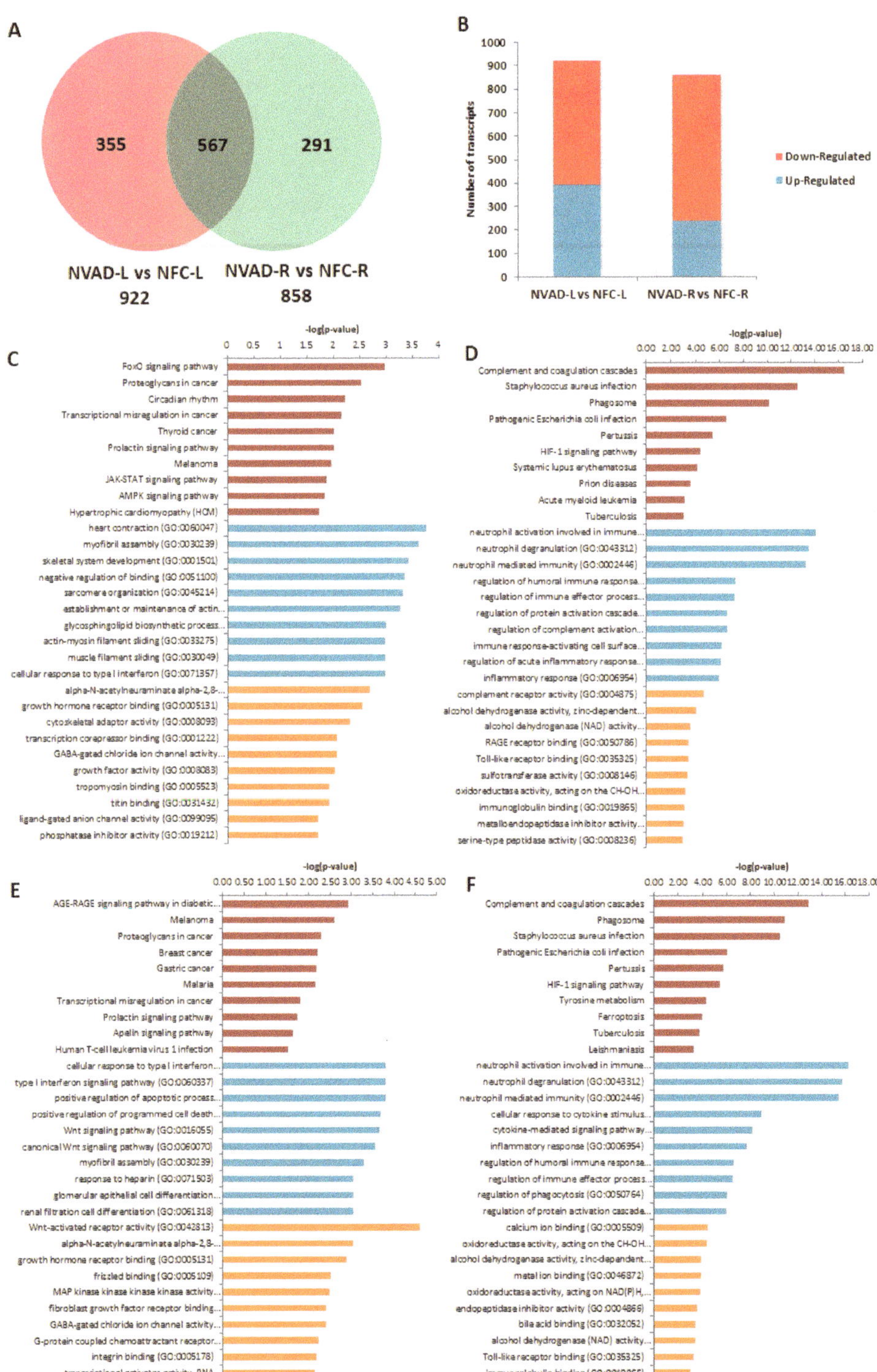

Figure 2. mRNA analysis of dilated cardiomyopathy (DCM) hearts without left ventricle assist device (NVAD) versus non-failing controls (NFC). (**A**) Number of significant differentially expressed transcripts

in the left (L) and right (R) ventricles. (**B**) Number of significantly up- and down-regulated transcripts in the L and R ventricles. Gene ontology (GO) analysis for (**C,E**) up-regulated and (**D,F**) down-regulated differentially expressed genes in the (**C,D**) L and (**E,F**) R ventricles. The top ten Kyoto Encyclopedia of Genes and Genomes (KEGG) pathways (red), GO terms for biological processes (blue) and GO terms for molecular functions are shown (yellow).

In the RV, 238 up-regulated pathological transcripts were the products of 157 genes (Figure 2E). Similar to up-regulated genes in the LV, the most significantly enriched KEGG pathways and BPs were involved in cellular physiological events. Genes involved in the apelin signalling pathway, which functions in key processes such as angiogenesis, cardiovascular function, cell proliferation, and energy regulation, were also enriched. MFs related to transcription, ion channels, cellular growth and metabolism were enriched. Genes involved in myocardial structure and contractility were not as enriched compared to what was seen in the LV. On the other hand, 620 down-regulated pathological transcripts were the products of 383 genes (Figure 2F). The enriched KEGG pathways, BPs and MFs of these genes were found to be similar to those for down-regulated genes in the LV (immune response, oxygen homeostasis). The lists of genes from all enrichment analyses, at least for the top 10 results for each category, are presented in Supplementary Material Table S1.

2.4. Gene Expression Changes in the LV and RV Post LVAD Implantation

The LV and RV pathological transcripts found above were then tracked to see how their expressions changed upon LVAD implantation, as demonstrated by the VAD vs. NFC and VAD vs. NVAD comparisons (Figure 3). Results from the VAD vs. NFC comparisons revealed that of the 922 pathological transcripts identified in the LV, 617 (67%) showed a decrease in FC with LVAD use (compared to the FC in NVAD vs. NFC), and 130 (14%) showed no significant changes when compared to the NFC (Figure 3A–C). Of the 858 pathological transcripts identified in the RV, 493 (57%) showed a decrease in FC with LVAD use compared to without LVAD use, and 157 (18%) showed no changes compared to the NFC (Figure 4A–C).

Comparisons between VAD and NVAD revealed significantly normalized transcripts (from pathological transcript lists) post-LVAD use in both the LV and RV. A transcript is considered normalized if the FC in the VAD vs. NVAD comparison shows the opposite direction of change (e.g., up-regulated) compared to the direction of change in the NVAD vs. NFC comparison (e.g., down-regulated). In the LV, 205 pathological transcripts (corresponding to 133 genes) were significantly normalized (Figure 3C). Enrichment analysis of these genes suggests that they function in immunity-related processes, oxygen homeostasis, as well as in the regulation of the ECM (Figure 3D). In the RV, 116 pathological transcripts (corresponding to 79 genes) were significantly normalized (Figure 4C). Enrichment analysis of these genes suggests that they also function in immunity (Figure 4D). Our results are consistent with increased apoptosis and cell death of cardiomyocytes and up-regulation of inflammatory cascade as known drivers of advanced HF.

Besides the normalization of pathological genes identified from our NVAD vs. NFC comparison, LVAD use also induced additional changes in gene expression as shown in the VAD vs. NVAD comparison. In the LV, 56 transcripts (corresponding to 36 genes) were significantly altered post-LVAD use. Enrichment analysis showed that the most enriched processes were involved in ECM remodelling, immune response, cellular physiological processes and angiogenesis (Figure 3E). In the RV, 23 transcripts (corresponding to 16 genes) were significantly altered post-LVAD use. Enrichment analysis showed that the majority of these genes were involved in immune system-related processes (Figure 4E). The gene lists for all the enrichment analyses in this section, at least for the top 10 results for each category, are presented in Supplementary Material Table S2.

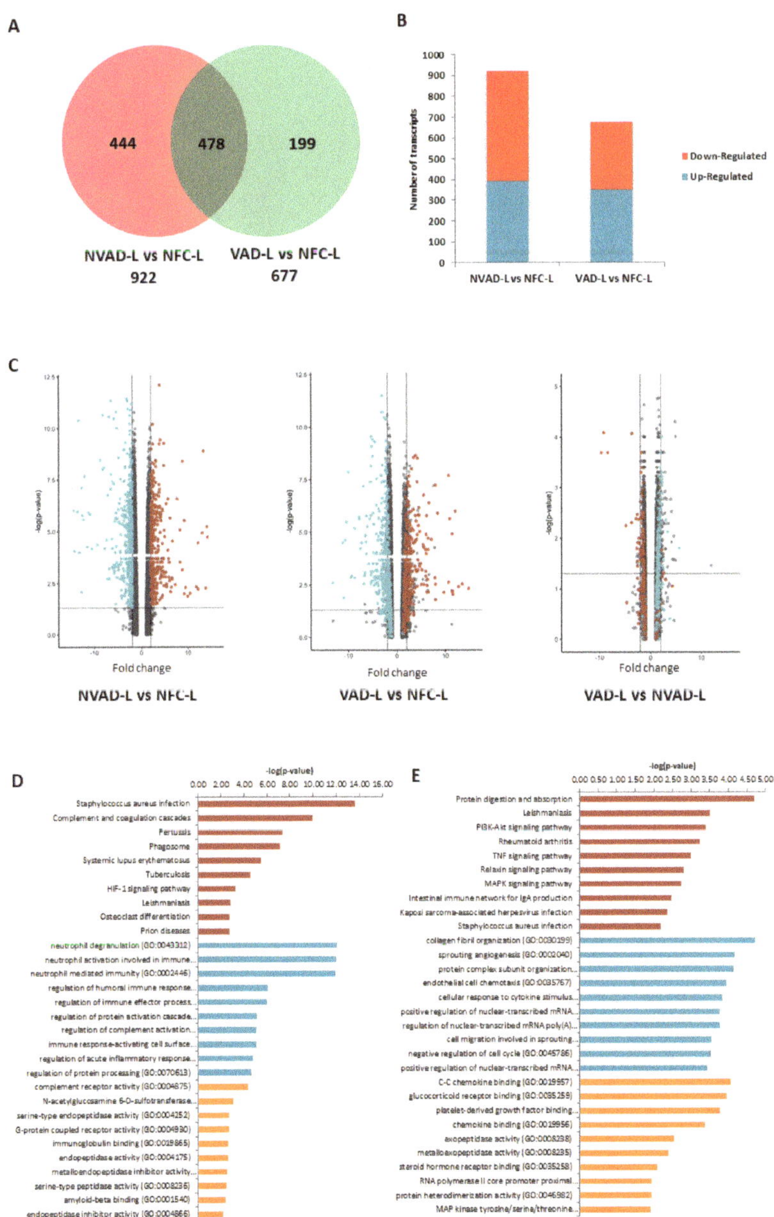

Figure 3. mRNA analysis of left ventricles (LV) from dilated cardiomyopathy (DCM) hearts with (VAD) and without left ventricle assist device (NVAD), using the non-failing controls (NFC) as reference. (**A**) Number of significant differentially expressed transcripts in the VAD and NVAD groups. (**B**) Number of significantly up- and down-regulated transcripts in the VAD and NVAD groups. (**C**) Volcano plots of microarray data from the VAD and NVAD groups. The LV DCM signature genes are marked (red: up-regulated, blue: down-regulated in NVAD group); grey vertical lines indicate two-fold fold-change values in either direction. Gene ontology (GO) analysis for (**D**) normalized genes in the LV following LVAD use and (**E**) additional differentially expressed LV genes

following LVAD use. The top ten Kyoto Encyclopedia of Genes and Genomes (KEGG) pathways (red), GO terms for biological processes (blue) and GO terms for molecular functions are shown (yellow).

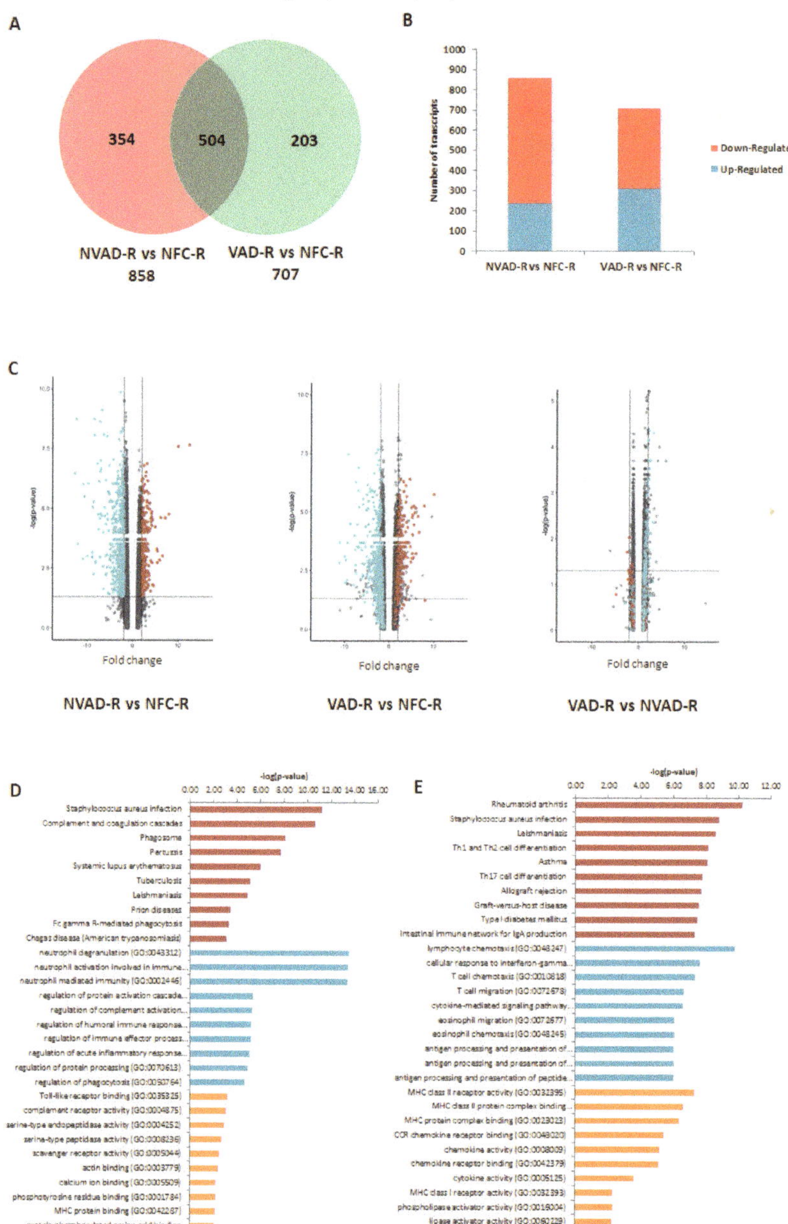

Figure 4. mRNA analysis of right ventricles (RV) from dilated cardiomyopathy (DCM) hearts with (VAD) and without left ventricle assist device (NVAD), using the non-failing controls (NFC) as reference. (**A**) Number of significant differentially expressed transcripts in the VAD and NVAD groups. (**B**) Number of significantly up- and down-regulated transcripts in the VAD and NVAD groups. (**C**) Volcano plots of

microarray data from the VAD and NVAD groups. The RV DCM signature genes are marked (red: up-regulated, blue: down-regulated in NVAD group); grey vertical lines indicate two-fold fold-change values in either direction. Gene ontology (GO) analysis for (**D**) normalized genes in the RV following LVAD use and (**E**) additional differentially expressed RV genes following LVAD use. The top ten Kyoto Encyclopedia of Genes and Genomes (KEGG) pathways (red), GO terms for biological processes (blue) and GO terms for molecular functions are shown (yellow).

2.5. Pathological miRNAs of DCM Hearts

The miRNA expression profile of the NVAD group was compared to that of the NFC group to identify miRNAs altered in DCM, which was termed pathological miRNAs. The target transcripts (and corresponding genes) of these pathological miRNAs were then compared to the pathological transcript/gene lists from our earlier mRNA analyses. This would determine whether the changes in mRNA expression we previously observed could be attributed to changes at the miRNA level. In the LV, 39 miRNAs were differentially expressed (Figure 5A), 16 up-regulated and 23 down-regulated (Figure 5B), in the NVAD compared to the NFC group. Of these 39 miRNAs, 18 regulate transcripts and genes that were altered in our earlier NVAD vs. NFC comparison (Table 2). Enrichment analysis of these genes showed that they are involved in immunity, oxygen homeostasis, cardiac muscle contraction, and cellular processes such as energy regulation, growth, proliferation, and apoptosis (Figure 5C). In the RV, 19 miRNAs were differentially expressed (Figure 5A), 12 up-regulated and seven down-regulated (Figure 5B), in the NVAD compared to the NFC group. Of these 19 miRNAs, 13 regulate transcripts and genes that were altered in our earlier NVAD vs. NFC comparison (Table 3). Enrichment analysis of these genes showed that they are involved in immunity, adrenergic signalling in cardiomyocytes, cardiac muscle contraction, and cellular processes such as energy regulation, growth, proliferation and apoptosis (Figure 5D).

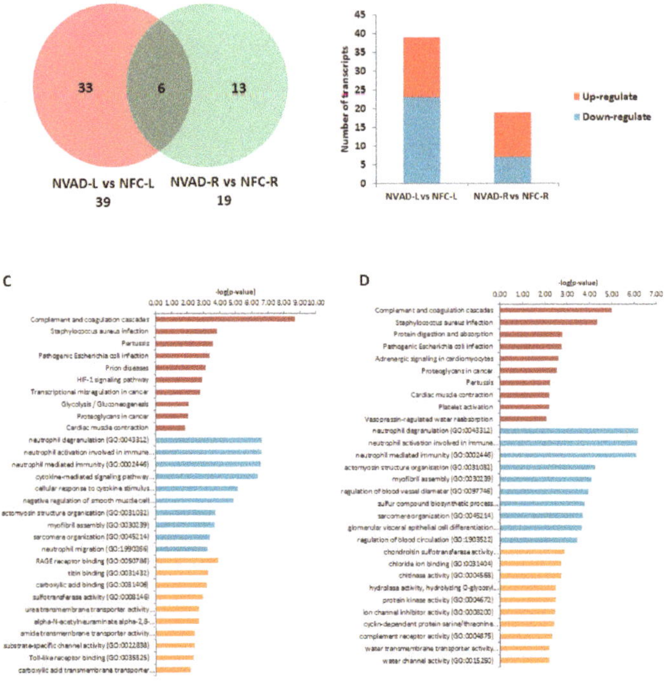

Figure 5. miRNA analysis of dilated cardiomyopathy (DCM) hearts without left ventricle assist device (NVAD) versus non-failing controls (NFC). (**A**) Number of significant differentially expressed

miRNAs in the left (L) and right (R) ventricles. (**B**) Number of significantly up- and down-regulated miRNAs in the L and R ventricles. Gene ontology (GO) analysis for target genes of the concordant miRNAs in the (**C**) L and (**D**) R ventricles. The top ten Kyoto Encyclopedia of Genes and Genomes (KEGG) pathways (red), GO terms for biological processes (blue) and GO terms for molecular functions are shown (yellow).

Table 2. Pathological miRNAs in the LV of DCM hearts and their corresponding target genes.

ID	Fold Change	p-Value	Target Transcripts from Corresponding mRNA Comparisons	
			Number of Transcripts	**Target Genes**
hsa-miR-451	16.22	4.30×10^{-8}	43	CCL18, FRZB, ACTC1, EIF1AY, C3AR1, RAMP1, AQP4, PLP2, F5, SLC35F1, KCNJ8, SLC25A27, LY96, STK38L, LDHA, TNFAIP8, EGLN3, ATP5F1, GPRASP1, UTY, SSPN, CTH, SORL1, GMNN, PTPRB
hsa-miR-182	15.33	1.03×10^{-6}	41	IGF1R, DNER, BCAT1, CREB3L1, RARRES1, RNASE6, TPST2, HIF3A, EMB, MGST1, FCN3, MARK3, FIGF, MLF1, SLITRK4, BDNF, HS3ST1, SAMSN1, GABRB1, RSAD2, AASS, MYH6, ZMYND12, PKD1L2, TM2D3, DCN
hsa-miR-495	2.87	1.14×10^{-5}	30	C3, CITED2, DNER, ALOX5AP, ACTC1, BCAT1, CRYM, AQP4, HSPA2, FAM46B, LARP6, CADM1, ATP5F1, GPR22, LRRTM4, RSAD2, AASS, DUSP27, AGTPBP1, PCDH20, HCLS1, TUBA1C, METTL7A
hsa-miR-135b	−3.95	0.0003	30	SERPINA3, CD53, GADD45A, FCER1G, RNASE6, IFIT2, ADH1A, FGF1, TIMP4, ANP32E, LILRB2, EFCAB2, RNASE2, CYP4Z1, ITIH5, IL1RL1, GABRB1, MAEL, RASAL2, HAS2, SLCO4A1, AGTPBP1, PTPRB
hsa-miR-374b	2.35	0.0005	25	F13A1, GADD45A, EGR1, CHST2, ASPN, CCL18, BIRC3, STK38L, ANKRD1, CTH, GOLGA8A, TMED5, ATP2A2, CNKSR2, PTPRB
hsa-miR-218	4.42	0.0017	41	PER3, FRZB, CRYM, PLCE1, IFI44L, DNAJB5, RABGAP1L, HAPLN1, SMTNL2, HEY2, ANKRD1, MLF1, CA2, ANK1, OGN, GABRB1, AASS, DUSP27, PHACTR1, GMNN, ST8SIA5, ASS1, UCHL1, RHOBTB1, DCN, EDIL3
hsa-miR-208a	−2.58	0.0018	24	SH3BGRL3, HMOX1, NPTX2, EMB, IFI44L, MGST1, TIMP4, LRRC1, MLF1, GPRASP1, C6, PDCD4, AGTPBP1, CPAMD8, HK2, BLM, TUBA1C
hsa-miR-373	5.76	0.0026	27	CD24, MAP4, NINJ2, PLCE1, MYBL1, C2orf40, GPR34, CD68, HS3ST1, CSAD, CNN1, ITIH5, GPRASP1, CHRFAM7A, SLMAP, SCUBE2, PRRT2, PKD1L2, UCHL1, IFI44
hsa-miR-628-5p	2.59	0.0037	32	CTSC, SMOC2, SLCO2A1, SLC1A2, BCL6, PLCE1, REPS2, OGDHL, PTN, DPY19L2, TYRP1, KCNIP2, C4A, MYOC, SCUBE2, GMNN, TSPAN13, BLM, PECAM1, TYROBP, ATP2A2

Table 2. *Cont.*

ID	Fold Change	*p*-Value	Target Transcripts from Corresponding mRNA Comparisons	
			Number of Transcripts	Target Genes
hsa-miR-431	−2.33	0.0055	20	*CD53, SLC7A8, NPR3, LARP6, ITGAM, MYBPC1, STAT4, ITIH5, WISP2, GSG1L, CLEC10A, PHACTR1, ARPC3, CPAMD8, DDAH1*
hsa-miR-224	2.82	0.0058	30	*HLA-C, PACSIN3, TPST2, CYP2J2, HIF3A, GRB14, OGDHL, TMEM45A, HSD17B11, ARL4C, CHST9, CA2, GSG1L, C6, CTH, ARPC3, FHL1, HMOX2, ASS1, PTPRB*
hsa-miR-95	4.09	0.0069	36	*SERPINA3, C1QB, TUBB2A, PENK, TBXAS1, PLK2, CA14, TTC9, IL20RA, EDARADD, ST8SIA2, TFPI, CHST9, EGLN3, IL1RL1, GPRASP1, GSG1L, RSAD2, MGAT4C, ENO1, GMNN, PTPRB*
hsa-miR-940	−2.26	0.0078	17	*SH3BGRL3, SERPINA3, S100A9, NES, SERPINA1, HIF3A, GCHFR, CENPA, CYP4B1, CAPG, TNFRSF12A, IFI30*
hsa-miR-601	−2.29	0.0134	23	*PIM1, CD109, PLA2G2A, C1QC, SLC6A6, CYP2J2, GCHFR, CENPA, C1orf162, LILRB2, NR4A3, GRK5, TPM3, CA2, HMOX2*
hsa-miR-329	2.22	0.0148	25	*CITED2, HIF3A, AHNAK, KCNJ8, DPY19L2, CADM1, CD68, LRRTM4, PHACTR1, PRDX6, CEBPD, CAMK1D, NNMT, SLMAP, PKD1L2, AQP3, OBSCN*
hsa-miR-187	−6.09	0.0186	33	*SH3BGRL3, CHGB, S100A4, IGF1R, ALOX5AP, CYP2J2, PLP2, FAM58A, PRIMA1, MGST1, DPY19L2, HFE2, LILRB2, STK38L, STAT4, WISP2, GSG1L, DUSP27, PHACTR1, MXRA5, S100A8, CPAMD8, UCHL1, OBSCN*
hsa-miR-10b	2.55	0.0238	36	*APOD, FCER1G, RAMP1, DIO2, TMEM97, CA14, MYBL1, FAM46B, SMTNL2, GRK5, CD68, TNFRSF11B, LCN6, CNN1, ANK1, ATP5F1, SGPP2, C10orf71, PRDX6, SLMAP, SCUBE2, AGTPBP1, PKD1L2, OBSCN*
hsa-miR-223	2.99	0.0311	51	*GADD45A, PER3, AKR1C1, RRAS2, UBR1, PLCE1, RABGAP1L, MYO5A, TDRD9, GPR34, C1orf162, VIT, SLC14A1, EFCAB2, TFPI, ANK1, GPR22, OGN, MAEL, FKBP5, DUSP27, HVCN1, FHL1, CNKSR2*
hsa-miR-4269	2.38	2.70E-06	-	-
hsa-miR-299-5p	3.58	9.71×10^{-5}	-	-
hsa-miR-4270	−2.12	0.0001	-	-
hsa-miR-4539	−2.06	0.0003	-	-
hsa-miR-1825	−2.04	0.0007	-	-
hsa-miR-3187-3p	−2.4	0.001	-	-
ENSG00000202498	−2.28	0.0013	-	-
ENSG00000202498_x	−2.31	0.0016	-	-
hsa-miR-3910	−2.03	0.0045	-	-
hsa-miR-4687-3p	−2.22	0.0057	-	-
hsa-miR-4741	−2.08	0.0071	-	-
hsa-miR-548x	−2	0.0077	-	-
hsa-miR-4689	−2.28	0.0107	-	-
hsa-miR-3128	−2.3	0.0109	-	-
hsa-miR-4793-3p	−2.69	0.011	-	-
hsa-miR-3201	−2.11	0.0114	-	-

Table 2. *Cont.*

ID	Fold Change	*p*-Value	Target Transcripts from Corresponding mRNA Comparisons	
			Number of Transcripts	**Target Genes**
hsa-miR-103b	−2.58	0.015	-	-
hsa-miR-1226	4.09	0.0158	-	-
hsa-miR-4458	2.21	0.0198	-	-
hsa-miR-4521	−3.93	0.0275	-	-
hsa-miR-1226	−2.12	0.0304	-	-

Table 3. Pathological miRNAs in the RV of DCM hearts and their corresponding target genes.

ID	Fold Change	*p*-Value	Target Transcripts from Corresponding mRNA Comparisons	
			Number of Transcripts	**Target Genes**
hsa-miR-182	19.89	6.64×10^{-6}	48	*ANXA1, IL1R1, DNER, BCAT1, CREB3L1, RARRES1, RNASE6, PC, EMB, MGST1, APOB, FCN3, ACE2, MARK3, OLFM1, FIGF, CLEC4G, KCNN3, KCNMB2, PTGER3, SAMSN1, GABRB1, RSAD2, MYH6, ATP1B3, ZMYND12, MLF1, PKD1L2, CHI3L2, TM2D3, DCN*
hsa-miR-124	−2.16	0.0002	38	*CDKN1A, IQGAP1, IL1R1, CD59, RDH10, PENK, BCAT1, NAP1L3, GCHFR, AGTR1, KLF15, OLFM1, IRAK3, PAPSS2, CADM1, RNASE2, TRIM45, RGS4, RBM47, CYP4B1, NID1, SLCO4A1, ARPC1B, GOLGA8A, CPAMD8, TWIST2, C1S, TSC22D1*
hsa-miR-451	7.13	0.0004	38	*KLF6, HCK, CCL18, FRZB, ACTC1, C3AR1, RAMP1, AQP4, F5, CDKN2B, LY96, STK38L, B3GNT7, FYB, SCGB1D2, TNFAIP8, GPRASP1, MAN1A1, SSPN, CTH, SAT1, CTHRC1, ARPC1B, UTY*
hsa-miR-181a-2	2.34	0.0006	20	*VAMP8, RBP4, SOX4, IER3, ACLY, DIO2, ABCG2, C1orf162, CD163, SUSD4, ANKRD1, TPO, PHACTR1, NNMT, AQP3*
hsa-miR-373	6.81	0.0009	25	*CD24, KLF6, ENSA, NINJ2, PLCE1, UAP1, MYBL1, GPR34, CD68, MYPN, CNN1, ITIH5, GPRASP1, RASSF2, CHRFAM7A, IFI44, SCUBE2, PKD1L2*
hsa-miR-138	−2.39	0.0013	28	*LYZ, EIF4EBP1, FCER1G, CSTA, C3AR1, ADH1A, DOCK2, UAP1, APBB1IP, FIGF, VIT, TPM4, ITGAM, LCN6, SYNPO2L, TMEM74, STON1, PHACTR1, PKD1L2, C1S, DCN*
hsa-miR-431	−4.17	0.0016	24	*CD53, IGFBP6, HCK, SLC7A8, NPR3, LARP6, BMPR1B, HPR, ITGAM, MYBPC1, HP, ITIH5, TPO, CLEC10A, PHACTR1, CPAMD8, COL12A1*
hsa-miR-92a-1	−2.77	0.0025	29	*TIMP1, SH3BGRL3, SERPINA3, SEC14L1, AIF1, PLA2G2A, FCER1G, RAMP1, UAP1, METTL7B, LY96, GPX3, DPYSL4, ITIH5, PHACTR1, SAT1, CTHRC1, DUSP4, TKT, TYROBP*
hsa-miR-21	−2.87	0.0034	22	*SMOC2, IGFBP6, ALOX5AP, PENK, BCAT1, NQO2, BIRC3, CENPA, TDRD9, LARP6, TNFSF12, RGS4, C5AR1, ITIH5, MAP3K8, PKD1L2, ASRGL1*

Table 3. *Cont.*

ID	Fold Change	*p*-Value	Target Transcripts from Corresponding mRNA Comparisons	
			Number of Transcripts	**Target Genes**
hsa-miR-10b	2.64	0.0051	26	*ENSA, FCER1G, RAMP1, DIO2, CA14, HS6ST2, UAP1, MYBL1, SMTNL2, GRK5, CD68, SCN3A, LCN6, CNN1, C10orf71, SRGN, PRDX6, ARPC1B, SCUBE2, AGTPBP1, PKD1L2*
hsa-miR-95	2.86	0.0193	38	*SERPINA3, EMP1, C1QB, TUBB2A, PENK, TBXAS1, CA14, ACE2, EDARADD, ST8SIA2, CHST9, SHMT1, IL1RL1, GPRASP1, RSAD2, STON1, MGAT4C, SAT1, ENO1, CHI3L2, CHI3L1, GADD45B*
hsa-miR-217	3.1	0.0487	38	*CHST7, ENSA, HMOX1, MFAP5, KCNS3, FNDC1, CENPA, DOCK2, COL21A1, GFRA1, FCGR2A, FIGF, STRBP, CD163, HSPA4L, MXI1, MYPN, FGF10, STON1, DUSP27, MGAT4C, SAT1, CAMK1D, CPAMD8, PKD1L2*
hsa-miR-216a	3.02	0.0488	25	*KLF6, ENSA, FCER1G, NPR3, IFI44L, PITPNC1, STK17B, HFE2, VIT, CADM1, ITIH5, PTGER3, CTH, CYP4B1, ZMYND12, CCDC109B, HCLS1, CHI3L2*
HBII-52-32_x	4.76	0.0028	-	-
hsa-miR-1972	2.57	0.0046	-	-
hsa-miR-3065-3p	−3.04	0.0068	-	-
hsa-miR-4524	2.05	0.0101	-	-
hsa-miR-1247	−2.56	0.0187	-	-
hsa-miR-4461	2.3	0.0321	-	-

2.6. miRNA Expression Changes in the LV and RV Post LVAD Implantation

Similar to our previous analysis for the pathological mRNAs, the LV and RV pathological miRNAs (from NVAD vs. NFC) were assessed to examine how their expression levels changed upon LVAD implantation. The results from the VAD vs. NFC comparison revealed that of the 39 pathological miRNAs identified in the LV, 21 (54%) showed a decrease in FC with LVAD use compared to without LVAD use, and eight (21%) showed no change when compared to the NFC (Figure 6A–D; Table 4). On the other hand, of the 19 pathological miRNAs identified in the RV, 5 (26%) showed a decrease in FC with LVAD use compared to without LVAD use, and seven (37%) showed no significant change when compared to the NFC (Table 5). Comparisons between the VAD and NVAD groups showed that majority of pathological miRNAs were not normalized after LVAD implantation. One miRNA (hsa_miR_4458) in the LV and three miRNAs (hsa_miR_21*, hsa_miR_1972 and hsa_miR_4461) in the RV showed normalization. However, LVAD use also induced changes in the expression of other miRNAs (not from our "pathological" list) in both the LV (11 miRNAs) and RV (nine miRNAs). Enrichment analysis of the target genes of these miRNAs showed that the majority of them function in immunity and cellular process such as growth, proliferation and apoptosis (Figure 6E,F).

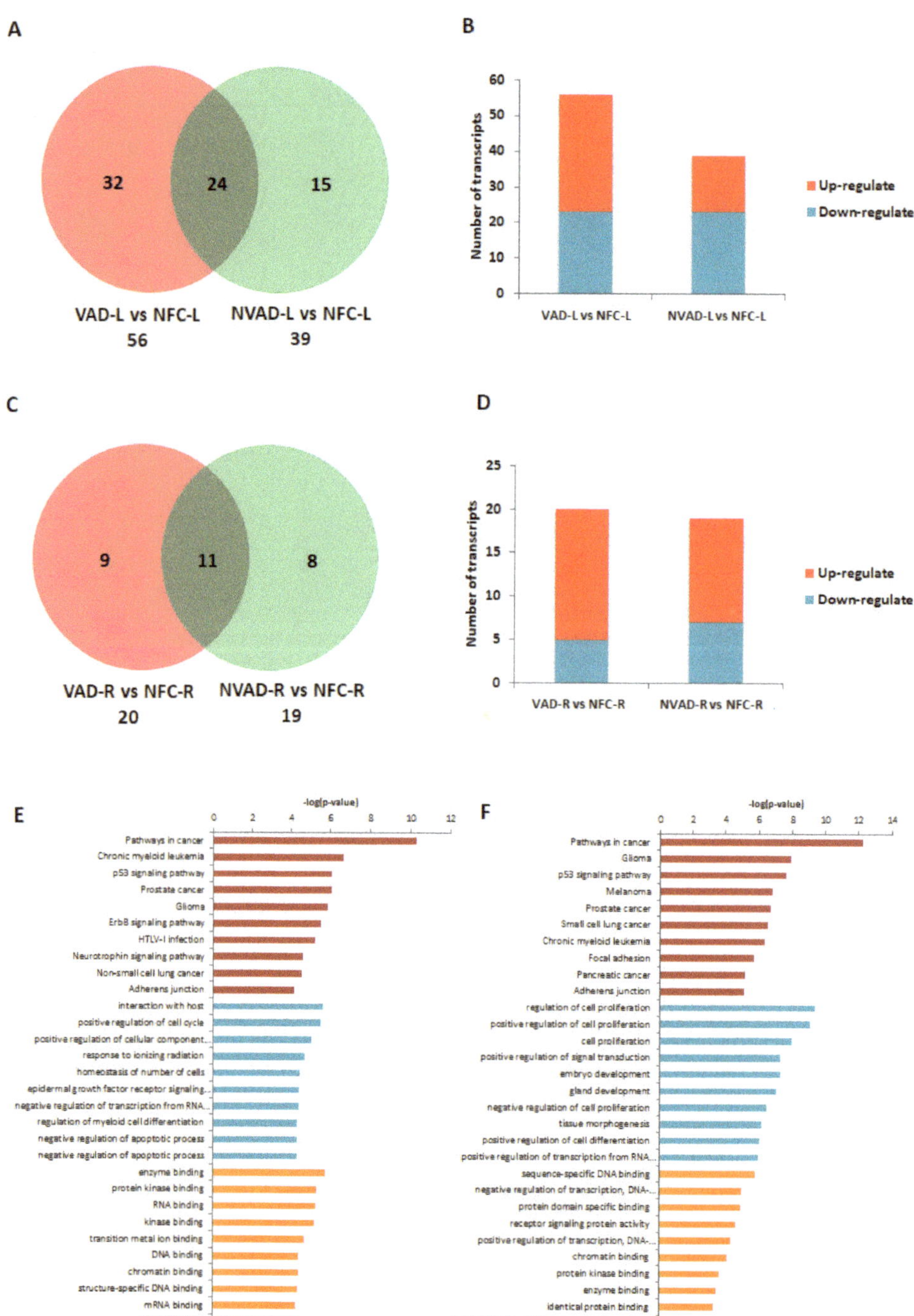

Figure 6. miRNA analysis of left (L) and right (R) ventricles from dilated cardiomyopathy (DCM) hearts with (VAD) and without left ventricle assist device (NVAD), using the non-failing controls (NFC) as reference. Number of significant differentially expressed miRNAs in the (**A**) L ventricle and

(**C**) R ventricle of VAD and NVAD groups. (**B**) Number of significantly up- and down-regulated miRNAs in the (**C**) L ventricle and (**D**) R ventricle of VAD and NVAD groups. Gene ontology (GO) analysis for target genes of the differentially expressed miRNAs in the (**E**) L and (**F**) R ventricles following LVAD use. The top ten Kyoto Encyclopedia of Genes and Genomes (KEGG) pathways (red), GO terms for biological processes (blue) and GO terms for molecular functions are shown (yellow).

Table 4. Changes in pathological miRNA expression post-LVAD implantation in the LV.

ID	FC in NVAD vs. NFC	FC in VAD vs. NFC	*p*-Value in VAD vs. NFC
Insignificant change compared to NFC			
hsa-miR-4458_st	2.21	1.73	0.5994
hsa-miR-4793-3p_st	−2.69	−1.96	0.3478
hsa-miR-187_st	−6.09	−1.68	0.3441
hsa-miR-373_st	5.76	1.48	0.1852
hsa-miR-103b_st	−2.58	−1.81	0.096
hsa-miR-95_st	4.09	2.58	0.0677
hsa-miR-431_st	−2.33	−1.76	0.0659
hsa-miR-1226_st	4.09	2.2	0.064
Decreased in FC			
hsa-miR-182_st	15.33	9.68	3.58×10^{-6}
hsa-miR-451_st	16.22	10.96	3.10×10^{-8}
hsa-miR-3187-3p_st	−2.4	−1.37	0.0139
hsa-miR-548x_st	−2	−1.36	0.0076
hsa-miR-299-5p_st	3.58	2.95	0.0019
hsa-miR-4539_st	−2.06	−1.58	0.03
hsa-miR-628-5p_st	2.59	2.17	0.0047
hsa-miR-1226-star_st	−2.12	−1.76	0.0171
hsa-miR-4687-3p_st	−2.22	−1.91	0.0144
ENSG00000202498_x_st	−2.31	−2.08	0.0044
hsa-miR-223_st	2.99	2.8	0.0225
hsa-miR-10b_st	2.55	2.38	0.0145
hsa-miR-374b_st	2.35	2.21	0.0015
ENSG00000202498_st	−2.28	−2.16	0.0012
hsa-miR-4689_st	−2.28	−2.16	0.046
hsa-miR-1825_st	−2.04	−1.93	0.0033
hsa-miR-601_st	−2.29	−2.19	0.0158
hsa-miR-940_st	−2.26	−2.19	0.0282
hsa-miR-4741_st	−2.08	−2.02	0.005
hsa-miR-3910_st	−2.03	−1.99	0.0172
hsa-miR-4270_st	−2.12	−2.09	9.94×10^{-5}

Table 5. Changes in pathological miRNA expression post-LVAD implantation in the RV.

ID	FC in NVAD vs. NFC	FC in VAD vs. NFC	*p*-Value in VAD vs. NFC
Insignificant change compared to NFC			
hsa-miR-1972_st	2.57	1.22	0.8742
hsa-miR-4461_st	2.3	1.14	0.7984
hsa-miR-21-star_st	−2.87	−1.27	0.647
hsa-miR-4524-star_st	2.05	1.32	0.3365
hsa-miR-373_st	6.81	1.66	0.0801
hsa-miR-431_st	−4.17	−3.62	0.0738
hsa-miR-1247_st	−2.56	−2.33	0.0583
Decreased in FC			
hsa-miR-182_st	19.89	11.3	1.75×10^{-5}
hsa-miR-451_st	7.13	5.16	0.0002
HBII-52-32_x_st	4.76	3.48	0.0029
hsa-miR-181a-2-star_st	2.34	1.91	0.0098
hsa-miR-10b_st	2.64	2.53	0.0068

3. Discussion

The main findings of our study are three-fold: (a) transcriptomic signatures of end-stage DCM hearts, (b) gene expression (mRNA) changes post-LVAD use, and (c) gene regulation (miRNA) changes post-LVAD use (Figure 7). The majority of differentially expressed genes in this study were down-regulated, suggesting a general loss-of-function model of pathogenesis and only the most significantly enriched transcripts in our dataset were considered.

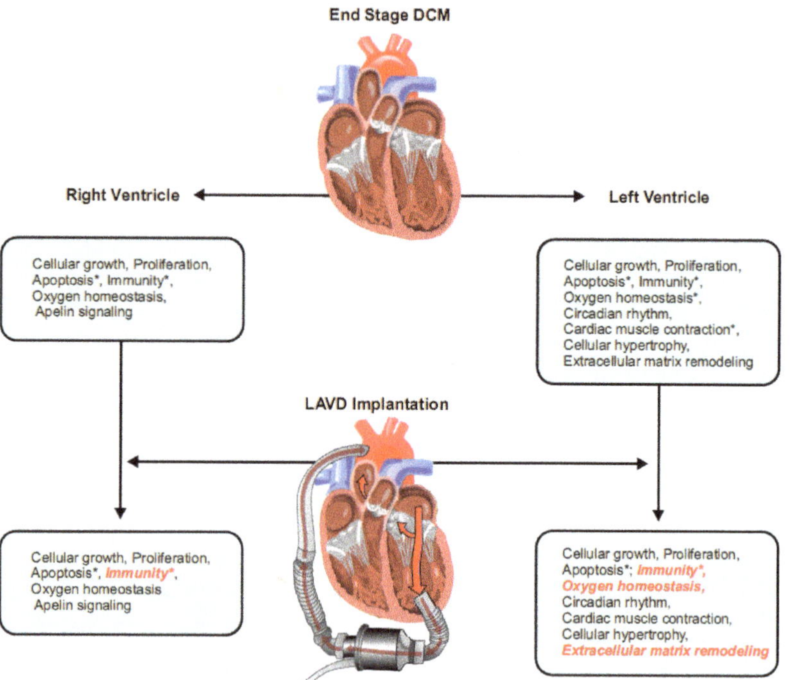

Figure 7. Summary of the findings from this study illustrating the seminal impact of LVAD therapy on reverse remodelling in the left ventricle and right ventricle. Processes that were normalized are italicized in red, and an asterisk (*) indicates pathways where the concerned genes also had corresponding changes at the miRNA level. DCM, dilated cardiomyopathy; LVAD, left ventricular assist device.

3.1. Pathological Genes Common in Both the LV and RV

Cellular processes such as growth, proliferation, and apoptosis were significantly enriched among up-regulated genes in both the LV and RV. Genes that functioned in critical signalling pathways such as FoxO, JAK-STAT, AMPK, and Wnt were all up-regulated in end-stage DCM hearts. Genes involved in cell survival, cell cycle regulation, and energy metabolism are differentially expressed between DCM and healthy control hearts [15,16]. Cardiomyocyte apoptosis is common in patients with end-stage cardiomyopathy and has been reported by several studies [16,17]. Unlike most cell types, healthy postnatal cardiomyocytes are relatively non-proliferative and any cell loss could be detrimental to myocardial structure and function. Up-regulating genes and pathways involved in cellular growth and proliferation may, therefore, serve as a compensatory mechanism in the failing myocardium in the face of increased cell death.

On the other hand, genes involved in the immune response were most significantly enriched among the down-regulated genes. These functional gene classes included components of the complement system, neutrophil-mediated immunity, inflammatory response, and the humoral immune response. Activation of the immune system in heart failure is a widely known phenomenon [18,19]. The pro-inflammatory response in DCM is most likely a response to cardiomyocyte damage and often triggers cardiac fibrosis [18]. Furthermore, this inflammatory response activates the humoral immune system, which produces autoantibodies that can cause further tissue damage. Immune response-related genes displayed the most pronounced regulated genes in end-stage DCM [20]. Although down-regulation of genes involved in immune response was observed, our study confirms alterations in the immune system related genes and emphasizes the robust nature of the immune response in the DCM environment. Further characterization of how the immune system changes in the course of DCM may have implications in the treatment of DCM at various stages in its progression.

Genes involved in the HIF-1 pathway were also differentially expressed in both chambers of the DCM heart. HIF-1 controls oxygen delivery and utilization by regulating angiogenesis, vascular remodelling, and metabolic processes [21,22]. HIF-1 plays a protective role in the pathophysiology of ischemic heart disease and pressure-overload heart failure. HIF-1 activation has also been observed in association with the ischemic environment in DCM and over-expression of this pathway can lead to the development of cardiomyopathy [23,24]. We showed that HIF-1 was down-regulated in both LV and RV of end-stage DCM hearts suggesting that disturbance of oxygen homeostasis may play a role in the pathogenesis of end-stage DCM.

3.2. LV- and RV-Specific Pathological Genes

In the LV, the regulation of circadian rhythm is one of the most significantly enriched processes amongst up-regulated genes. Molecular circadian clocks exist in all cardiovascular cell types and various cardiovascular processes such as endothelial function, thrombus formation, blood pressure, and heart rate are under the regulation of the circadian clock [25]. Disruption of this rhythm results in cardiovascular diseases including heart failure, myocardial infarction, and arrhythmias. Disruption of the circadian rhythm leads to altered sarcomeric structure, cardiac fibrosis, and eccentric hypertrophy of myocardial walls, which eventually result in LV dilation and systolic dysfunction [26,27]. Our study is the first to show an association between dysregulated circadian rhythm and DCM at the transcript level in human hearts. Genes involved in cardiac muscle contraction were likewise enriched among up-regulated genes in the LV. Various processes such as myofibril assembly, sarcomere organization, or actin–myosin filament sliding were all up-regulated. LV transcriptomic and proteomic profiling in human end-stage DCM hearts found a large number of up-regulated pathways and processes belonging to the cardiomyocyte contractility family [15]. Up-regulation of genes involved in the cardiomyocyte compartment may serve to compensate for the impaired contractility observed in DCM. The cardiomyocyte hypertrophy pathway was also up-regulated in the LV of DCM hearts which is supported by our histological data. Our finding provides molecular evidence that pathological hypertrophy is a hallmark feature of DCM.

ECM remodelling was enriched among the down-regulated genes in the LV. Remodelling of the ECM, particularly altered collagen homeostasis, has an important role in the pathogenesis of DCM [28–30]. In general, the rate of collagen turnover is controlled by the balance between matrix metalloproteinase (MMP) and tissue inhibitors of matrix metalloproteinase (TIMP) activity in the ECM [28–30]. However, TIMP expression in DCM hearts with some studies showing increased levels and others showing decreased levels compared to healthy controls suggesting a complex interplay between MMPs and TIMPs in the pathogenesis of DCM [28–31]. Our study showed decreased expression in *TIMP1*, *TIMP3*, and *TIMP4*, which potentially leads to decreased inhibition of MMPs and increased collagen turnover which was corroborated by our histological data.

In the RV, genes involved in the apelin signalling pathway were enriched among the up-regulated genes. The apelin/apelin receptor system functions in various cardiovascular processes such as vascular homeostasis, angiogenesis, cardiomyocyte contractility, cardiac hypertrophic response, and even in early cardiac development [32,33]. Our data for the first time, report significantly up-regulated apelin signalling pathway in the RV of end-stage DCM hearts. Whether this is an attempt of the failing heart to deal with the hypoxic environment and myocardial injury or up-regulation of the apelin pathway actually contributes to the deterioration of the end-stage DCM heart remains to be determined. The dynamic expression pattern of this pathway during disease development warrants further investigations. Our network analysis illustrated unique gene interactions at multiple levels in the LV and RV in patients treated medically and in those with LVAD and is consistent with an emerging view shaped by pre-clinical models that RV and LV have distinct embryological origins, workload and remodelling [34].

3.3. Gene Expression Post-LVAD Use

Upon LVAD implantation, genes involved in the immune response showed normalization (up-regulation) towards the healthy state in both the LV and RV. Proteomic analysis revealed increased abundance of innate immune response-related proteins, but a decreased abundance of complement system proteins [35]. However, it remains unknown if the up-regulation in immune response genes observed in this study is due to the unloading effect of LVAD on the ventricles or a triggered immune response by the insertion of a foreign object into the LV. Given the long duration of LVAD support in our study (median: 156 days, IQR: 131–268 days), perhaps the initial acute immune activation post-LVAD implantation plays less of a role in the observed up-regulation of immune response genes at the transcriptomic level. Moreover, our LV samples were taken at least 1 cm away from the insertion site of the LV cannula.

In the LV, genes involved in ECM remodelling were normalized after LVAD use. Both *TIMP1* and *TIMP3* (down-regulated in end-stage DCM without LVAD) were up-regulated post LVAD implantation. Besides gene normalization, LVAD use induced additional transcriptomic changes in the LV of the mechanically unloaded heart. Genes encoding for the alpha chain of collagen types 1, 3, and 14 (*COL1A1, COL3A1, COL14A1*) were all up-regulated in DCM hearts with LVAD compared to DCM hearts without LVAD. The effect of LVAD on the degree of collagen formation and fibrosis in the DCM heart is variable with some studies showing decreased collagen content post-LVAD, while others found increased collagen cross-linking with subsequent increased myocardial stiffness [36–38]. Although we observed an up-regulation of collagen expression, significant enrichment of TIMPs was also noted. TIMPs are suppressor of fibrosis and are known to decrease the deposition of excess collagen [28,39]. This complex interplay between the determinants of ECM remodelling can explain the overall reduction in fibrosis observed on staining. Our findings confirm the complex regulation of the ECM with up-regulation of TIMPs leading to increased inhibition of MMP activity and subsequently decreased ECM remodelling in the setting of reduced overall myocardial fibrosis.

Genes involved in the HIF-1 pathway for oxygen homeostasis were normalized (up-regulation) in the LV post-LVAD. Other genes related to angiogenesis were also up-regulated after LVAD implantation, which makes sense since the HIF-1 pathway was found to be a master regulator of angiogenesis [21,40]. Interestingly, the phosphatidylinositol 3-kinase (PI3K) pathway was up-regulated post-LVAD insertion and is known to play key roles in numerous cellular processes including angiogenesis and cytoskeletal remodelling [41,42]. Levels of heme oxygenase-1 (HO-1), a stress protein whose presence is induced by hypoxia, is decreased in DCM after LVAD use, suggesting improved myocardial hypoxia [43]. In our study, the normalization of genes involved in the HIF-1 pathway may contribute to the improved myocardial oxygen homeostasis following the unloading effect of LVAD. Pathways such as PI3K, MAPK, and TNFα were up-regulated after LVAD implantation. These pathways are important in numerous physiological processes such as the regulation of cellular metabolism, growth, proliferation, and apoptosis, which are critical mediators of the reverse remodelling from mechanical unloading of the failing heart via LVAD use [44,45].

3.4. Gene Regulation Post-LVAD Use

miRNA expression showed a trend towards normalization post-LVAD use, and four miRNAs (1 in the LV, 3 in the RV) were significantly normalized following LVAD implantation. In the LV, miR-4458 was significantly normalized following the unloading effect of LVAD. miR-4458 is a relatively new miRNA identified in several carcinomas [46,47]. At the cellular level, miR-4458 was shown to suppress cell proliferation and promote apoptosis in breast cancer and haemangioma [47,48]. miR-4458 functions as a negative modulator in cardiac hypertrophy and inhibition of this miRNA exacerbates cardiac hypertrophy. Our study is the first to demonstrate that miR-4458 is regulated in end-stage DCM with its expression normalizing after LVAD use. Our histological data support this finding, showing that cardiomyocyte CSA was normalized following LVAD implantation.

In the RV, 3 miRNAs (miR-21*, miR-1972, miR-4461) were significantly normalized post-LVAD use. While miR-1972 and miR-4461 have not been studied in the context of cardiovascular disease, miR-21* has been implicated in adverse myocardial remodelling [49,50]. Recent studies on miRNAs revealed the abundance of many miRNA* (miRNAstar), previously believed to undergo intracellular degradation [50]. MiR-21* induce hypertrophy of cardiomyocytes and inhibition of miR-21* alleviated cardiac hypertrophy in a model of angiotensin II-induced heart disease [51]. In contrast, another study revealed the intrinsic anti-hypertrophic function of miR-21* in cardiomyocytes [52]. Our study supports miR-21* role in the pathogenesis of heart failure and we found that LVAD use normalized the expression of this miRNA. LVAD use also induced changes in the expression of other miRNAs whose target genes function in the immune response and cellular physiological processes. While some studies found many miRNAs normalizing after LVAD use, others showed minimal changes [53]. Findings from our study suggest that changes in gene expression after LVAD implementation is partly attributable to miRNA regulation, and LVAD use has a more pronounced effect on miRNA normalization in the RV compared to the LV.

3.5. Strengths and Limitations

The limitations of our study include those of all descriptive studies using microarray or RNA sequencing. Microarray technology relies on a pre-determined set of probes so our study may not capture the full transcriptomic profiles of these hearts. However, proof of concept regarding gene expression and regulation generated in DCM hearts with or without LVAD should be further evaluated using next-generation sequencing [54]. Previous studies have examined molecular signatures in DCM hearts after LVAD use; however, the pre-LVAD samples were obtained from the apex region, while the post-LVAD samples were from the left ventricular free wall (LVFW) or septum. This regional variability between pre-and post-LVAD tissues may affect the evaluation of gene expression and regulation. The LVFW was consistently used in our study to control this regional variation. However, inter-individual variability may still play a role in our study since the pre-VAD and post-VAD samples were from different patients (as opposed to from the same people at the time of LVAD insertion and transplantation). The clinical background including mutations is an important modifier of myocardial gene expression. Future studies should expand analysis to these important variables to provide information on gene expression in the diseased human hearts. Another strength of our study is our ability to assess the transcriptomic changes in both the LV and RV of these hearts; in contrast, the majority of previous studies only looked at changes in the LV. We also studied both mRNA and miRNA expression and were, therefore, able to evaluate the effects of LVAD use on both gene expression and gene regulation, as well as assess whether changes in miRNA expression were accompanied by corresponding changes in mRNA expression.

4. Materials and Methods

4.1. Study Patients and Protocol

The study protocol (Pro00011739) was reviewed and approved by the Health Research Ethics Board at the University of Alberta. The inclusion criteria were broad and included consecutive patients having heart transplantation at the Mazankowski Alberta Heart Institute and we were able to consent 99% of our patients. The only exclusion criterion was lack of written consent. All patients who participated in the study provided written informed consent in accordance with the Declaration of Helsinki. Clinical information for patients with DCM was collected via review of medical records from electronic databases maintained by Alberta Health Services based on the Human Explanted Heart Program (HELP) at the University of Alberta [4,54]. Available information on non-failing donor hearts was provided by the Human Organ Procurement and Exchange (HOPE) program in Northern Alberta.

4.2. Heart Tissue Procurement

LV and RV free-wall (LVFW and RVFW) tissue from two groups of explanted failing human hearts, DCM with LVAD ($n = 8$, male:female [M:F] = 7:1) or without LVAD ($n = 8$, M:F = 7:1) were collected via the HELP program. Non-failing control human hearts ($n = 6$, M:F = 4:2) were collected via the HOPE program. All hearts received cold cardioplegia, promptly explanted, preserved in cold saline, and kept on ice. Transmural myocardial tissue samples from the mid LVFW and RVFW were collected within 5 min of explantation and completed within 15 min, frozen in liquid nitrogen and stored at $-80\,^{\circ}\mathrm{C}$ for subsequent use [4].

4.3. Histology
4.3.1. Picrosirius Red Staining

Formalin-fixed LVFW and RVFW were embedded in paraffin and sectioned with 5-μm thickness. Sections were stained using picrosirius red (PSR; Sigma, Oakville, ON, Canada) following standard procedure. Samples were visualized at $100\times$ magnification using the Olympus IX-8 fluorescence microscope. Images were taken in at least 10 fields of view per sample to cover the whole slide. Myocardial collagen content was quantified using Metamorph Basic (Version 7.7.0.0). Values obtained across all fields of view were averaged for each sample. Image acquisition and myocardial collagen content quantification were conducted in a double-blinded fashion.

4.3.2. Wheat Germ Agglutinin Staining

OCT-embedded frozen blocks of LVFW and RVFW were sectioned with 5 μm thickness. Sections were stained using Oregon Green 488–conjugated wheat germ agglutinin (Invitrogen, Burlington, ON, Canada) following standard procedure. Samples were visualized at $200\times$ magnification using the Olympus IX-8 fluorescence microscope. Images were taken in at least 10 random fields of view per sample. Cardiomyocyte cross-sectional area was measured by tracing the area congruent to the wheat germ agglutinin-stained cardiomyocyte membrane using Metamorph Basic (Version 7.7.0.0). Values obtained across all fields of view were averaged for each sample. Image acquisition and cardiomyocyte cross-sectional area quantification were conducted in a double-blinded fashion.

4.4. RNA Processing and Microarray

RNA extraction was performed using an RNeasy Mini Kit (Qiagen, Toronto, ON, Canada) according to the manufacturer's instructions. The purity of the total RNA was determined from the ratio of absorbance readings at 260 and 280 nm, with an A260/280 ratio between 1.8 and 2.0 indicating acceptable purity. RNA samples were stored at $-80\,^{\circ}\mathrm{C}$ for microarray analysis. BioAnalyzer was used to assess RNA sample quality. All RNA analysed had an RNA Integrity Number (RIN) of 7.5–10. Microarray analysis was performed using the PrimeView Human Gene Expression Array (Thermo Fisher, Ottawa, ON, Canada) and GeneChip miRNA 3.0 Array (Thermo Fisher), according to the manufacturer's protocol. mRNA expression analysis was performed using TaqMan reverse transcription polymerase chain reaction as before, using primers and probes obtained from Invitrogen [28]. mRNA levels for each gene were normalized to 18S levels for corresponding sample.

4.5. Data Generation and Analysis

Transcriptomic data, both mRNA and miRNA, were normalized and analysed using Transcriptome Analysis Console (Version 4.0.1) (Thermo Fisher). One-Way Between-Subject ANOVA was used to calculate p-values, and a p-value of less than 0.05 was considered significant. Cut-off fold change (FC) for differentially expressed transcripts was set at 2 (≥ 2 for up regulation or ≤ -2 for down-regulation). For analysis of transcript expression normalization after treatment with LVAD, only the p-value cut-off was applied in order to capture all effects of treatment on transcript expression.

To identify significantly enriched functional pathways and gene ontology (GO) terms related to biological processes and molecular functions, differentially expressed genes were analysed using Enrichr (Available online: https://amp.pharm.mssm.edu/Enrichr/ (accessed on 7 January 2022)). The top ten most significantly enriched Kyoto Encyclopedia of Genes and Genomes (KEGG) pathways, GO terms related to biological process (BP), and GO terms related to molecular function (MF) were reported for each enrichment analysis. miRNet (Available online: https://www.mirnet.ca/miRNet/home.xhtml (accessed on 7 January 2022)) was used to predict miRNA target genes and their enrichment analysis. The top ten most significantly enriched KEGG pathways, GO terms related to BP, and GO terms related to MF were reported for the target genes, as previously performed. miRNA–mRNA interactions were analysed using Transcriptome Analysis Console (Version 4.0.1) (Thermo Fisher).

4.6. Statistical Analysis

Parametric continuous variables were expressed as means with their respective standard deviations (SDs). Non-parametric continuous variables were expressed as medians with their respective interquartile ranges (IQRs). Categorical variables were expressed as the total number in each category and the corresponding percentage of the study group they represented. Data distribution was assessed by the Kolmogorov–Smirnov test, D'Agostino–Pearson omnibus normality test, and Shapiro–Wilk normality test. All statistical analyses were performed using SPSS Statistics version 25 (IBM, New York, NY, USA). ANOVA followed by post hoc Tukey's multiple comparison test, student's t-test, two-tailed Mann–Whitney test, and two-sided Fisher's exact test and were used for parametric continuous, non-parametric continuous, and categorical data, respectively.

5. Conclusions

Our study used microarray technology to determine the transcriptomic signatures of end-stage DCM hearts as well as the effects of LVAD use on these signatures. We identified major physiological processes contributing to the pathogenesis of end-stage DCM. Our results also suggest that the unloading effect of LVAD normalizes the expression of various genes and miRNAs towards healthy levels and that it also induces additional transcriptomic changes in the expression of both mRNAs and regulatory miRNAs. Right and left ventricular remodelling are distinct and we have identified multiple possible therapeutic targets for HF driven by LV and/or RV failure.

Supplementary Materials: The following supporting information can be downloaded at: https://www.mdpi.com/article/10.3390/ijms23042050/s1.

Author Contributions: Conceptualization: M.P., S.S., R.B. and G.Y.O.; methodology: M.P., S.S., R.B. and K.S.F.; software: M.P., S.S. and K.S.F.; validation: M.P., S.S. and K.S.F.; formal analysis: M.P., S.S., R.B. and K.S.F.; investigation: M.P., S.S. and R.B.; resources: D.K., J.C.M., K.S.F., P.F.H. and G.Y.O.; data curation: G.Y.O.; writing—original draft preparation: M.P., S.S., R.B. and G.Y.O.; writing: review and editing: M.P., S.S., R.B. and G.Y.O.; supervision: K.S.F., P.F.H. and G.Y.O.; project administration: D.K., J.C.M., P.F.H. and G.Y.O.; funding acquisition: P.F.H. and G.Y.O. All authors have read and agreed to the published version of the manuscript.

Funding: Our study is funded through the University of Alberta Hospital Foundation (UHF), Canadian Institute of Health Research (CIHR) PJT-425751, Alberta Innovates-Health Solutions (AI-

HS), and Heart and Stroke Foundation (HSF) G-18-0022161. Dr Oudit is supported by a Tier II Canada Research Chair in Heart failure through the Government of Canada (Ottawa, ON, Canada).

Institutional Review Board Statement: The study was conducted in accordance with the Declaration of Helsinki and approved by the Health Research Ethics Board at the University of Alberta (Pro00011739, 16 November 2021).

Informed Consent Statement: Informed consent was obtained from all subjects involved in the study.

Data Availability Statement: Data available upon request.

Acknowledgments: We would like to thank all the patients for participating in this study.

Conflicts of Interest: The authors declare no conflict of interest.

References

1. Yogasundaram, H.; Alhumaid, W.; Dzwiniel, T.; Christian, S.; Oudit, G.Y. Cardiomyopathies and genetic testing in heart failure: Role in defining phenotype-targeted approaches and management. *Can. J. Cardiol.* **2021**, *37*, 547–559. [CrossRef]
2. Dec, G.W.; Fuster, V. Idiopathic dilated cardiomyopathy. *N. Engl. J. Med.* **1994**, *331*, 1564–1575. [CrossRef]
3. Caplan, A.L. Finding a solution to the organ shortage. *Can. Med. Assoc. J.* **2016**, *188*, 1182–1183. [CrossRef]
4. Zhang, H.; Viveiros, A.; Nikhanj, A.; Nguyen, Q.; Wang, K.; Wang, W.; Freed, D.H.; Mullen, J.C.; MacArthur, R.; Kim, D.H.; et al. The human explanted heart program: A translational bridge for cardiovascular medicine. *Biochim. Biophys. Acta -Mol. Basis Dis.* **2021**, *1867*, 165995–166009. [CrossRef] [PubMed]
5. Jacob, K.A.; Buijsrogge, M.P.; Ramjankhan, F.Z. Left ventricular assist devices for advanced heart failure. *N. Engl. J. Med.* **2017**, *376*, 1893–1894. [CrossRef] [PubMed]
6. Heerdt, P.M.; Holmes, J.W.; Cai, B.; Barbone, A.; Madigan, J.D.; Reiken, S.; Lee, D.L.; Oz, M.C.; Marks, A.R.; Burkhoff, D. Chronic unloading by left ventricular assist device reverses contractile dysfunction and alters gene expression in end-stage heart failure. *Circulation* **2000**, *102*, 2713–2719. [CrossRef] [PubMed]
7. Burkhoff, D.; Topkara, V.K.; Sayer, G.; Uriel, N. Reverse remodeling with left ventricular assist devices. *Circ. Res.* **2021**, *128*, 1594–1612. [CrossRef]
8. Rose, E.A.; Gelijns, A.C.; Moskowitz, A.J.; Heitjan, D.F.; Stevenson, L.W.; Dembitsky, W.; Long, J.W.; Ascheim, D.D.; Tierney, A.R.; Levitan, R.G.; et al. Long-term use of a left ventricular assist device for end-stage heart failure. *N. Engl. J. Med.* **2001**, *345*, 1435–1443. [CrossRef]
9. Rogers, J.G.; Butler, J.; Lansman, S.L.; Gass, A.; Portner, P.M.; Pasque, M.K.; Pierson, R.N., 3rd; Investigators, I.N. Chronic mechanical circulatory support for inotrope-dependent heart failure patients who are not transplant candidates: Results of the intrepid trial. *J. Am. Coll. Cardiol.* **2007**, *50*, 741–747. [CrossRef]
10. Owens, A.T.; Jessup, M. Should left ventricular assist device be standard of care for patients with refractory heart failure who are not transplantation candidates? Left ventricular assist devices should not be standard of care for transplantation-ineligible patients. *Circulation* **2012**, *126*, 3088–3094. [CrossRef]
11. Slaughter, M.S.; Rogers, J.G.; Milano, C.A.; Russell, S.D.; Conte, J.V.; Feldman, D.; Sun, B.; Tatooles, A.J.; Delgado, R.M., 3rd; Long, J.W.; et al. Advanced heart failure treated with continuous-flow left ventricular assist device. *N. Engl. J. Med.* **2009**, *361*, 2241–2251. [CrossRef] [PubMed]
12. Kirklin, J.K.; Naftel, D.C.; Pagani, F.D.; Kormos, R.L.; Stevenson, L.W.; Blume, E.D.; Miller, M.A.; Timothy Baldwin, J.; Young, J.B. Sixth intermacs annual report: A 10,000-patient database. *J. Heart Lung Transplant.* **2014**, *33*, 555–564. [CrossRef] [PubMed]
13. Friedberg, M.K.; Redington, A.N. Right versus left ventricular failure: Differences, similarities, and interactions. *Circulation* **2014**, *129*, 1033–1044. [CrossRef] [PubMed]
14. Frankfurter, C.; Molinero, M.; Vishram-Nielsen, J.K.K.; Foroutan, F.; Mak, S.; Rao, V.; Billia, F.; Orchanian-Cheff, A.; Alba, A.C. Predicting the risk of right ventricular failure in patients undergoing left ventricular assist device implantation: A systematic review. *Circ. Heart Fail.* **2020**, *13*, e006994. [CrossRef] [PubMed]
15. Colak, D.; Alaiya, A.A.; Kaya, N.; Muiya, N.P.; AlHarazi, O.; Shinwari, Z.; Andres, E.; Dzimiri, N. Integrated left ventricular global transcriptome and proteome profiling in human end-stage dilated cardiomyopathy. *PLoS ONE* **2016**, *11*, e0162669. [CrossRef] [PubMed]
16. Narula, J.; Haider, N.; Virmani, R.; DiSalvo, T.G.; Kolodgie, F.D.; Hajjar, R.J.; Schmidt, U.; Semigran, M.J.; Dec, G.W.; Khaw, B.A. Apoptosis in myocytes in end-stage heart failure. *N. Engl. J. Med.* **1996**, *335*, 1182–1189. [CrossRef] [PubMed]
17. Zorc, M.; Vraspir-Porenta, O.; Zorc-Pleskovic, R.; Radovanovic, N.; Petrovic, D. Apoptosis of myocytes and proliferation markers as prognostic factors in end-stage dilated cardiomyopathy. *Cardiovasc. Pathol.* **2003**, *12*, 36–39. [CrossRef]
18. Frantz, S.; Falcao-Pires, I.; Balligand, J.L.; Bauersachs, J.; Brutsaert, D.; Ciccarelli, M.; Dawson, D.; de Windt, L.J.; Giacca, M.; Hamdani, N.; et al. The innate immune system in chronic cardiomyopathy: A european society of cardiology (esc) scientific statement from the working group on myocardial function of the esc. *Eur. J. Heart Fail.* **2018**, *20*, 445–459. [CrossRef]
19. Adamo, L.; Rocha-Resende, C.; Prabhu, S.D.; Mann, D.L. Reappraising the role of inflammation in heart failure. *Nat. Rev. Cardiol.* **2020**, *17*, 269–285. [CrossRef]

20. Barth, A.S.; Kuner, R.; Buness, A.; Ruschhaupt, M.; Merk, S.; Zwermann, L.; Kaab, S.; Kreuzer, E.; Steinbeck, G.; Mansmann, U.; et al. Identification of a common gene expression signature in dilated cardiomyopathy across independent microarray studies. *J. Am. Coll. Cardiol.* **2006**, *48*, 1610–1617. [CrossRef]
21. Semenza, G.L. Hypoxia-inducible factor 1 and cardiovascular disease. *Annu. Rev. Physiol.* **2014**, *76*, 39–56. [CrossRef] [PubMed]
22. Kido, M.; Du, L.; Sullivan, C.C.; Li, X.; Deutsch, R.; Jamieson, S.W.; Thistlethwaite, P.A. Hypoxia-inducible factor 1-alpha reduces infarction and attenuates progression of cardiac dysfunction after myocardial infarction in the mouse. *J. Am. Coll. Cardiol.* **2005**, *46*, 2116–2124. [CrossRef] [PubMed]
23. Holscher, M.; Schafer, K.; Krull, S.; Farhat, K.; Hesse, A.; Silter, M.; Lin, Y.; Pichler, B.J.; Thistlethwaite, P.; El-Armouche, A.; et al. Unfavourable consequences of chronic cardiac hif-1alpha stabilization. *Cardiovasc. Res.* **2012**, *94*, 77–86. [CrossRef] [PubMed]
24. Moslehi, J.; Minamishima, Y.A.; Shi, J.; Neuberg, D.; Charytan, D.M.; Padera, R.F.; Signoretti, S.; Liao, R.; Kaelin, W.G., Jr. Loss of hypoxia-inducible factor prolyl hydroxylase activity in cardiomyocytes phenocopies ischemic cardiomyopathy. *Circulation* **2010**, *122*, 1004–1016. [CrossRef] [PubMed]
25. Crnko, S.; Du Pre, B.C.; Sluijter, J.P.G.; Van Laake, L.W. Circadian rhythms and the molecular clock in cardiovascular biology and disease. *Nat. Rev. Cardiol.* **2019**, *16*, 437–447. [CrossRef] [PubMed]
26. Alibhai, F.J.; LaMarre, J.; Reitz, C.J.; Tsimakouridze, E.V.; Kroetsch, J.T.; Bolz, S.S.; Shulman, A.; Steinberg, S.; Burris, T.P.; Oudit, G.Y.; et al. Disrupting the key circadian regulator clock leads to age-dependent cardiovascular disease. *J. Mol. Cell. Cardiol.* **2017**, *105*, 24–37. [CrossRef]
27. Martino, T.A.; Oudit, G.Y.; Herzenberg, A.M.; Tata, N.; Koletar, M.M.; Kabir, G.M.; Belsham, D.D.; Backx, P.H.; Ralph, M.R.; Sole, M.J. Circadian rhythm disorganization produces profound cardiovascular and renal disease in hamsters. *Am. J. Physiol. Regul. Integr. Comp. Physiol.* **2008**, *294*, R1675–R1683. [CrossRef]
28. Sakamuri, S.S.; Takawale, A.; Basu, R.; Fedak, P.W.; Freed, D.; Sergi, C.; Oudit, G.Y.; Kassiri, Z. Differential impact of mechanical unloading on structural and nonstructural components of the extracellular matrix in advanced human heart failure. *Transl. Res.* **2016**, *172*, 30–44. [CrossRef]
29. Takawale, A.; Sakamuri, S.S.; Kassiri, Z. Extracellular matrix communication and turnover in cardiac physiology and pathology. *Compr. Physiol.* **2015**, *5*, 687–719. [CrossRef]
30. Jana, S.; Zhang, H.; Lopaschuk, G.D.; Freed, D.H.; Sergi, C.; Kantor, P.F.; Oudit, G.Y.; Kassiri, Z. Disparate remodeling of the extracellular matrix and proteoglycans in failing pediatric versus adult hearts. *J. Am. Heart Assoc.* **2018**, *7*, e010427. [CrossRef]
31. Spinale, F.G. Myocardial matrix remodeling and the matrix metalloproteinases: Influence on cardiac form and function. *Physiol. Rev.* **2007**, *87*, 1285–1342. [CrossRef] [PubMed]
32. Zhong, J.C.; Zhang, Z.Z.; Wang, W.; McKinnie, S.M.K.; Vederas, J.C.; Oudit, G.Y. Targeting the apelin pathway as a novel therapeutic approach for cardiovascular diseases. *Biochim. Biophys. Acta* **2017**, *1863*, 1942–1950. [CrossRef] [PubMed]
33. de Oliveira, A.A.; Vergara, A.; Wang, X.; Vederas, J.C.; Oudit, G.Y. Apelin pathway in cardiovascular, kidney, and metabolic diseases: Therapeutic role of apelin analogs and apelin receptor agonists. *Peptides* **2021**, *147*, 170697. [CrossRef]
34. Voelkel, N.F.; Quaife, R.A.; Leinwand, L.A.; Barst, R.J.; McGoon, M.D.; Meldrum, D.R.; Dupuis, J.; Long, C.S.; Rubin, L.J.; Smart, F.W.; et al. Right ventricular function and failure: Report of a national heart, lung, and blood institute working group on cellular and molecular mechanisms of right heart failure. *Circulation* **2006**, *114*, 1883–1891. [CrossRef] [PubMed]
35. Shahinian, J.H.; Mayer, B.; Tholen, S.; Brehm, K.; Biniossek, M.L.; Fullgraf, H.; Kiefer, S.; Heizmann, U.; Heilmann, C.; Ruter, F.; et al. Proteomics highlights decrease of matricellular proteins in left ventricular assist device therapydagger. *Eur. J. Cardiothorac. Surg.* **2017**, *51*, 1063–1071. [CrossRef]
36. Maybaum, S.; Mancini, D.; Xydas, S.; Starling, R.C.; Aaronson, K.; Pagani, F.D.; Miller, L.W.; Margulies, K.; McRee, S.; Frazier, O.H.; et al. Cardiac improvement during mechanical circulatory support: A prospective multicenter study of the lvad working group. *Circulation* **2007**, *115*, 2497–2505. [CrossRef]
37. Bruckner, B.A.; Stetson, S.J.; Farmer, J.A.; Radovancevic, B.; Frazier, O.H.; Noon, G.P.; Entman, M.L.; Torre-Amione, G.; Youker, K.A. The implications for cardiac recovery of left ventricular assist device support on myocardial collagen content. *Am. J. Surg.* **2000**, *180*, 498–501. [CrossRef]
38. Bruckner, B.A.; Stetson, S.J.; Perez-Verdia, A.; Youker, K.A.; Radovancevic, B.; Connelly, J.H.; Koerner, M.M.; Entman, M.E.; Frazier, O.H.; Noon, G.P.; et al. Regression of fibrosis and hypertrophy in failing myocardium following mechanical circulatory support. *J. Heart Lung Transplant.* **2001**, *20*, 457–464. [CrossRef]
39. Fan, D.; Takawale, A.; Lee, J.; Kassiri, Z. Cardiac fibroblasts, fibrosis and extracellular matrix remodeling in heart disease. *Fibrogenes. Tissue Repair* **2012**, *5*, 15. [CrossRef]
40. Knutson, A.K.; Williams, A.L.; Boisvert, W.A.; Shohet, R.V. Hif in the heart: Development, metabolism, ischemia, and atherosclerosis. *J. Clin. Investig.* **2021**, *131*, e137557. [CrossRef]
41. Oudit, G.Y.; Penninger, J.M. Cardiac regulation by phosphoinositide 3-kinases and pten. *Cardiovasc. Res.* **2009**, *82*, 250–260. [CrossRef] [PubMed]
42. Patel, V.B.; Zhabyeyev, P.; Chen, X.; Wang, F.; Paul, M.; Fan, D.; McLean, B.A.; Basu, R.; Zhang, P.; Shah, S.; et al. Pi3kalpha-regulated gelsolin activity is a critical determinant of cardiac cytoskeletal remodeling and heart disease. *Nat. Commun.* **2018**, *9*, 5390. [CrossRef] [PubMed]

43. Grabellus, F.; Schmid, C.; Levkau, B.; Breukelmann, D.; Halloran, P.F.; August, C.; Takeda, N.; Takeda, A.; Wilhelm, M.; Deng, M.C.; et al. Reduction of hypoxia-inducible heme oxygenase-1 in the myocardium after left ventricular mechanical support. *J. Pathol.* **2002**, *197*, 230–237. [CrossRef] [PubMed]
44. Chen, Y.; Park, S.; Li, Y.; Missov, E.; Hou, M.; Han, X.; Hall, J.L.; Miller, L.W.; Bache, R.J. Alterations of gene expression in failing myocardium following left ventricular assist device support. *Physiol. Genom.* **2003**, *14*, 251–260. [CrossRef]
45. Heineke, J.; Molkentin, J.D. Regulation of cardiac hypertrophy by intracellular signalling pathways. *Nat. Rev. Mol. Cell Biol.* **2006**, *7*, 589–600. [CrossRef]
46. Yang, M.; Zhang, J.; Jin, X.; Li, C.; Zhou, G.; Feng, J. Nrf1-enhanced mir-4458 alleviates cardiac hypertrophy through releasing ttp-inhibited tfam. *Vitr. Cell. Dev. Biol. Anim.* **2020**, *56*, 120–128. [CrossRef]
47. Liu, X.; Wang, J.; Zhang, G. Mir-4458 regulates cell proliferation and apoptosis through targeting socs1 in triple-negative breast cancer. *J. Cell Biochem.* **2019**, *120*, 12943–12948. [CrossRef]
48. Wu, M.; Tang, Y.; Hu, G.; Yang, C.; Ye, K.; Liu, X. Mir-4458 directly targets igf1r to inhibit cell proliferation and promote apoptosis in hemangioma. *Exp. Ther. Med.* **2020**, *19*, 3017–3023. [CrossRef]
49. Kura, B.; Kalocayova, B.; Devaux, Y.; Bartekova, M. Potential clinical implications of mir-1 and mir-21 in heart disease and cardioprotection. *Int. J. Mol. Sci.* **2020**, *21*, 700. [CrossRef]
50. Duygu, B.; Da Costa Martins, P.A. Mir-21: A star player in cardiac hypertrophy. *Cardiovasc. Res.* **2015**, *105*, 235–237. [CrossRef]
51. Bang, C.; Batkai, S.; Dangwal, S.; Gupta, S.K.; Foinquinos, A.; Holzmann, A.; Just, A.; Remke, J.; Zimmer, K.; Zeug, A.; et al. Cardiac fibroblast-derived microrna passenger strand-enriched exosomes mediate cardiomyocyte hypertrophy. *J. Clin. Investig.* **2014**, *124*, 2136–2146. [CrossRef] [PubMed]
52. Yan, M.; Chen, C.; Gong, W.; Yin, Z.; Zhou, L.; Chaugai, S.; Wang, D.W. Mir-21-3p regulates cardiac hypertrophic response by targeting histone deacetylase-8. *Cardiovasc. Res.* **2015**, *105*, 340–352. [CrossRef] [PubMed]
53. Yang, K.C.; Yamada, K.A.; Patel, A.Y.; Topkara, V.K.; George, I.; Cheema, F.H.; Ewald, G.A.; Mann, D.L.; Nerbonne, J.M. Deep rna sequencing reveals dynamic regulation of myocardial noncoding rnas in failing human heart and remodeling with mechanical circulatory support. *Circulation* **2014**, *129*, 1009–1021. [CrossRef] [PubMed]
54. Litvinukova, M.; Talavera-Lopez, C.; Maatz, H.; Reichart, D.; Worth, C.L.; Lindberg, E.L.; Kanda, M.; Polanski, K.; Heinig, M.; Lee, M.; et al. Cells of the adult human heart. *Nature* **2020**, *588*, 466–472. [CrossRef] [PubMed]

Article

Cardiomyogenic Differentiation Potential of Human Dilated Myocardium-Derived Mesenchymal Stem/Stromal Cells: The Impact of HDAC Inhibitor SAHA and Biomimetic Matrices

Rokas Miksiunas [1], Ruta Aldonyte [1], Agne Vailionyte [2,3], Tadas Jelinskas [2,3], Romuald Eimont [2,3], Gintare Stankeviciene [2,3], Vytautas Cepla [2,3], Ramunas Valiokas [2,3], Kestutis Rucinskas [4], Vilius Janusauskas [4], Siegfried Labeit [5,6] and Daiva Bironaite [1,*]

[1] Department of Regenerative Medicine, State Research Institute Centre for Innovative Medicine, Santariskiu Str. 5, LT-08460 Vilnius, Lithuania; rokas.miksiunas@gmail.com (R.M.); aldonyte.ruta@gmail.com (R.A.)

[2] Department of Nanoengineering, Center for Physical Sciences and Technology, Savanoriu Ave. 231, LT-02300 Vilnius, Lithuania; agne@ferentis.eu (A.V.); tadas@ferentis.eu (T.J.); romuald@ferentis.eu (R.E.); gintare@ferentis.eu (G.S.); vytautas@ferentis.eu (V.C.); ramunas@ferentis.eu (R.V.)

[3] Ferentis UAB, Savanoriu Ave. 231, LT-02300 Vilnius, Lithuania

[4] Clinic of Cardiovascular Diseases, Institute of Clinical Medicine, Faculty of Medicine, Vilnius University, M. K. Ciurlionio Str. 21/27, LT-03101 Vilnius, Lithuania; Kestutis.Rucinskas@santa.lt (K.R.); Vilius.Janusauskas@santa.lt (V.J.)

[5] Medical Faculty Mannheim, University of Heidelberg, 68169 Mannheim, Germany; labeit@medma.de

[6] Myomedix GmbH, 69151 Neckargemünd, Germany

* Correspondence: daibironai@gmail.com or daiva.bironaite@imcentras.lt

Citation: Miksiunas, R.; Aldonyte, R.; Vailionyte, A.; Jelinskas, T.; Eimont, R.; Stankeviciene, G.; Cepla, V.; Valiokas, R.; Rucinskas, K.; Janusauskas, V.; et al. Cardiomyogenic Differentiation Potential of Human Dilated Myocardium-Derived Mesenchymal Stem/Stromal Cells: The Impact of HDAC Inhibitor SAHA and Biomimetic Matrices. *Int. J. Mol. Sci.* **2021**, *22*, 12702. https://doi.org/10.3390/ijms222312702

Academic Editors: Nicole Wagner and Kay-Dietrich Wagner

Received: 28 October 2021
Accepted: 20 November 2021
Published: 24 November 2021

Publisher's Note: MDPI stays neutral with regard to jurisdictional claims in published maps and institutional affiliations.

Abstract: Dilated cardiomyopathy (DCM) is the most common type of nonischemic cardiomyopathy characterized by left ventricular or biventricular dilation and impaired contraction leading to heart failure and even patients' death. Therefore, it is important to search for new cardiac tissue regenerating tools. Human mesenchymal stem/stromal cells (hmMSCs) were isolated from post-surgery healthy and DCM myocardial biopsies and their differentiation to the cardiomyogenic direction has been investigated in vitro. Dilated hmMSCs were slightly bigger in size, grew slower, but had almost the same levels of MSC-typical surface markers as healthy hmMSCs. Histone deacetylase (HDAC) activity in dilated hmMSCs was 1.5-fold higher than in healthy ones, which was suppressed by class I and II HDAC inhibitor suberoylanilide hydroxamic acid (SAHA) showing activation of cardiomyogenic differentiation-related genes alpha-cardiac actin (*ACTC1*) and cardiac troponin T (*TNNT2*). Both types of hmMSCs cultivated on collagen I hydrogels with hyaluronic acid (HA) or 2-methacryloyloxyethyl phosphorylcholine (MPC) and exposed to SAHA significantly downregulated focal adhesion kinase (*PTK2*) and activated *ACTC1* and *TNNT2*. Longitudinal cultivation of dilated hmMSC also upregulated alpha-cardiac actin. Thus, HDAC inhibitor SAHA, in combination with collagen I-based hydrogels, can tilt the dilated myocardium hmMSC toward cardiomyogenic direction in vitro with further possible therapeutic application in vivo.

Keywords: cardiomyogenic differentiation; cardiac mesenchymal stromal cell; hydrogels; histone deacetylase inhibitors

1. Introduction

Dilated cardiomyopathy (DCM) refers to heterogeneous myocardial disorders characterized by ventricle dilation and depressed myocardial performance [1]. It starts in the left ventricle and initiates chamber dilation, wall stretching, and thinning. Changed volume of the left ventricle further spreads to the right ventricle and then to the atria subsequently leading to the dilation of the whole heart [2]. Dilated cardiomyopathy and its further complications (heart failure, arrhythmias, thrombosis) are among the leading global causes of human mortality [3]. Current DCM treatment strategies are not always successful and often lead to heart transplantation. Therefore, new technologies modifying heart cells and/or extracellular environment to restore and/or regenerate damaged heart tissue remain relevant these days.

Studies investigating the effects of direct stem cell injection into heart muscle, epicardial patch or systemic cell deliver revealed embryonic, induced pluripotent, hematopoietic, adult mesenchymal stem/stromal (MSC), and cardiac progenitors being mostly employed for the cardiac regeneration studies both in vitro and in vivo [4–6]. However, stem cell transplantation-based DCM therapies have met several obstacles, such as cell-related tumorigenicity and immunogenicity, poor retention/engraftment, and insufficient tissue targeting [6]. The application of sophisticated tissue-engineered and cell-secreted products mimicking natural heart environment was also shown to improve survival of transplanted cells and subsequently human heart functioning [7–9]. However, the successful therapeutic engraftment of stem cells or extracellular environment-based components in the heart is still low due to the lack of more detail mechanistic investigations. Thus, new biosystems and/or biotechniques intracellularly and extracellularly stimulating cardiomyogenic differentiation processes are needed in order to improve failed heart functioning.

Histone deacetylases (HDACs) catalyze the removal of acetyl groups from lysine residues in the NH_2 terminal tails of core histones, resulting in a more closed chromatin structure and repression of genes activity [10]. In addition to the histones, HDACs also control activity of various non-histone proteins such as tubulin, importin, Hsp90, and hypoxia factor-1 [11]. Altogether, 18 HDACs are recently known and grouped into the main four classes based on their structural and functional characteristics [12]. The class I HDACs (1, 2, 3, and 8) mainly control intracellular functions such as cell cycle, viability, proliferation and death, whereas class II HDACs (4, 5, 6, 7, 9, and 10) are more related to the tissue specific functions [12]. Class III HDACs include the Sirtuin (SIRT) family of seven proteins that are dependent on the NAD^+ and do not contain zinc as do other HDACs [13]. Class IV HDAC include only HDAC11, which biological function has still not been fully investigated [14]. HDACs play a fundamental role in regulation of cell proliferation, survival, and differentiation, while impaired HDAC functions are widely involved in many pathophysiological processes such as cancer, neurodegeneration or inflammations-related disorders [15–17]. However, the role of HDACs and their inhibitors in DCM pathophysiology has not been fully investigated.

Cardiomyogenic differentiation in vitro can be induced by various factors inducing intracellular and extracellular changes. Accumulating data suggest that cells cultivated in conditions mimicking natural environment in vivo experience less stress and obtain more relevant phenotype [18]. It was also shown that adult tissue-derived MSC, when cultured for a long period on plastic surface, change cell shape and responses to the external stimuli [19]. Moreover, the plastic surface does not mimic the environment in which cells grow in vivo: the extracellular matrix (ECM) chemistry, cell-to-ECM contacts and other paracrine factors are changed. Extensive studies indicate that ECM components and integrin receptors-related signaling are very important regulators of myocardial hypertrophy, dilated cardiomyopathy, and even heart failure [20–22]. One of the major proteins involved in the ECM-cell signaling cascade is the non-receptor protein tyrosine kinase (FAK), which can be activated by growth factors or through the integrin-mediated cell attachment to the extracellular surfaces [23]. FAK activation has been shown also to be strongly involved in the regulation of myocyte hypertrophy model systems in vitro [24,25] and in vivo [26,27] and might be an important element regulating intracellular signaling systems and cardiac gene expression.

Since the main component of heart tissue ECM is collagen I (~80%), which has a tunable mechanical elasticity and ability to crosslink with various additives, it makes collagen I a very attractive biomaterial for cardiomyogenic differentiation studies [28]. Hyaluronic acid (HA) is another most abundant non-proteoglycan, a polysaccharide component of the heart ECM, which in combination with collagen I was successfully used for the rat heart tissue engineering purposes [29]. Recent study also showed that polymer

2-methacryloyloxyethyl phosphorylcholine (MPC), beside its structural similarity to the cell membrane phospholipids, has anti-inflammatory properties that were used to improve survival of long-term corneal implants [30]. Moreover, the corneal implants with MPC showed faster eye nerve regeneration and recovery than implants without MPC. Beside the chemical composition of the heart ECM mimicking biomatrices, their mechanical properties as well as surface topography can also regulate cell–cell, cell–ECM interactions changing an intracellular signaling and stimulating myogenic differentiation in vitro [31–33]. However, the impact of hybrid collagen I, HA, and MPC hydrogels and linear surface topography on the expression of cardiomyogenic differentiation-related genes in human dilated myocardium-derived hmMSC has not been investigated.

Therefore, the main goal of this study was to investigate abilities to tilt the dilated myocardium-derived hmMSC to the cardiomyogenic differentiation. We have also investigated the levels of MSC origin biomarkers, growth intensity, and HDAC activity in dilated myocardium derived-MSC comparing to the hmMSC derived from the normal functioning myocardium in vitro. The HDAC activity in dilated hmMSC was significantly higher compared to the healthy ones, which has been suppressed by class I and II HDAC inhibitor SAHA (also known as Vorinostat) subsequently activating expression of two main cardiomyogenic differentiation-related genes alpha cardiac actin (*ACTC1*) and cardiac troponin T (*TNNT2*). The impact of hybrid collagen I hydrogels with or without hyaluronic acid and polymer 2-methacryloyloxyethyl phosphorylcholine on the expression of *ACTC1* and *TNNT2* and possible role of FAK as well as longitudinal cell growth mode has been investigated. Data of this study show that HDAC inhibitor SAHA alone or in combination with hybrid collagen I-based hydrogels positively affect dilated hmMSC differentiation to cardiomyogenic direction in vitro.

2. Results

2.1. Cell Isolation and Morphology

Human healthy and pathological (DCM myocardium-derived) hmMSC were isolated by explant outgrowth method and had typical stromal cell-like morphology (Figure 1A). hmMSC isolated from dilated human myocardium (Figure 1C) were bigger in shape and grew flatter on the plastic surface than hmMSC isolated from the healthy myocardium (Figure 1B). The size of healthy and dilated heart myocardium-derived hmMSC was calculated by evaluating cell attachment surface (μm^2) (Figure 1D).

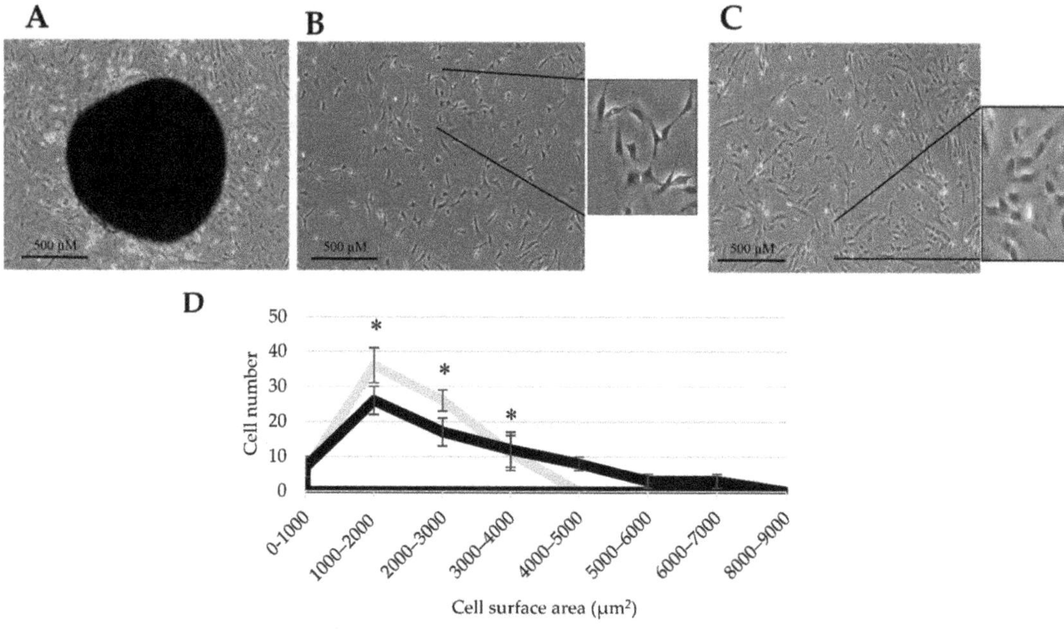

Figure 1. Morphology and size of healthy and pathological/dilated myocardium-derived hmMSC. (**A**) hmMSC outgrowth from human left ventricle myocardium biopsy material. (**B**) Healthy hmMSC in culture (2 weeks). (**C**) Pathological hmMSC in culture (2 weeks). (**D**) Cell size evaluation according to the cell attachment surface (μm^2) using ImageJ software. Data are shown as mean value $\pm$ SD. * Data are significant at $p \leq 0.05$ from not measuring not less than 60 cells ($n = 60$) of three patients from each group and comparing healthy and pathological cells. Statistical data were calculated using Excel software.

2.2. Characterization of Healthy and Dilated Myocardium-Derived hmMSC

Beside the morphology, we have also identified the MSC origin of healthy and pathological/dilated hmMSC by measuring levels of typical surface markers and differentiation potential to the main MSC directions (osteo, chondro, and adipo). The investigation of MSC typical biomarkers allows to evaluate the initial status of the healthy and dilated hmMSC and their further capabilities.

MSCs-typical surface markers were assessed by flow cytometer (Figure 2A). Healthy hmMSC were positive for CD29 (~98%), CD44 (~99%), CD90 (~87%), CD105 (~86%), and CD73 (~92%) and were nearly negative for CD45 (~1.1%), CD14 (~1.1%). Dilated myocardium-derived hmMSC (pathological) exhibited similar levels of surface markers: CD29 (~97%), CD44 (~90%) CD90 (~70%), CD105 (~85%), and CD73 (~78%).

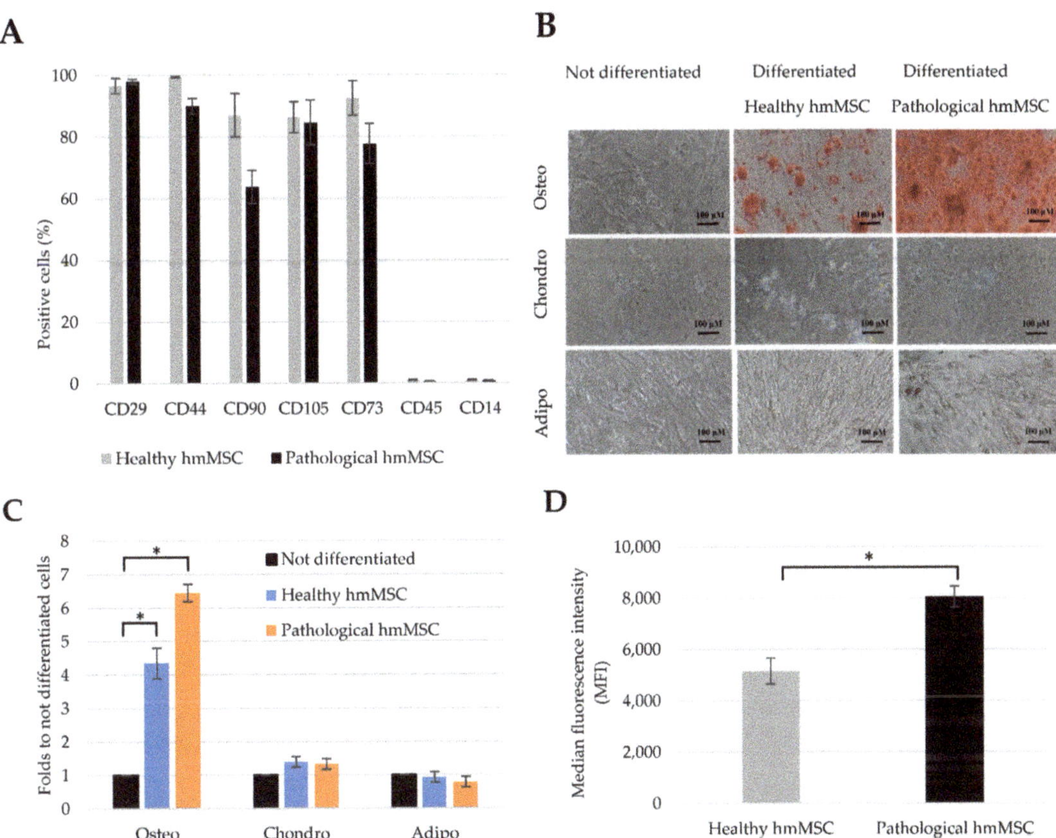

Figure 2. The identification of human healthy and pathological/dilated myocardium-derived hmMSC. (**A**) The levels of MSC-typical surface biomarkers on healthy and pathological hmMSC. (**B**) Qualitative evaluation of hmMSC differentiation towards osteo-, chondro-, and adipogenic directions (**C**) Spectrophotometric evaluation of osteo-, chondro-, and adipogenic differentiation. (**D**) Intracellular calcium level measured by Cal-520 dye. Data are shown as mean value ± SD. * Data are significant at $p \leq 0.05$ level from three repeats ($n = 3$) of three patients from each group comparing osteo differentiated and not differentiated cells or intracellular calcium in healthy and pathological hmMSC using Excel software.

The qualitative evaluation of differentiation potential of both types of the hmMSC to MSC-typical directions varied: cells showed osteo (strong), chondro (very weak), and adipo (almost not detected) (Figure 2B). Quantitative (spectrophotometric) evaluation of differentiation potential showed similar results (Figure 2C). Interestingly, both types of the hmMSC showed significantly increased differentiation to the osteo direction compared to the chondro and adipo (Figure 2C). The osteogenic differentiation was more prominent in the pathological hmMSC than in healthy MSC that could be related to the changed intracellular calcium level during DCM degeneration (Figure 2D). There is also a possibility that strong osteogenic differentiation of both types of the hmMSC is age-related sign: both cell types were isolated from elder men (55–65 years old) myocardiums that often show increased calcium deposits causing calcification and stiffening of the valve cusps in vivo. This part needs more detailed investigations.

2.3. Proliferation and Cardiomyogenic Differentiation of hmMSC on ECM Components-Precoated Plastic Surfaces

In order to understand, which type of hydrogel will be the best for the further cardiomyogenic differentiation studies, we investigated the proliferation of both types of the hmMSC on natural collagen I (0.2%) and fibronectin (2 ng/mL) precoated and not precoated plastic surfaces for nine days. Results indicate that healthy hmMSC proliferated faster than pathological hmMSC on all types of surfaces (Figure 3). The proliferation rate of both types of the cells was highest on natural collagen I precoated surface, compared to the plain plastic or fibronectin (Figure 3). Both types of the hmMSC showed exponential growth mode.

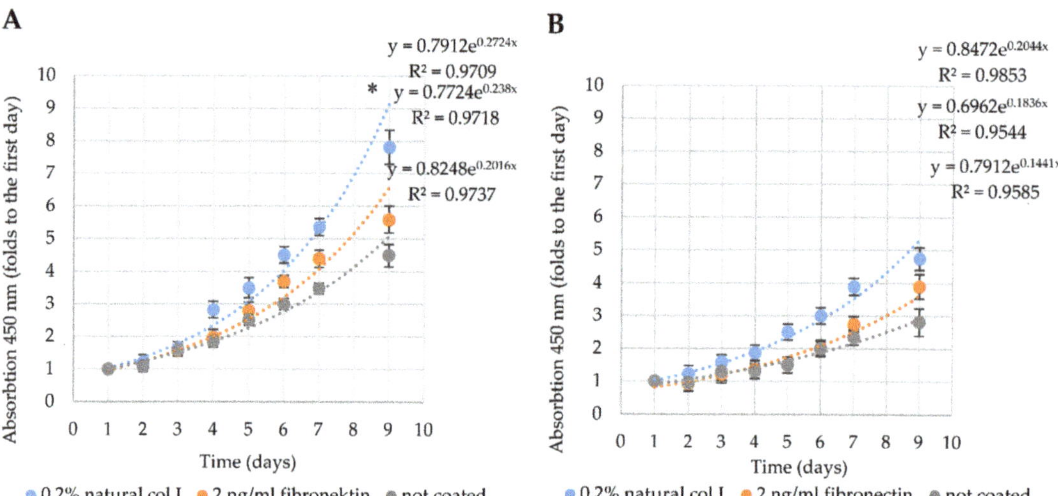

Figure 3. Proliferation of healthy and pathological/dilated myocardium-derived hmMSC on differently precoated and not precoated plastic surfaces. (**A**) Proliferation of healthy hmMSC. (**B**) Proliferation of pathological/dilated hmMSC. Both types of hmMSC were cultivated on surfaces precoated with 0.2% of natural collagen I, 2 ng/mL of fibronectin, and not precoated plastic surface for 9 days. Cell proliferation was measured by the CCK-8 kit. * Data are presented as mean of absorption (450 nm) $\pm$ SD and are significant at $p \leq 0.05$ comparing cells grown on natural collagen I precoated plastic surface to not precoated or precoated with fibronectin. Not less than three repeats ($n = 3$) of three patients from each group were measured.

Further, we investigated the total HDAC activity (Figure 4A) and the effect of HDAC inhibitor SAHA on it (Figure 4B). The HDAC activity was significantly higher in the pathological cells (almost 50%) compared to the healthy hmMSC (Figure 4A), and was significantly suppressed by 1 μM SAHA during three days of incubation (Figure 4B). In parallel, we investigated the impact of 1 μM SAHA on the expression of most typical cardiomyogenic differentiation-related genes in vitro alpha-cardiac actin (*ACTC1*) (Figure 4C) and cardiac troponin T (*TNNT2*) (Figure 4D) in healthy and pathological hmMSC growing on plastic surface. The expression of *ACTC1* and *TNNT2* in both types of hmMSCs exposed to 1 μM SAHA for 14 days was upregulated (Figure 4C,D). HDAC inhibitor SAHA most significantly affected cardiac gene expression in pathological hmMSC. Data show that dilated myocardium-derived hmMSC still retained ability to differentiate to the cardio myogenic direction, which can be stimulated in vitro by epigenetic regulator SAHA.

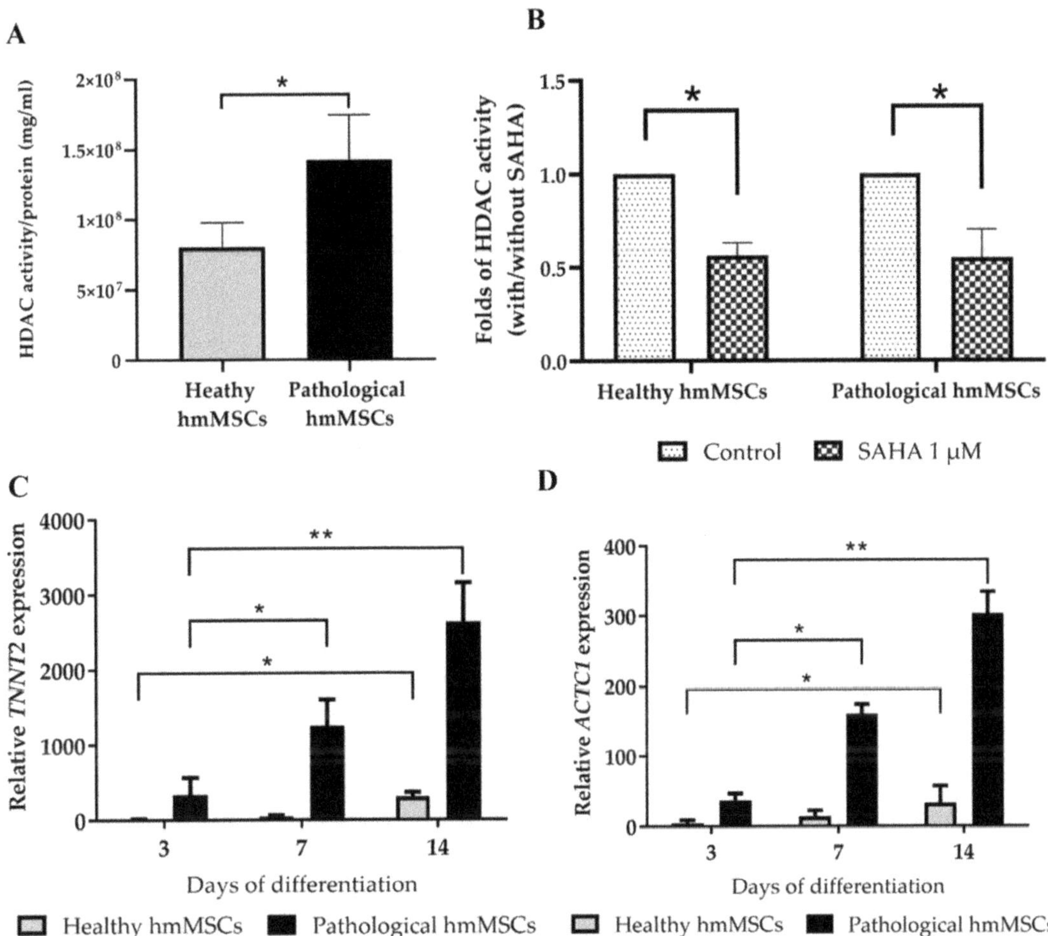

Figure 4. The effect of HDAC inhibitor SAHA on HDAC activity and cardiomyogenic differentiation of healthy and pathological/dilated myocardium-derived hmMSC. (**A**) HDAC activity in the healthy and pathological hmMSC. (**B**) The effect of SAHA on HDAC activity during 3 days of incubation. Expression of cardiomyogenic differentiation-specific genes: alpha-cardiac actin (*ACTC1*) (**C**) and cardiac troponin T (*TNNT2*). (**D**) The relative cardiac gene expression is shown as $2^{-\Delta Ctx}$2,000,000 and was normalized to the *ACTB* gene. Data are shown as mean $\pm$ SD and are significant at * $p \leq 0.05$, ** $p \leq 0.01$ levels from not less than three repeats ($n = 3$) of three patients from each group as calculated using the GraphPad Prism 6 software.

2.4. The Attachment and Growth of Healthy and Dilated Myocardium-Derived hmMSC on Hybrid Collagen I Hydrogels

Since both types of hmMSC showed better proliferation on natural collagen I-precoated plastic surface compared to the fibronectin or plain plastic surface, we investigated hmMSC attachment and survival on collagen type I (Col)-based hydrogels supplemented with HA and MPC during 24 h of cultivation (Figure 5). The healthy and pathological hmMSCs showed better adherence to the Col hydrogels and aligned growth mode (Figure 5A, arrows) compared to the cells attached to the Col-MPC and Col-HA hydrogels (Figure 5). Overall, healthy hmMSC slightly better adhered to the all types of hybrid hydrogels compared to the pathological hmMSC (Figure 5A,B).

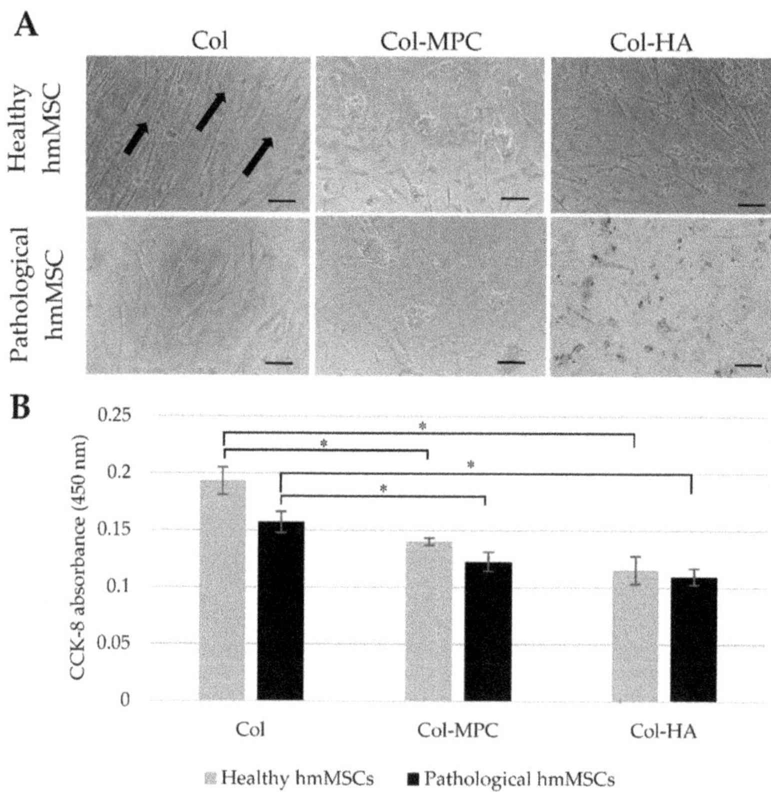

Figure 5. Attachment and survival of healthy and pathological/dilated myocardium-derived hmMSC on collagen I-based hydrogelss. (**A**) Healthy and pathological hmMSCs were allowed to attach to the collagen type I (Col) hydrogels, and collagen type I hydrogels with hyaluronic acid (Col-HA) and 2-methacryloyloxyethyl phosphorylcholine (Col-MPC) for 24 h. (**B**) Assessment of cell viability by CCK-8 absorption after 24 h in culture. * Significantly better attachment of hmMSC on Col hydrogels compared to the Col-MPC and Col-HA. Data are shown as absorption mean ± SD and are significant at $p \leq 0.05$ from not less than tree repeats (n = 3) of two patients from each group. Scale bars are 100 μm. Arrows show linear growth mode of healthy hmMSC on Col hydrogel.

Further, we have investigated expression of cardiomyogenic differentiation genes *ATCT1* and *TNNT2* in healthy and dilated hmMSC cultured on hybrid hydrogels with and without HDAC inhibitor SAHA for 14 days. For the cardiomyogenic differentiation, both types of the hmMSC were cultured to complete confluence before adding SAHA.

2.5. The Differentiation to Cardiomyogenic Direction of Healthy and Pathological hmMSC on Hybrid Collagen Type I Hydrogels

Since both types of hmMSC better attached and grew on Col hydrogel than on Col-MPC and Col-HA, we further investigated the impact of collagen I-based hydrogels on SAHA-stimulated cardiomyogenic differentiation of hmMSCs. In order to investigate the impact of hydrogels, healthy and pathological/dilated hmMSC were grown to monolayer on Col, Col-HA, and Col-MPC hydrogels and exposed to 1 μM SAHA for 3, 7, and 14 days (Figure 6). The differentiation of hmMSCs to cardiomyogenic direction was estimated by the expression of the *ACTC1* and *TNNT2* genes and compared to the cells cultivated on plastic.

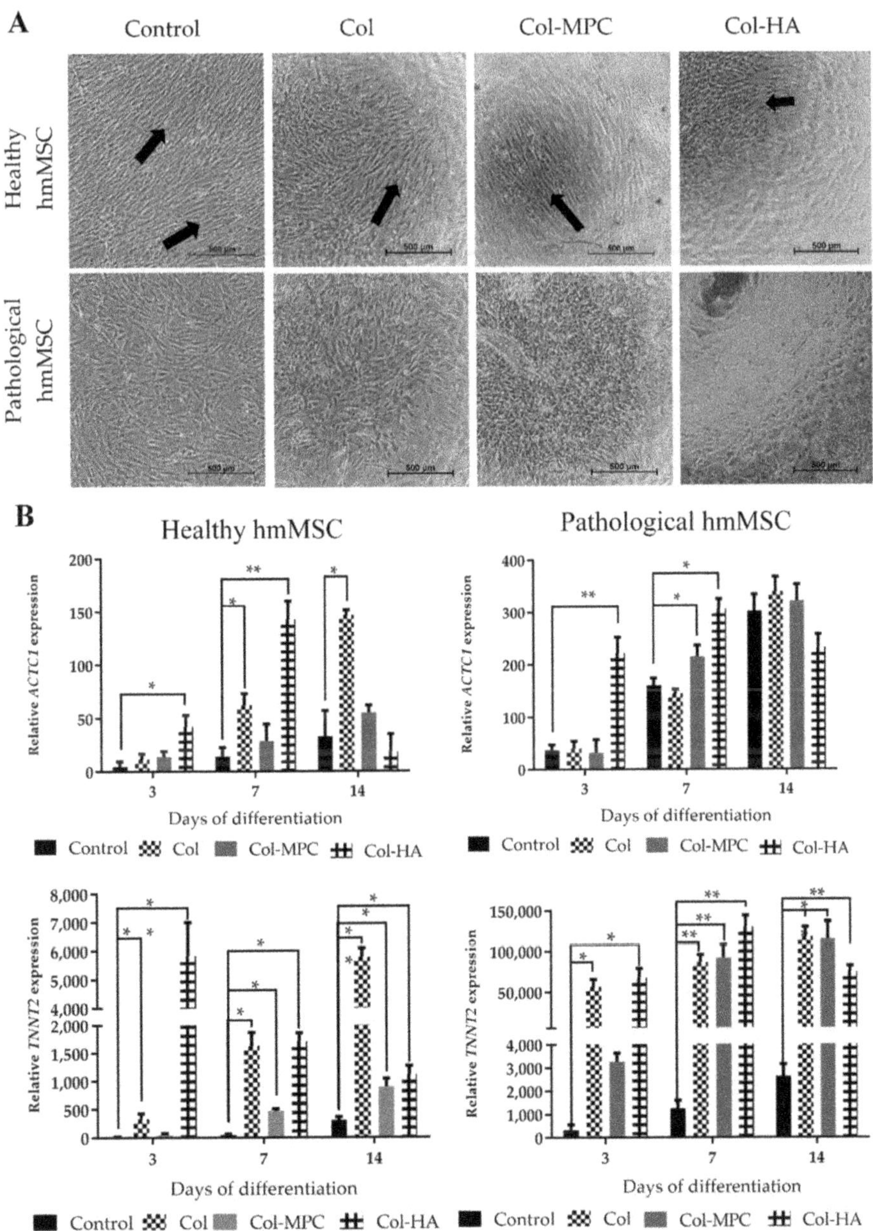

Figure 6. Expression of alpha-cardiac actin (*ACTC1*) and cardiac troponin T (*TNNT2*) genes in healthy and pathological/dilated myocardium-derived hmMSC during exposure to 1 µM SAHA for 3, 7, and 14 days. (**A**) Morphology of healthy and pathological hmMSC during 14 days of exposure to 1 µM SAHA on Col, Col-MPC, Col-HA hydrogels and plastic (Control). (**B**) Expression of *ACTC1* (*upper panels*) and *TNNT2* (*lower panels*) in healthy and pathological hmMSC grown on Col, Col-MPC, Col-HA hydrogels and plastic (Control). Relative cardiac gene expression is shown as $2^{-\Delta Ct} \times 2,000,000$ and was normalized to the *ACTB* gene. Data of 3 replicates ($n = 3$) of two patients in each group are shown with ±SD. Data are significant at * $p < 0.05$, ** $p < 0.01$ comparing expression of *ACTC1* and *TNNT2* cultivated on the hydrogels with the cells cultivated on plastic. The arrows indicate the longitudinal growth pattern of healthy hmMSC.

The healthy hmMSC maintained more linear growth pattern on hybrid collagen I-based hydrogels compared to the pathological hmMSC during 14 days of cultivation (Figure 6A, arrows). All employed hydrogels showed different impact on *ACTC1* expression over the time: the highest *ACTC1* expression in both types of hmMSC cultivated on Col-HA was at the seventh day (142.5 $\pm$ 17.7 and 306.6 $\pm$ 18.8-fold, for the healthy and pathological cells, respectively), whereas hmMSC cultivated on Col and Col-MPC hydrogels most intensively expressed *ACTC1* at the 14th day (146.5 $\pm$ 4.8 and 55.1 $\pm$ 7.2-fold for the healthy, and 339.2 $\pm$ 28.2 and 322.1 $\pm$ 31.1-fold for the pathological cells, respectively) compared to the cells cultivated on plastic (Figure 6B, upper panels). The lowest *ACTC1* activation was in the cells cultivated on plastic (control).

A similar tendency was observed in *TNNT2* expression: the level of *TNNT2* in healthy and pathological hmMSC cultivated on Col-HA matrices was strongly upregulated at the third (5830.6 $\pm$ 1174.6 and 67812.3 $\pm$ 11048.2-fold, respectively) and seventh (1708.1 $\pm$ 152.7 and 130072.9 $\pm$ 14245.2-fold, respectively) day, whereas cultivated on Col and Col-MPC hydrogels was stably increasing up to the 14th day compared to the cells cultivated on plastic (Figure 6B, lower panels). The lowest stimulation of *TNNT2* expression, similar to *ACTC1*, was in the cells cultivated on plastic (Figure 6B, lower panels). In general, the activation of *ACTC1* and *TNNT2* was stronger in pathological hmMSCs, compared to the healthy ones, suggesting retained regeneration potential of dilated myocardium.

2.6. The Expression of Focal Adhesion Kinase (FAK) in Healthy and Pathological hmMSC Grown on Hybrid Collagen I Hydrogels

In addition to *ACTC1* and *TNNT2*, the expression of FAK gene (*PTK2*) in the hmMSCs grown on the hybrid hydrogels has been estimated. FAK is a non-receptor protein tyrosine kinase participating in cell-ECM adherence signal transferring and stress responses. Data show that expression of FAK kinase gene *PTK2* was lower in both types of the hmMSC cultured on hybrid collagen I-based hydrogels compared to the plastic (Figure 7). In addition, the expression of *PTK2* in healthy hmMSC cultured on plastic was higher compared to the pathological hmMSC (Figure 7) suggesting healthy hmMSC being more sensitive to the adherence stress than pathological. Moreover, the lower expression of *PTK2* in both types of hmMSC grown on hydrogels can be explained by around 100-fold lower stiffness of all tested hydrogels: Col is ~140 kPa, Col-MPC is ~95, Col-HA is ~110 compared to the plastic (10,000 kPa) [31]. Reduced adherence-caused stress as well as hydrogel composition in the cells cultivated on hydrogels positively affected their cardiomyogenic differentiation-related gene expression.

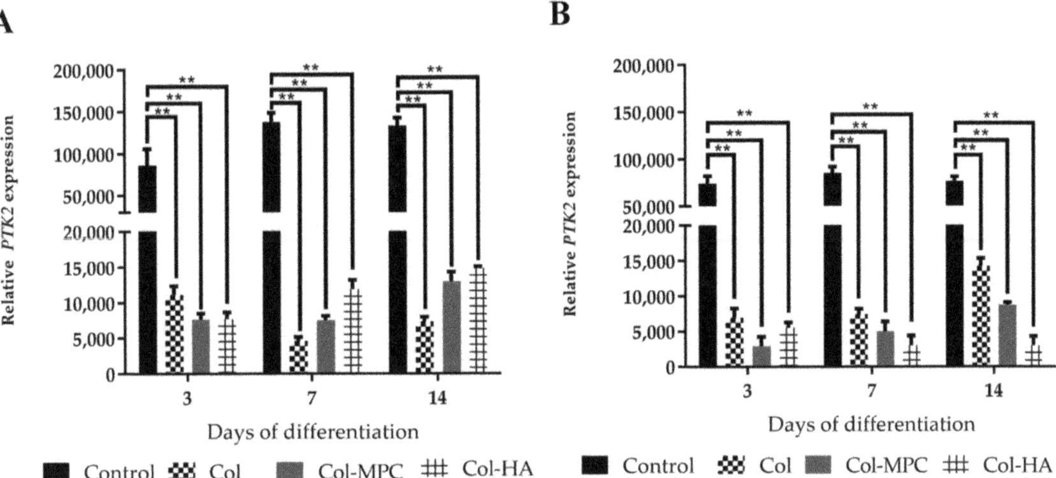

Figure 7. Expression of focal adhesion kinase gene *(PTK2)* in healthy and pathological/dilated myocardium-derived hmMSC cultivated on plastic and Col, Col-MPC and Col-HA hydrogels, and exposed to 1 µM SAHA for 3, 7, and 14 days. (**A**) Expression of focal adhesion kinase *(PTK2)* gene in the healthy hmMSC. (**B**) Expression of focal adhesion kinase gene *PTK2* in the pathological hmMSC. Relative *PTK2* gene expression is shown as $2^{-\Delta Ctx}2{,}000{,}000$ and was normalized to the *ACTB* gene. Data are shown as mean $\pm$ SD of three replicates ($n = 3$) from two patients of each group and are significant at ** $p < 0.01$ levels comparing expression of *PTK2* in the cells cultivated on the hydrogels with the cells cultivated on plastic.

2.7. The Impact of Longitudinal hmMSC Culturing on the Cardiomyogenic Differentiation-Related Proteins

In order to investigate the impact of linear pattern of hmMSC growth on the level of alpha-cardiac actin, the different width of fibronectin bioprinted stripes were used. The fibronectin lines were bioprinted on 1 cm diameter glass precoated with the low cell adherence material as shown in Figure 8 and described in the method part. The hmMSC cultivated on such biochips can attach only to the various topography of bioprinted fibronectin places.

Previous micrographs of this study (Figures 5, 6 and 8A, scheme) showed that healthy myocardium-derived hmMSCs grow in more neatly longitudinal manner than dilated myocardium-derived hmMSC. Both types of the hmMSC were cultivated on 20 ± 0.3 µm and 45 ± 1 µm wide fibronectin stripes-printed biochips (Figure 8B,C) for three days in IMDM medium without an additional stimuli and the level of alpha-cardiac actin was determined immunocytochemically (Figure 8C,D). In parallel, both types of hmMSC were cultivated on uniformly precoated fibronectin surface (Figure 8C, upper panels). The level of alpha-cardiac actin was strongly upregulated on 45 ± 1 µm fibronectin stripes compared to the 20 ± 0.3 µm stripes or entire coating (Figure 8D). Data show that artificially oriented longitudinal cell cultivation manner can increase the level of alpha-cardiac actin in both types of hmMSC without an additional stimulus. In addition, the expression of alpha-cardiac actin gene *ACTC1* in hmMSC on fibronectin lines was also evaluated by qPCR method using cell-to-Ct kit (Invitrogen™ TaqMan™ Gene Expression Cells-to-CT™ Kit), but the number of hmMSC was too low.

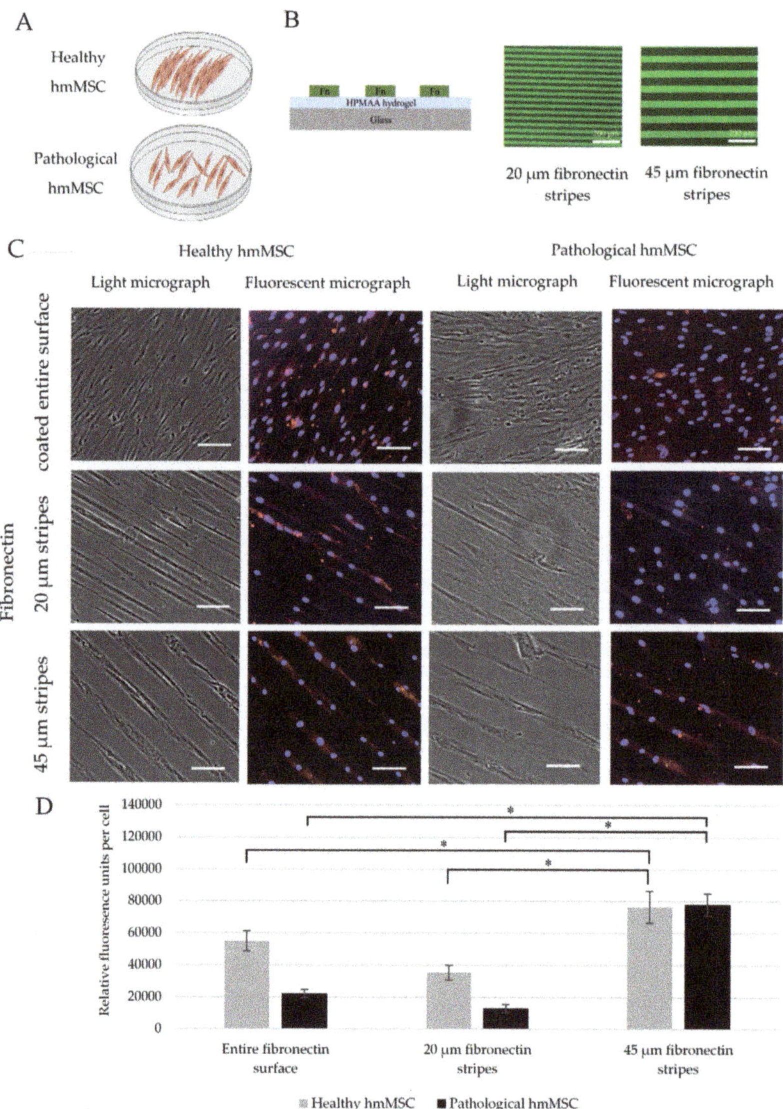

Figure 8. The level of alpha-cardiac actin in healthy and pathological/dilated myocardium-derived hmMSC cultured on uniformly-coated or linearly-printed fibronectin surfaces. (**A**) Schematic visualization of growth patterns of healthy and pathological hmMSC. (**B**) The schematic presentation of 20 ± 0.3 µm and 45 ± 1 µm fibronectin stripes. Fn—fibronectin, HPMAA—hydrophobic gel containing methyl methacrylic acid (MAA), 2-hydroxyethyl methacrylate (HEMA), and PEG methacrylate (PEG10MA). (**C**) Immunocytochemical micrographs of alpha-cardiac actin in healthy and pathological hmMSC. Upper panels—cells grown on uniformly fibronectin (2 ng/mL) coated surface, middle panel—on 20 ± 0.3 µm fibronectin-printed stripes, bottom panel—on 45 ± 1 µm µm fibronectin-printed stripes. Nuclei were stained with DAPI (blue). Scale is 100 µm. (**D**) Quantitative calculation of alpha-cardiac actin expression was evaluated by the fluorescence intensity (red) using ImageJ program and expressed as relative fluorescence units per cell. * Data are shown as mean of fluorescence intensity $\pm$ SD and are significant at $p \leq 0.05$ compared fluorescence of the hmMSC on the 45 µm strips with the hmMSC fluorescence on 20 µm and entire fibronectin coating. Not less than 3 micrographs ($n = 3$) of the cells from three different patients of each group were measured.

3. Discussion

It has long been thought that adult human heart is terminally differentiated organ without self-regenerating capabilities and only functions of cardiomyocytes (CM) were investigated [34]. However, it was shown that heart is composed of various types of the cells and even maintain slow regeneration potential [35,36]. It has been also shown that human cardiac regeneration may take place from the pre-existing CM [37,38], cardiospheres [39], or c-Kit$^+$ cardiac progenitor cells residing in the heart tissue [40,41]. In addition, the MSC isolated from different types of adult tissues can have specific features: human fetal heart MSC was shown to have cardiac immunophenotyping and differentiation potential to CM, endothelial, or smooth muscle cells [42–44]. Human heart myocardium-derived MSCs can be used to investigate the human heart pathologies and/or possible ways of heart regeneration [45,46]. Other authors also suggested that adult human heart tissue-derived MSC have better cardio regenerating capabilities compared to the bone marrow or other types of adult MSC [47]. So far, it is a huge demand of cardiac tissue regenerating biomodels investigating the endogenous heart repair mechanisms, intrinsic and extrinsic signaling pathways, epigenetic, ECM-based and other molecular mechanisms [48]. In this study, the hmMSC were isolated from human healthy and DCM myocardium biopsies and their MSC origin [49], proliferation and ability to differentiate to cardiomyogenic direction combining intracellular (HDAC inhibitor SAHA) and extracellular biomatrices (hybrid collagen I-based hydrogels) has been investigated.

For a long time, HDACs were used as a promising tool for the development of new HDAC inhibitors, as potential anticancer drugs [50]. Recent studies suggested different members of class I and class II HDACs being important in cardiovascular physiology and pathophysiology: the interaction of class IIa HDACs with myocyte enhancer factor-2 (MEF-2) was shown to be an essential regulator of mouse cardiac hypertrophy [51,52]. The increased level of class I HDACs, particularly HDAC2, was directly related to the cardiac hypertrophy [53] or caused fatal cardiac arrhythmias [54]. Therefore, the investigation of HDACs activity and their inhibitors in heart diseases, particularly DCM, can be of promising therapeutic interests [55]. In this study, the dilated myocardium-derived hmMSC showed different properties compared to the healthy hmMSC: they were bigger in size, slower proliferated, showed more pronounced osteogenic differentiation than chondro and adipo, but had similar MSC-typical surface markers [56,57]. The strong osteogenic differentiation of both types of the hmMSC might be an age-related sign: both cell types were isolated from elder men (55–65 years old) myocardiums that often show increased calcium deposits causing calcification and stiffening of the valve cusps in vivo [45]. Moreover, the dilated hmMSC had significantly (1.5-fold) higher level of HDACs activity, which could negatively affect dilated myocardium. This finding encouraged to investigate the effects of HDAC inhibitors on dilated myocardium-derived hmMSC.

Data show, suberoylanilide hydroxamic acid SAHA (Vorinostat), a class I and II pan-histone deacetylase inhibitor, in addition to suppressing HDAC activity in pathological/dilated hmMSC, at the same time stimulating expression of cardiomyogenic differentiation-related genes. In previous studies, SAHA has been clinically approved for the treatment of cutaneous T-cell lymphoma (CTCL) and was shown to exert anticancer activities in various other types of tumors [58]. In addition, recent our publication showed SAHA suppressing *HDAC1* and *HDAC2* expression in dilated myocardium-derived MSC, improved mitochondrial activity and energetic status of human dilated myocardium-derived hmMSC [46]. Similar findings also confirmed HDAC1 and HDAC2 being the main regulators of histone acetylation [59]. Other HDACs can also negatively affect cardiac functioning: overexpression of *HDAC4* in mouse cardiomyocytes reduced functional recovery after ischemia/reperfusion injury, while excessive activity of HDAC6 in diabetic rats increased their vulnerability to ischemia/reperfusion injury [60,61]. On the other hand, the HDACs inhibitors slowed down fibrosis in hypertensive rats, improved mouse left ventricle end-diastolic pressure and cardiovascular functioning [62,63]. Despite the positive effect of HDACs inhibitors on animal hearts, their effect on human heart cells, particularly derived from the dilated myocardium, has not been investigated. Data of this study show HDAC inhibitor SAHA, in addition to the suppression of HDAC activity, stimulating expression of cardio myogenic differentiation-related genes *ACTC1* and *TNNT2*, particularly in dilated hmMSC. Further, the new type of hybrid collagen I-based hydrogels with HA and MPC have been developed to enhance SAHA effect on cardiomyogenic gene expression in hmMSC.

Collagen is the most abundant structural component of various tissues such as skin, tendon, lung, vasculature, and other in mammals [64]. In the heart, the extracellular collagen matrix supports cardiomyocytes and coronary microcirculation, ventricle diastolic functioning, vasculature system and transmits generated force to pump blood [65]. Fibrillar collagen type I and III are the major components of the heart extracellular matrix (ECM) constituting around 80% and 11%, respectively [66]. As natural collagen I is a mechanical-load flexible material, the main its role in various tissues is to maintain a tissue tensile and mechanical compression properties [67]. Injectable collagen hydrogels [68], as well as decellularized intact heart tissue ECM [69], have been used to stabilize ventricle, limit adverse remodeling, and improve cardiac functioning after myocardial infarction in animal modelling systems. However, as heart is a highly physically loaded organ, the hydrogels used for the mechanical studies in vitro or implanted in vivo should withstand high compression and/or compaction [70]. As biological function of collagen lies predominantly in its mechanical properties, its use for the heart regeneration purposes is the most reliable [28]. Therefore, due to the abundance of collagen I in the heart, biodegradability and mechanical-load elasticity, the hybrid type of collagen I-based scaffolds can be a promising therapeutic tool with the wide applicability in heart engineering field.

Beside the collagens, HA is the most abundant non-proteoglycan polysaccharide component of the heart ECM. In the heart, HA is involved in cardiac development at embryonic stage, healing processes, and also participates in pathological conditions such as atherosclerosis and myocardial infarction [71]. Since healthy and pathological cells had similarly high levels (99% and 90%, respectively) of CD44, a known HA receptor, data suggest that HA component in Col-HA hydrogel might be a dominating factor for the quick (3–7 days) activation of cardiac differentiation-related genes *ACTC1* and *TNNT2* in both types of the hmMSC, while on Col and Col-MPC it required longer (up to 14 days) of incubation time. There is also a possibility that HA, due to its hydrophilicity, could interrupt tight collagen packaging and improve cell binding to the hydrogel and subsequent quicker gene expression [72]. In addition, the combination of collagen I hydrogels with MPC, a cell membrane mimicking polymer [73], can regulate cell attachment [74], and was shown to be successfully used in corneal tissue regeneration [30,75]. Recently, MPC has been also shown to have anti-inflammatory feature that is important for the long-term tissue regeneration purposes [30]. Despite the fact that various composition of hydrogel biomatrices have been employed to mimic heart ECM [76], increase vascularization [77], deliver small molecules [78], or conveniently transplanted cells [79], the interest in complex cardiac tissue engineering techniques is constantly growing.

In addition, the positive impact of hydrogels on increased expression of cardiomyogenic differentiation genes *ACTC1* and *TNNT2* can be also related to the lower hydrogels stiffness: (Col = 140.28 ± 7.20 kPa, Col-MPC = 96.09 ± 13.19 and Col-HA = 113.02 ± 35.05 kPa compared to the plastic surface (10,000 kPa), while most tissues in our body are much softer (1–50 kPa) [31]. The softer hydrogel surfaces compared to the plastic decreased expression of cell adhesion stress-related focal adhesion kinase (*PTK2*) leading to the higher expression of cardiomyogenic differentiation genes. The stiffness of hydrogels used in this study differed little from each other, therefore, we assume that the composition of hydrogels also could have a significant impact on cardiac gene expression. Mechanical signals from the extracellular surfaces to the cells are usually transferred through the integrin-based adhesion and distinct molecules, such as FAK, which are particularly important for the further intracellular signal transfer [80]. It was shown that FAK, a non-receptor protein tyrosine kinase, also regulates cell adhesion-related responses such as cell proliferation, differentiation, and surface composition [23,31]. In addition, studies in vivo showed higher FAK level in the volume-overloaded human heart, which inhibition prevented load-induced cardiac hypertrophy in mice [81,82]. It was also shown that inhibition of FAK attenuated fibrosis in post-myocardial infarction mice model that could have a promising pharmaceutical strategy [83]. Data of this study also suggest that FAK expression in hmMSC grown on hydrogels compared to the hmMSC on plastic was downregulated that positively influenced cardiac gene expression.

Finally, cardiac regenerating strategies can be expanded via a hydrogel substrate surface topography and/or chemistry. Broad spectrum of ECM-based proteins or their derivative peptides can be incorporated into hydrogels or printed on their surface targeting special cells or tissues [84]. However, the linear cell growth pattern has been mostly investigated in the pro-myogenic techniques of skeletal myoblast such as murine myoblast cell line C2C12 and showing improved myotube formation by linear growth mode [85]. The oriented myoblasts growth also increased myotube fusion index and enhanced response to electrical pulse stimulation [86]. In another study, the linear nanofiber scaffolds promoted axial growth and enhanced cardiac differentiation of induced pluripotent stem cells (iPSC)-derived cardiomyocytes by stimulating cardiac troponin T expression [87]. Although axial growth on nanofiber scaffolds were shown to promote cardiac differentiation, there are no data showing the effect of aligned growth mode on human dilated myocardium-derived hmMSC. In this study, the specially bioengineered biochips with different breadth of fibronectin lines showed that both healthy and dilated hmMSC naturally stretching on 45 ± 1 μm strips can upregulate alpha-cardiac actin without any additional stimulus.

Though, data of this study show that dilated myocardium-derived hmMSC were bigger in size, slower proliferated, but had almost the same levels of MSC-typical cell surface markers compared to the healthy hmMSC. In addition, dilated myocardium-derived hmMSC had significantly higher HDAC level compared to the healthy myocardium cells, which can be suppressed by HDAC inhibitor SAHA leading to the increased expression of cardiomyogenic differentiation-related genes such as alpha cardiac actin and troponin T. Data also show that healthy and dilated myocardium-derived hmMSC cultured on collagen I-based hybrid hydrogels and exposed to SAHA more intensively expressed cardiomyogenic differentiation-related genes compared to the cell grown on plastic. The artificially oriented linear culturing of dilated hmMSC also stimulated the level of alpha-cardiac actin. Overall, the combination of intracellular stimuli such as epigenetic modulator SAHA with the extracellular (heart ECM-based hydrogels or other biomatrices) could be a suitable tool for targeted regulation of cardiomyogenic regeneration mechanisms and better understanding of DCM therapeutic and/or regenerating possibilities.

4. Materials and Methods

4.1. Isolation and Cultivation of hmMSC

Human healthy and dilated myocardium-derived primary hmMSC were isolated from biopsy material after obtaining patient's (55–65 year-old men) written informed consent. The idiopathic dilated cardiomyopathy (NYHA group III) has been identified by the cardiologists according to the weak ventricle functioning (ejection fraction < 20%), diastolic diameter higher than 5.5 cm and other clinical parameters. Cell isolated from dilated myocardium were named as pathological or dilated hmMSC, while from the normally contracting myocardium, they were named healthy hmMSC. Not less than three patients' samples of each group were used. Tissue fragments were washed in PBS (Sigma Aldrich, St. Louis, MO, USA), non-muscle tissue was removed, and fragments were cut into small pieces (2–3 mm in diameter). Biopsy material was pretreated with trypsin/EDTA (Thermo Fisher Scientific, Waltham, MA, USA), for 10 min, placed on fibronectin-coated (PeproTech, Rocky Hill, NJ, USA) 6-well surface and submerged in IMDM medium (Thermo Fisher Scientific, Waltham, MA, USA) containing 20% FBS (Thermo Fisher Scientific, Waltham, MA, USA) and antibiotics (Thermo Fisher Scientific, Waltham, MA, USA). Cell migration was allowed for 7–14 days until cells reached confluence of outgrowing. Cells were cultivated at 37 °C and 5% CO_2 in a humidified atmosphere. The medium was changed twice a week. The confluent layer of cell outgrowth was trypsinized, washed with PBS and transferred to the 0.2% gelatine-coated cell culture flaks with IMDM, 10% of FBS, and antibiotics. hmMSC were used up to fifth passage number.

4.2. Evaluation of hmMSC Morphology

The size of healthy and pathological hmMSC was measured when cells reached a 50% confluence to better determine individual cell boundaries. Cell dimension was measured with the ImageJ software and analyzed using Microsoft Excel. The attachment area of individual hmMSC was marked and expressed as μm^2. The data are presented as cell size mean $\pm$ SD evaluating not less than 60 cells from three patients' biopsies of both types.

4.3. Identification of hmMSC Surface Markers

The MSC origin of human healthy and dilated myocardium-derived MSC (hmMSC) has been investigated as described elsewhere [43,46]. Briefly, cell growth media was aspired, washed with PBS and exposed to 0.25% trypsin-EDTA (Thermo Fisher Scientific, Waltham, MA, USA) for 1–2 min. After that, cells were suspended in 1% BSA (Sigma Aldrich, St. Louis, MO, USA) in PBS, counted, diluted to 0.5×10^6 cells/mL, transferred into special flow cytometer tubes and incubated with specific fluorochromes-conjugated antibodies on ice for 30 min. Next, cells were washed with 1% BSA in PBS, centrifuged at 600 g for 5 min and resuspended in 1% BSA in PBS. Isotype controls were prepared. The cell samples were analyzed on BD FACSAriaTM IIU flow cytometer using the BD FACSDiva software (BD Biosciences, San Jose, CA, USA). Flow cytometry data were presented as a percent of cells having markers (for cell surface markers) and as fluorescence intensity (MFI) for other flow measurements.

The antibodies used to evaluate hmMSC surface markers were: CD29 (Integrin beta-1-ImmunoglobulinG1-Allophycocyanin (1A-219-T100, Exbio, Praha, Czech Republic)), CD44 (homing cell adhesion molecule-ImmunoglobulinG2b-Fluorescein isothiocyanate (555478, BD Biosciences, San Jose, CA, USA)), CD90 (thymocyte differentiation antigen 1-ImmunoglobulinG1-Fluorescein isothiocyanate (328108, BioLegend, San Diego, CA, USA)), CD105 (endoglin-ImmunoglobulinG1-Allophycocyanin (MHCD10505, Thermo Fisher Scientific, Waltham, MA, USA)), CD73 (ecto-5'-nucleotidase-ImmunoglobulinG1-Fluorescein isothiocyanate (561254, BD Biosciences, San Jose, CA, USA)), CD45 (protein tyrosine phosphatase, receptor type, C -ImmunoglobulinG2a-Fluorescein isothiocyanate (sc-70686, Santa Cruz Biotechnology, Dallas, TX, USA)), and CD14 (macrophage protein, which binds lipopolysaccharide-ImmunoglobulinG2a-Allophycocyanin (BioLegend, San Diego, CA, USA)).

4.4. Differentiation of hmMSC to Adipo-, Osteo-, and Chondrogenic Directions

Capabilities of hmMSC to differentiate toward at least one of MSC-typical directions such as adipo-, osteo-, or chondrogenic were investigated as described earlier with some modifications [88].

For the adipogenic differentiation, cells were grown to full confluence and the growth medium was changed to adipogenic differentiation medium (DMEM medium with 1 g/L glucose (Thermo Fisher Scientific, Waltham, MA, USA), 20% FBS (Biochrom GmbH, Berlin, Germany), 1 µM dexamethasone (Sigma Aldrich, St. Louis, MO, USA), 0.5 µM IBMX (MyBioSource, San Diego, CA, USA), 60 µM indomethacin (Sigma Aldrich, St. Louis, MO, USA)), which was changed twice a week. Differentiation was performed for 21 days. Adipogenic differentiation was identified by staining of fat (triglycerides and other lipids) droplets with Oil Red: the cells were fixed with glutaraldehyde and incubated with 12 mM Oil Red dye (Carl Roth GmbH, Karlsruhe, Germany). Red fat droplets were photographed under a light microscope, dye was extracted with 100% isopropanol (Eurochemicals, Vilnius, Lithuania) and the absorbance was measured at 520 nm. Cells directed to the adipogenic differentiation were compared to the control cells grown in normal growth medium.

For the osteogenic differentiation studies, cells were grown to 80% of confluence, the medium was changed to osteogenic differentiation medium (DMEM with 4.5 g glucose (Thermo Fisher Scientific, Waltham, MA, USA), 10% FBS (Biochrom GmbH, Berlin, Germany), 0.1 μM dexamethasone (Sigma Aldrich, St. Louis, MO, USA), 50 μg/mL L-ascorbic acid (Santa Cruz Biotechnology, Dallas, TX, USA), 10 mM β-glycerophosphate (Santa Cruz Biotechnology, Dallas, TX, USA), and 3 mM NaH_2PO_4 (Sigma Aldrich, St. Louis, MO, USA). Cell differentiation was performed for 21 days with medium changes twice a week. After 21 days of differentiation, the cells were fixed with cold ethanol, stained with 40 mM Alizarin Red (Carl Roth GmbH, Karlsruhe, Germany) (stains the calcified bodies) dye at room temperature (RT), photographed under a light microscope. The dye was extracted with 10% cetylpyridinium chloride (Sigma Aldrich, St. Louis, MO, USA) and absorbance was measured at 562 nm. Differentiated cells were compared to the cells grown in normal growth medium.

For the chondrogenic differentiation the cells were grown to the complete monolayer and growth medium was changed to the chondrogenic differentiation medium (DMEM media (Thermo Fisher Scientific, Waltham, MA, USA) with 0.1 μM dexamethasone (Sigma Aldrich, St. Louis, MO, USA), 0.17 mM ascorbic acid (Santa Cruz Biotechnology, Dallas, TX, USA), 0.35 mM proline (Carl Roth GmbH, Karlsruhe, Germany), 1x insulin-transferrin-selenium (Thermo Fisher Scientific, Waltham, MA, USA), and 10 ng/mL TGF-β3 (Thermo Fisher Scientific, Waltham, MA, USA) for 21 days with medium changed twice a week. After 21 days, cells were fixed with 95% methanol (Sigma Aldrich, St. Louis, MO, USA) and stained with 3% Alcian Blue dye (Carl Roth GmbH, Karlsruhe, Germany). Alcian blue dye acts as a large cationic molecule with several positive charges that bind to the negative polysaccharides on glycosaminoglycans and stain them blue. Alcian blue dye also stains acidic mucous mucins (glycoproteins that enter the mucous membranes of the glands). This paint belongs to the group of multivalent dyes that are soluble in water. The blue color is due to the presence of copper in the molecule. After incubation, the cells were photographed, incubated with DMSO (Sigma Aldrich, St. Louis, MO, USA), and absorbance was measured at 678 nm. The chondrogenic differentiation with TGF-β3 were compared with chondrogenic differentiation without TGF-β3.

4.5. Proliferation of Healthy and DCM Myocardium-Derived hmMSC

Proliferation was measured using the CCK-8 (Cell Counting Kit-8) kit. Cells with CCK-8 reagent were incubated at 37 °C for 3 h and absorption was measured spectrophotometrically at 450 nm. Tetrazolium salt-based assays provide a reliable and convenient platform to assess cell viability, proliferation and/or to test toxicity. Tetrazolium salt WST-8 (2-(2-methoxy-4-nitrophenyl)-3-(4-nitrophenyl)-5-(2,4-disulfophenyl)-2H-tetrazolium, monosodium salt) in cell counting kit (CCK-8) is reduced by cellular dehydrogenases to give an orange color product (formazan) that is soluble in cell culture medium and absorption can be measured at 450 nm. Viability was determined at indicated time points according to the manufacturers' instructions (Dojindo, Kumamoto, JPN). Cell attachment to the gels 24 h after the seeding has been also measured by WST-8 reagent.

4.6. Evaluation of Calcium Concentration with Flow Cytometry

Cells (10^5/well) were seeded in 6-well plate. Then, trypsinized cells were washed with PBS and incubated with 1 μM of calcium specific fluorescent dye Cal-520 (Interchim, Montlucon, France) at 37 °C for 30 min. After staining, cells were washed twice with PBS and 1% BSA and suspended in 1% BSA. Flow cytometry analysis has been performed with BD FACSAria II (BD Biosciences, San Jose, CA, USA).

4.7. Detection of HDAC Activity

The cells were washed twice with PBS, lifted with trypsin-EDTA solution, washed with PBS, and lysed with lysing buffer (50 mM Tris-HCl, pH 7.5, 5% glycerol, 0.3% Triton X-100). Protein concentration was determined with the Pierce™ Modified Lowry Protein

Assay Kit. Cell samples were diluted with HDAC substrate buffer containing fluorescent HDAC substrate. HDAC activity was evaluated measuring fluorescence of universal HDAC substrate ((S)-tert-Butyl (6-acetamido-1-((4-methyl-2-oxo-2H-chromen-7-yl)amino)-1-oxohexan-2-yl)carbamate (BOC-Ac-Lys-AMC) diluted in HDAC substrate buffer (50 mM Tris- HCl, pH 8.1, 250 µM EDTA, 250 mM NaCl, 10% glycerol). The reaction was carried out at 30 °C for 30 min, where the HDACs cleave the acetyl group from the peptide substrate.

4.8. Preparation of Hydrogels

Collagen I-based hydrogels of 8.5% (w/w) (Col) were produced as previously described [89] using 4-(4,6-dimethoxy-1,3,5-triazin-2-yl)-4-methylmorpholinium chloride (DMTMM) (Merck KGaA, Darmstadt, Germany) as crosslinking agent. Shortly, 600 mg of collagen I (NMP collagen PS, Nippon Meatpackers, Ibaraki, Japan) aqueous solution of 12% (w/w) was mixed with 0.625 M 2-(N-morpholino)ethanesulfonic acid (MES) buffer (Merck KGaA, Darmstadt, Germany) within a glass syringe mixing system at RT. The calculated amount of a crosslinker was mixed in. The molar ratio of DMTMM to ε-amine groups of lysine in collagen molecule (Col-NH$_2$) was 1:1. After thoroughly mixing, hydrogel solution was casted between two glass plates to form 500 µm thickness sheet. The hydrogel was then left to stay overnight in 100% humidity at RT.

To enhance stability against enzymatic digestion 6.5% (w/w) of collagen I hydrogels, the interpenetrating with 2-Methacryloyloxyethyl phosphorylcholine (MPC) polymeric network were fabricated as described earlier [90]. Briefly, 12% (w/w) of collagen solution was buffered with MES in a syringe mixing system, then MPC solution (Merck KGaA, Darmstadt, Germany) in MES was added, Col:MPC (w/w) ratio was 2:1. After mixing thoroughly, poly(ethylene glycol) diacrylate (PEGDA) (Merck KGaA, Darmstadt, Germany) was added (PEGDA:MPC (w/w) 1:3). Next, calculated volumes of 4% (w/w) ammonium persulfate (APS) (Merck KGaA, Darmstadt, Germany) (APS:MPC (w/w) 0.03:1) and 2% (w/v) N,N,N′,N′-tetramethylethane-1,2-diamine (TEMED) (Merck KGaA, Darmstadt, Germany) (APS:TEMED (w/w) 1:0.77) solutions in MES were mixed in. Then, a calculated amount of DMTMM (DMTMM:Col-NH$_2$ (mol:mol) 1:1) was added. Hydrogel solution was mixed thoroughly and casted into 500 µm thickness sheets as described above. The final concentrations of Col and MPC in the hydrogel were 6.5% (w/w) and 3% (w/w), respectively.

For further enhancement of biointeractive properties of Col hydrogel, a new formulation of Col with hyaluronic acid (Merck KGaA, Darmstadt, Germany) was developed. The solution of 12% (w/w) Col was dispensed in a syringe mixing system as described above and a calculated amount of 1% (v/v) HA aqueous solution was mixed in. The final concentrations of Col and HA in the hydrogel were 8.5% (w/w) and 0.05% (w/w), respectively.

After crosslinking, mechanically robust and 500 µm thick Col, Col-HA and Col-MPC hydrogels sheets were trephined into 10 mm diameter cell culture substrate disks. Prior to use, hydrogel disks were kept refrigerated in PBS with traces of chloroform to maintain sterility. Tissue culture performance of hydrogel substrates was compared to conventional tissue culture plastic. All hydrogel parameters are shown in Supplementary Materials.

4.9. Preparation of Fibronectine Lines on Glass Chips

For the cell culture aligned growth experiments, in-house-made biochips were used. Biochips were prepared as described earlier [91]. Briefly, the biochips were 10 mm diameter and 130 µm thickness glass substrates coated with 34 ± 9 nm (in air) with 34 ± 9 nm (in air) poly(ethylene glycol) methacrylate hydrogel layer (PEGMA) (Merck KGaA, Darmstadt, Germany). Two different width fibronectins from Yo Proteins AB (Huddinge, Sweden) and bovine plasma fibronectin labelled with HiLyte™ Fluor 488 (FN-HiLyte™) (Cytoskeleton, Denver, CO, USA) lines 20 ± 0.3 µm and 45 ± 1 µm width with 20 ± 0.7 µm and 53 ± 1.2 µm spaces in between, respectively, were used. The fibronectin was microcontact printed on glass substrates (Merck KGaA, Darmstadt, Germany) poly(ethylene glycol) methacrylate hydrogel layer. The patterned surfaces were monitored using Olympus BX51 upright microscope (Olympus, Tokyo, Japan) equipped with 10xNA 0.3 air objective.

4.10. Cultivation of hmMSC on Hydrogels and Biochips

Prior to the cell cultivation, all hydrogels and biochips were washed (twice) with PBS in 37 °C and 5% CO_2 incubator for 30 min. Next, all substrates were primed by submerging in cell culture medium for 30 min. The medium was then aspirated, substrates were left to dry at RT for about 10 min and cells were plated on the top of hydrogel and left to attach in 37 °C and 5% CO_2 incubator for 24 h. Hydrogels were used to evaluate cell viability and cardiomyogenic differentiation for 14 days, while the expression of alpha cardiac actin was investigated on fibronectin stripes without any additional stimuli.

4.11. Cardiomyogenic Differentiation of Healthy and Dilated Myocardium-Derived hmMSC

For the cardiomyogenic differentiation, confluent hmMSC on hydrogels and on plastic were exposed to differentiation medium DMEM/F12 with 2% FBS, antibiotics and 1 μM of SAHA (Sigma Aldrich, St. Louis, MO, USA), and inhibitor of class I and II histone deacetylase (HDAC) inhibitor SAHA was added for 3, 7, and 14 days. Differentiation medium was changed every 2 days. Control cells were grown in DMEM/F12 medium (Thermo Fisher Scientific, Waltham, MA, USA) with 2% FBS and antibiotics without SAHA. Cardiomyogenic differentiation was evaluated by expression of alpha-cardiac actin and cardiac troponin T (cTnT) at gene and protein levels.

4.12. Gene Expression Levels

Cells grown on hydrogels and exposed to 1 μM SAHA for 14 days were washed, lysed in QIAzol® reagent (Qiagen GmbH, Hilden, Germany) and centrifuged at 14,000× g, 4 °C for 10 min. For one time-point gene expression measurement cells were collected from ten hydrogels. Supernatant was mixed with 200 μL of chloroform (Sigma Aldrich, St. Louis, MO, USA), pipetted and incubated for 10 min at RT. Next, mix was centrifuged at 12,000× g for 15 min at 4 °C. Subsequently, RNA was collected from the upper phase, 0.5 mL of isopropanol was added and mix was incubated for 10 min at RT. After centrifugation at 12,000× g, 4 °C for 10 min RNA-containing pellets were collected and washed in ethanol (Sigma Aldrich, St. Louis, MO, USA) twice. Finally, RNA was resuspended in water and its concentration was assessed by SpectraDrop™ technology (Molecular Devices, San Jose, CA, USA). Complimentary DNA was synthesized by using the High Capacity cDNA Reverse Transcription Kit® (Thermo Fisher Scientific, Waltham, MA, USA) according to the manufacturer's suggestions. Produced cDNA was stored at −20 °C until used. PCR was run by using Maxima Probe qPCR Master Mix (2X), ROX Solution kit (Thermo Fisher Scientific, Waltham, MA, USA) in triplicates on the cycler AriaMx Real-Time PCR System® (Agilent Technologies, Santa Clara, CA, USA). Temperature regimes were as follows: (1) denaturation and activation of Taq polymerase for 10 min at 95 °C, (2) denaturation for 15 s at 95 °C (40 cycles), (3) primer recognition and amplification for 60 s at 60 °C (40 cycles). The gene expression was calculated using $2^{-\Delta Ct}×10,000,000$ equation, where ΔCt is a subtraction of target gene Ct from housekeeping gene Ct, Ct is a threshold cycle. For the evaluation of the statistical significance $2^{-\Delta Ct}×2,000,000$ values were used. Gene expression was detected with these primers (Thermo Fisher Scientific, Waltham, MA, USA): *ACTC1* (Hs01109515_m1), *TNNT2* (Hs00943911_m1), and *PTK2* (Hs01056457_m1).

4.13. Immunofluorescence Assay

Cell cultures were fixed in 4% paraformaldehyde at RT for 15 min, washed with PBS, permeabilized with 0.1% Triton X-100 for 15 min and blocked with 1% BSA (Sigma Aldrich, St. Louis, MO, USA) at RT for 30 min as described in [92]. Primary antibodies to alpha-cardiac actin (GTX101876, GeneTex, Irvine, CA, USA) were added at dilution 1:50 and incubated at RT for 1 h. Washing steps with PBS and incubation with Alexa 549-conjugated secondary antibody (1:500) followed. Cells were visualized under fluorescent microscope

Nikon Eclipse TE2000-U (Nikon, Tokyo, JPN). Cells were observed with the 4x objective after 14 days of differentiation on hydrogels and the 20 × objective was used in the experiments with fibronectin lines. All micrographs were done with Digital Sight DS-2MBWc camera attached to the Nikon Eclipse TE2000-U microscope. Red and UV filters were used to detect alpha-cardiac actin and DAPI staining, respectively. After staining, red (alpha-cardiac actin) and blue (DAPI) fluorescence was captured and analyzed with the ImageJ program. First, the average of cell surface area was determined by following formula:

Average cell surface area (μm^2) = total surface covered by cells in one micrograph/number of cell nucleus stained with DAPI.

Second, we determined an average of fluorescence intensity of 1 μm^2 background surface of fibronectin stripes. Thus, the average cell fluorescence was calculated by following formula:

Average of normalized fluorescence (relative fluorescence units per 1 μm^2) = cell fluorescence in 1 μm^2 surface − background fluorescence of fibronectin stripes in 1 μm^2 surface.

Third, we evaluated relative fluorescence per cell area (μm^2) by following formula:

Relative fluorescence per cell area = average of cell surface area (μm^2) × average of normalized fluorescence (relative fluorescence units per 1 μm^2).

4.14. Statistical Analysis

Statistical analysis was performed using Excel (Microsoft Corporation, Redmont, WA, USA) and the Graphpad Prism 6.01 (GraphPad Software, San Diego, CA, USA) software. The differences between measurements were analyzed for their statistical significance using the independent-samples Student two-sided *t*-test. Results from three patients' cells of both types with not less than three-five repeats for each measurement were presented, unless indicated otherwise. Data were expressed as mean ± SD and were significant at * $p \leq 0.05$ comparing with corresponding control samples.

4.15. Ethical Approval

The study was approved by the local Bioethics Committee (license No. 158200-14-741-257). All patients gave written informed consent to investigate heart samples. The investigation conforms to the principles outlined in the Declaration of Helsinki.

5. Conclusions

Overall, data from this study show that expression of cardiomyogenic differentiation-related genes such as cardiac actin and troponin T in dilated myocardium-derived hmMSC can be efficiently stimulated by combining epigenetic stimulus such as HDAC inhibitor SAHA and extracellular collagen I-based biomatrices or applying spatial/topographical clues. Data show that combining extra- and intracellular stimuli can increase expression of cardiomyogenic differentiation-related genes in human healthy and dilated myocardium-derived hmMSC in vitro. Findings of this study also show that collagen I-based hydrogel substrates better mimicked natural cardiac environment than plastic, decreased expression of cell adhesion stress-related focal adhesion kinase, and activated cardiac genes. In addition, the composition of collagen I-based hydrogels is not less important for the cardiac gene expression than stiffness and showed time-depended cardiac gene upregulation. Moreover, data from this study show that human dilated myocardium-derived hmMSC retained ability to upregulate cardiomyogenic differentiation-related genes, which can be purposefully stimulated by complex means in vitro. Further in vitro and in vivo studies are needed to assess the feasibility of this therapeutic approach.

Supplementary Materials: The following are available online at https://www.mdpi.com/article/10.3390/ijms222312702/s1.

Author Contributions: Conceptualization, D.B., R.A. and R.V.; methodology, D.B., R.A., R.M., A.V., T.J., R.E., G.S. and V.C.; software, S.L. and R.M.; validation, K.R., V.J. and R.M.; formal analysis, S.L. and R.V.; investigation, R.M.; resources, K.R. and V.J.; data curation, R.M.; writing—original draft preparation, D.B., R.A. and R.V.; writing—review and editing, D.B., R.A., R.V. and S.L.; visualization, R.M.; supervision, D.B.; project administration, D.B.; funding acquisition, D.B. and R.A. All authors have read and agreed to the published version of the manuscript.

Funding: This research was funded by Lithuanian Research Council, project No. S-MIP-17-13.

Institutional Review Board Statement: The study was conducted according to the guidelines of the Declaration of Helsinki, and approved by the Vilnius Regional Biomedical Research Ethics Committee (protocol No. 158200-14-741-257, date of approval 10 June 2014).

Informed Consent Statement: Informed consent was obtained from all subjects involved in the study.

Data Availability Statement: Data are contained within the article or Supplementary Material.

Acknowledgments: Authors are thankful to Saule Valiuniene and Jolita Buivydaviciute for technical assistance.

Conflicts of Interest: R.V. is a majority shareholder and the CEO, and R.A., A.V., T.J., G.S., R.E., and V.C. are employees of Ferentis UAB.

Abbreviations

ACTB	Actin beta
ACTC1	Alpha cardiac actin 1
CCK-8	Cell counting kit 8
APS	Ammonium persulfate
ECM	Extracellular matrix
DAPI	40,6-diamidino-2-phenylindole
DCM	Dilated cardiomyopathy
DMTMM	4-(4,6-dimethoxy-1,3,5-triazin-2-yl)-4-methylmorpholinium chloride
HA	Hyaluronic acid
HDAC	Histone deacetylases
hmMSC	Human myocardium-derived mesenchymal stem cell
MEF-2	Myocyte enhancer factor-2
MES	2-(N-morpholino)ethanesulfonic acid
MPC	2-methacryloyloxyethyl phosphorylcholine
MSC	Mesenchymal stem cells
PEGDA	poly(ethylene glycol) diacrylate
SAHA	Suberoylanilide hydroxamic acid
TEMED	N,N,N′,N′-tetramethylethane-1,2-diamine
TNNT2	Cardiac Muscle Troponin T

References

1. Richardson, P.; McKenna, W.; Bristow, M.; Maisch, B.; Mautner, B.; O'Connell, J.; Olsen, E.; Thiene, G.; Goodwin, J.; Gyarfas, I.; et al. Report of the 1995 World Health Organization/International Society and Federation of Cardiology Task Force on the Definition and Classification of Cardiomyopathies. *Circulation* **1996**, *93*, 841–842. [CrossRef] [PubMed]
2. Lakdawala, N.K.; Winterfield, J.R.; Funke, B.H. Dilated Cardiomyopathy. *Circ. Arrhythmia Electrophysiol.* **2013**, *6*, 228–237. [CrossRef] [PubMed]
3. Ambrosy, A.P.; Fonarow, G.C.; Butler, J.; Chioncel, O.; Greene, S.J.; Vaduganathan, M.; Nodari, S.; Lam, C.S.P.; Sato, N.; Shah, A.N.; et al. The global health and economic burden of hospitalizations for heart failure: Les-sons learned from hospitalized heart failure registries. *J. Am. Coll. Cardiol.* **2014**, *63*, 1123–1133. [CrossRef]
4. Cahill, T.J.; Choudhury, R.P.; Riley, P.R. Heart regeneration and repair after myocardial infarction: Transla-tional opportunities for novel therapeutics. *Nat. Rev. Drug Discov.* **2017**, *16*, 699–717. [CrossRef]
5. Telukuntla, K.S.; Suncion, V.Y.; Schulman, I.H.; Hare, J.M. The Advancing Field of Cell-Based Therapy: Insights and Lessons from Clinical Trials. *J. Am. Hear. Assoc.* **2013**, *2*, e000338. [CrossRef]

6. Tang, J.-N.; Cores, J.; Huang, K.; Cui, X.; Luo, L.; Zhang, J.-Y.; Li, T.-S.; Qian, L.; Cheng, K. Concise Review: Is Cardiac Cell Therapy Dead? Embarrassing Trial Outcomes and New Directions for the Future. *STEM CELLS Transl. Med.* **2018**, *7*, 354–359. [CrossRef] [PubMed]

7. Segers, V.F.; Lee, R.T. Biomaterials to Enhance Stem Cell Function in the Heart. *Circ. Res.* **2011**, *109*, 910–922. [CrossRef]

8. Heallen, T.R.; Martin, J.F. Heart repair via cardiomyocyte-secreted vesicles. *Nat. Biomed. Eng.* **2018**, *2*, 271–272. [CrossRef]

9. Liu, B.; Lee, B.W.; Nakanishi, K.; Villasante, A.; Williamson, R.; Metz, J.; Kim, J.; Kanai, M.; Bi, L.; Brown, K.; et al. Cardiac recovery via extended cell-free delivery of extracellular vesicles secreted by cardiomyocytes derived from induced pluripotent stem cells. *Nat. Biomed. Eng.* **2018**, *2*, 293–303. [CrossRef]

10. Taunton, J.; Hassig, C.A.; Schreiber, S.L. A Mammalian Histone Deacetylase Related to the Yeast Transcriptional Regulator Rpd3p. *Science* **1996**, *272*, 408–411. [CrossRef] [PubMed]

11. Glozak, M.A.; Sengupta, N.; Zhang, X.; Seto, E. Acetylation and deacetylation of non-histone proteins. *Gene* **2005**, *363*, 15–23. [CrossRef]

12. Yoon, S.; Eom, G.H. HDAC and HDAC Inhibitor: From Cancer to Cardiovascular Diseases. *Chonnam Med. J.* **2016**, *52*, 1–11. [CrossRef] [PubMed]

13. Faller, Y.D.A.D.V. Transcription Regulation by Class III Histone Deacetylases (HDACs)—Sirtuins. *Transl. Oncogenom.* **2008**, *1*, 53–65. [CrossRef]

14. Liu, S.-S.; Wu, F.; Jin, Y.-M.; Chang, W.Q.; Xu, T.-M. HDAC11: A rising star in epigenetics. *Biomed. Pharmacother.* **2020**, *131*, 110607. [CrossRef]

15. Glauben, R.; Batra, A.; Stroh, T.; Erben, U.; Fedke, I.; Lehr, H.A.; Leoni, F.; Mascagni, P.; Dinarello, C.A.; Zeitz, M.; et al. Histone deacetylases: Novel targets for prevention of colitis-associated cancer in mice. *Gut* **2008**, *57*, 613–622. [CrossRef]

16. Simões-Pires, C.; Zwick, V.; Nurisso, A.; Schenker, E.; Carrupt, P.A.; Cuendet, M. HDAC6 as a target for neu-rodegenerative diseases: What makes it different from the other HDACs? *Mol. Neurodegener.* **2013**, *8*, 7. [CrossRef]

17. Adcock, I.M. HDAC inhibitors as anti-inflammatory agents. *Br. J. Pharmacol.* **2007**, *150*, 829–831. [CrossRef] [PubMed]

18. Caliari, S.; Burdick, J.A. A practical guide to hydrogels for cell culture. *Nat. Methods* **2016**, *13*, 405–414. [CrossRef]

19. Izadpanah, R.; Kaushal, D.; Kriedt, C.; Tsien, F.; Patel, B.; Dufour, J.; Bunnell, B.A. Long-Term In Vitro Expansion Alters the Biology of Adult Mesenchymal Stem Cells. *Cancer Res.* **2008**, *68*, 4229–4238. [CrossRef] [PubMed]

20. Valencik, M.L.; McDonald, J.A. Cardiac expression of a gain-of-function α5-integrin results in perinatal lethality. *Am. J. Physiol. Circ. Physiol.* **2001**, *280*, H361–H367. [CrossRef]

21. Shai, S.-Y.; Harpf, A.E.; Babbitt, C.J.; Jordan, M.C.; Fishbein, M.C.; Chen, J.; Omura, M.; Leil, T.; Becker, K.D.; Jiang, M.; et al. Cardiac Myocyte-Specific Excision of the β1 Integrin Gene Results in Myocardial Fibrosis and Cardiac Failure. *Circ. Res.* **2002**, *90*, 458–464. [CrossRef]

22. Valencik, M.L.; Keller, R.S.; Loftus, J.C.; McDonald, J.A. A lethal perinatal cardiac phenotype resulting from altered integrin function in cardiomyocytes. *J. Card. Fail.* **2002**, *8*, 262–272. [CrossRef]

23. Parsons, J.T. Focal adhesion kinase: The first ten years. *J. Cell Sci.* **2003**, *116*, 1409–1416. [CrossRef]

24. Dorn, G.W.; Force, T. Protein kinase cascades in the regulation of cardiac hypertrophy. *J. Clin. Investig.* **2005**, *115*, 527–537. [CrossRef]

25. Torsoni, A.S.; Constancio, S.S.; NadruzJr, W.; Hanks, S.K.; Franchini, K.G. Focal Adhesion Kinase Is Activated and Mediates the Early Hypertrophic Response to Stretch in Cardiac Myocytes. *Circ. Res.* **2003**, *93*, 140–147. [CrossRef]

26. Peng, X.; Kraus, M.S.; Wei, H.; Shen, T.-L.; Pariaut, R.; Alcaraz, A.; Ji, G.; Cheng, L.; Yang, Q.; Kotlikoff, M.I.; et al. Inactivation of focal adhesion kinase in cardiomyocytes promotes eccentric cardiac hypertrophy and fibrosis in mice. *J. Clin. Investig.* **2005**, *116*, 217–227. [CrossRef]

27. DiMichele, L.A.; Doherty, J.T.; Rojas, M.; Beggs, H.E.; Reichardt, L.F.; Mack, C.P.; Taylor, J.M. Myocyte-Restricted Focal Adhesion Kinase Deletion Attenuates Pressure Overload–Induced Hypertrophy. *Circ. Res.* **2006**, *99*, 636–645. [CrossRef] [PubMed]

28. Wenger, M.P.; Bozec, L.; Horton, M.A.; Mesquida, P. Mechanical Properties of Collagen Fibrils. *Biophys. J.* **2007**, *93*, 1255–1263. [CrossRef] [PubMed]

29. Nazir, R.; Bruyneel, A.; Carr, C.; Czernuszka, J. Collagen type I and hyaluronic acid based hybrid scaffolds for heart valve tissue engineering. *Biopolymers* **2019**, *110*, e23278. [CrossRef]

30. Simpson, F.C.; McTiernan, C.D.; Islam, M.M.; Buznyk, O.; Lewis, P.N.; Meek, K.M.; Haagdorens, M.; Audiger, C.; Lesage, S.; Gueriot, F.-X.; et al. Collagen analogs with phosphorylcholine are inflammation-suppressing scaffolds for corneal regeneration from alkali burns in mini-pigs. *Commun. Biol.* **2021**, *4*, 1–15. [CrossRef] [PubMed]

31. Skardal, A.; Mack, D.; Atala, A.; Soker, S. Substrate elasticity controls cell proliferation, surface marker expression and motile phenotype in amniotic fluid-derived stem cells. *J. Mech. Behav. Biomed. Mater.* **2013**, *17*, 307–316. [CrossRef] [PubMed]

32. Newman, P.; Galenano-Niño, J.L.; Graney, P.; Razal, J.; Minett, A.I.; Ribas, J.; Ovalle-Robles, R.; Biro, M.; Zreiqat, H. Relationship between nanotopographical alignment and stem cell fate with live imaging and shape analysis. *Sci. Rep.* **2016**, *6*, 37909. [CrossRef]

33. Almonacid Suarez, A.M.; Zhou, Q.; Van Rijn, P.; Harmsen, M.C. Directional topography gradients drive optimum alignment and differentiation of human myoblasts. *J. Tissue Eng. Regen. Med.* **2019**, *13*, 2234–2245. [CrossRef]

34. Mitcheson, J.S.; Hancox, J.C.; Levi, A.J. Cultured adult cardiac myocytes: Future applications, culture meth-ods, morphological and electrophysiological properties. *Cardiovasc. Res.* **1998**, *39*, 280–300. [CrossRef]

35. Bergmann, O.; Bhardwaj, R.D.; Bernard, S.; Zdunek, S.; Barnabé-Heider, F.; Walsh, S.; Zupicich, J.; Alkass, K.; Buchholz, B.A.; Druid, H.; et al. Evidence for Cardiomyocyte Renewal in Humans. *Science* **2009**, *324*, 98–102. [CrossRef] [PubMed]

36. Bergmann, O.; Zdunek, S.; Felker, A.; Salehpour, M.; Alkass, K.; Bernard, S.; Sjostrom, S.L.; Szewczykowska, M.; Jackowska, T.; Dos Remedios, C.; et al. Dynamics of Cell Generation and Turnover in the Human Heart. *Cell* **2015**, *161*, 1566–1575. [CrossRef] [PubMed]

37. Beltrami, A.P.; Urbanek, K.; Kajstura, J.; Yan, S.; Finato, N.; Bussani, R.; Nadal-Ginard, B.; Silvestri, F.; Leri, A.; Beltrami, C.A.; et al. Evidence That Human Cardiac Myocytes Divide after Myocardial Infarction. *N. Engl. J. Med.* **2001**, *344*, 1750–1757. [CrossRef]

38. Senyo, S.; Steinhauser, M.L.; Pizzimenti, C.L.; Yang, V.K.; Cai, L.; Wang, M.; Wu, T.-D.; Guerquin-Kern, J.-L.; Lechene, C.P.; Lee, R.T. Mammalian heart renewal by pre-existing cardiomyocytes. *Nature* **2012**, *493*, 433–436. [CrossRef]

39. Davis, D.R.; Smith, R.R.; Marbán, E. Human Cardiospheres are a Source of Stem Cells with Cardiomyogenic Potential. *Stem Cells* **2010**, *28*, 903–904. [CrossRef]

40. Bearzi, C.; Rota, M.; Hosoda, T.; Tillmanns, J.; Nascimbene, A.; De Angelis, A.; Yasuzawa-Amano, S.; Trofimova, I.; Siggins, R.W.; LeCapitaine, N.; et al. Human cardiac stem cells. *Proc. Natl. Acad. Sci. USA* **2007**, *104*, 14068–14073. [CrossRef]

41. Kulvinskiene, I.; Aldonyte, R.; Miksiunas, R.; Mobasheri, A.; Bironaite, D. Biomatrices for Heart Regeneration and Cardiac Tissue Modelling In Vitro. In *Cell Biology and Translational Medicine*; Springer: Cham, Switzerland, 2020; Volume 10, pp. 43–77. [CrossRef]

42. Orbay, H.; Tobita, M.; Mizuno, H. Mesenchymal Stem Cells Isolated from Adipose and Other Tissues: Basic Biological Properties and Clinical Applications. *Stem Cells Int.* **2012**, *2012*, 1–9. [CrossRef] [PubMed]

43. Han, I.; Kwon, B.-S.; Park, H.-K.; Kim, K.S. Differentiation Potential of Mesenchymal Stem Cells Is Related to Their Intrinsic Mechanical Properties. *Int. Neurourol. J.* **2017**, *21*, S24–S31. [CrossRef] [PubMed]

44. Garikipati, V.N.S.; Singh, S.P.; Mohanram, Y.; Gupta, A.K.; Kapoor, D.; Nityanand, S. Isolation and characterization of mesenchymal stem cells from human fetus heart. *PLoS ONE* **2018**, *13*, e0192244. [CrossRef]

45. Zabirnyk, A.; Perez, M.D.M.; Blasco, M.; Stensløkken, K.-O.; Ferrer, M.D.; Salcedo, C.; Vaage, J. A Novel Ex Vivo Model of Aortic Valve Calcification. A Preliminary Report. *Front. Pharmacol.* **2020**, *11*, 568764. [CrossRef]

46. Miksiunas, R.; Rucinskas, K.; Janusauskas, V.; Labeit, S.; Bironaite, D. Histone Deacetylase Inhibitor Suberoylanilide Hydroxamic Acid Improves Energetic Status and Cardiomyogenic Differentiation of Human Dilated Myocardium-Derived Primary Mesenchymal Cells. *Int. J. Mol. Sci.* **2020**, *21*, 4845. [CrossRef]

47. Rossini, A.; Frati, C.; Lagrasta, C.; Graiani, G.; Scopece, A.; Cavalli, S.; Musso, E.; Baccarin, M.; Di Segni, M.; Fagnoni, F.; et al. Human cardiac and bone marrow stromal cells exhibit distinctive properties related to their origin. *Cardiovasc. Res.* **2010**, *89*, 650–660. [CrossRef] [PubMed]

48. Horrobin, D.F. Modern biomedical research: An internally self-consistent universe with little contact with medical reality? *Nat. Rev. Drug Discov.* **2003**, *2*, 151–154. [CrossRef] [PubMed]

49. Dominici, M.; Le Blanc, K.; Mueller, I.; Slaper-Cortenbach, I.; Marini, F.C.; Krause, D.S.; Deans, R.J.; Keating, A.; Prockop, D.J.; Horwitz, E.M. Minimal criteria for defining multipotent mesenchymal stromal cells. The International Society for Cellular Therapy position statement. *Cytotherapy* **2006**, *8*, 315–317. [CrossRef]

50. Kim, H.J.; Bae, S.C. Histone deacetylase inhibitors: Molecular mechanisms of action and clinical trials as anti-cancer drugs. *Am. J. Transl. Res.* **2011**, *3*, 166–179. [PubMed]

51. McKinsey, T.A. Isoform-selective HDAC inhibitors: Closing in on translational medicine for the heart. *J. Mol. Cell. Cardiol.* **2011**, *51*, 491–496. [CrossRef]

52. Lu, J.; McKinsey, T.A.; Nicol, R.L.; Olson, E.N. Signal-dependent activation of the MEF2 transcription factor by dissociation from histone deacetylases. *Proc. Natl. Acad. Sci. USA* **2000**, *97*, 4070–4075. [CrossRef]

53. Eom, G.H.; Kook, H. Role of histone deacetylase 2 and its posttranslational modifications in cardiac hyper-trophy. *BMB Rep.* **2015**, *48*, 131–138. [CrossRef] [PubMed]

54. Monteforte, N.; Napolitano, C.; Priori, S.G. Genética y arritmias: Aplicaciones diagnósticas y pronósticas. *Rev. Esp. Cardiol.* **2012**, *65*, 278–286. [CrossRef]

55. Bush, E.W.; McKinsey, T.A. Targeting histone deacetylases for heart failure. *Expert Opin. Ther. Targets* **2009**, *13*, 767–784. [CrossRef]

56. da Silva Meirelles, L.; Chagastelles, P.C.; Nardi, N.B. Mesenchymal stem cells reside in virtually all post-natal organs and tissues. *J. Cell Sci.* **2006**, *119*, 2204–2213. [CrossRef] [PubMed]

57. Miller, J.D.; Weiss, R.M.; Heistad, D.D. Calcific Aortic Valve Stenosis: Methods, Models, and Mechanisms. *Circ. Res.* **2011**, *108*, 1392–1412. [CrossRef]

58. Zhao, Y.; Yu, D.; Wu, H.; Liu, H.; Zhou, H.; Gu, R.; Zhang, R.; Zhang, S.; Wu, G. Anticancer activity of SAHA, a potent histone deacetylase inhibitor, in NCI-H460 human large-cell lung carcinoma cells in vitro and in vivo. *Int. J. Oncol.* **2013**, *44*, 451–458. [CrossRef]

59. Montgomery, R.L.; Davis, C.A.; Potthoff, M.J.; Haberland, M.; Fielitz, J.; Qi, X.; Hill, J.A.; Richardson, J.A.; Olson, E.N. Histone deacetylases 1 and 2 redundantly regulate cardiac morphogenesis, growth, and contractility. *Genes Dev.* **2007**, *21*, 1790–1802. [CrossRef]

60. Zhang, L.; Wang, H.; Zhao, Y.; Wang, J.; Dubielecka, P.M.; Zhuang, S.; Qin, G.; Chin, Y.E.; Kao, R.L.; Zhao, T.C. Myocyte-specific overexpressing HDAC4 promotes myocardial ischemia/reperfusion injury. *Mol. Med.* **2018**, *24*, 1–10. [CrossRef]

61. Leng, Y.; Wu, Y.; Lei, S.; Zhou, B.; Qiu, Z.; Wang, K.; Xia, Z. Inhibition of HDAC6 Activity Alleviates Myocardial Ischemia/Reperfusion Injury in Diabetic Rats: Potential Role of Peroxiredoxin 1 Acetylation and Redox Regulation. *Oxid. Med. Cell. Longev.* **2018**, *2018*, 1–15. [CrossRef] [PubMed]

62. Nural-Guvener, H.F.; Zakharova, L.; Nimlos, J.; Popovic, S.; Mastroeni, D.; Gaballa, M.A. HDAC class I inhibitor, Mocetinostat, reverses cardiac fibrosis in heart failure and diminishes CD90+ cardiac myofibroblast activation. *Fibrogenes. Tissue Repair* **2014**, *7*, 10. [CrossRef] [PubMed]

63. Iyer, A.; Fenning, A.; Lim, J.; Le, G.T.; Reid, R.C.; Halili, M.A.; Fairlie, D.P.; Brown, L. Antifibrotic activity of an inhibitor of histone deacetylases in DOCA-salt hypertensive rats. *Br. J. Pharmacol.* **2010**, *159*, 1408–1417. [CrossRef] [PubMed]

64. Kadler, K.E.; Baldock, C.; Bella, J.; Boot-Handford, R. Collagens at a glance. *J. Cell Sci.* **2007**, *120*, 1955–1958. [CrossRef]

65. Horn, M.; Trafford, A.W. Aging and the cardiac collagen matrix: Novel mediators of fibrotic remodelling. *J. Mol. Cell. Cardiol.* **2015**, *93*, 175–185. [CrossRef] [PubMed]

66. de Souza, R. Aging of myocardial collagen. *Biogerontology* **2002**, *3*, 325–335. [CrossRef] [PubMed]

67. Janicki, J.S. Myocardial collagen remodeling and left ventricular diastolic function. *Braz. J. Med Biol. Res.* **1992**, *25*, 975–982.

68. McLaughlin, S.; McNeill, B.; Podrebarac, J.; Hosoyama, K.; Sedlakova, V.; Cron, G.; Smyth, D.; Seymour, R.; Goel, K.; Liang, W.; et al. Injectable human recombinant collagen matrices limit adverse remodeling and improve cardiac function after myocardial infarction. *Nat. Commun.* **2019**, *10*, 1–14. [CrossRef]

69. Bai, R.; Tian, L.; Li, Y.; Zhang, J.; Wei, Y.; Jin, Z.; Liu, Z.; Liu, H. Combining ECM Hydrogels of Cardiac Bioactivity with Stem Cells of High Cardiomyogenic Potential for Myocardial Repair. *Stem Cells Int.* **2019**, *2019*, 6708435. [CrossRef]

70. Lane, B.A.; Harmon, K.A.; Goodwin, R.L.; Yost, M.J.; Shazly, T.; Eberth, J.F. Constitutive modeling of compressible type-I collagen hydrogels. *Med Eng. Phys.* **2018**, *53*, 39–48. [CrossRef]

71. Bonafè, F.; Govoni, M.; Giordano, E.; Caldarera, C.M.; Guarnieri, C.; Muscari, C. Hyaluronan and cardiac regeneration. *J. Biomed. Sci.* **2014**, *21*, 1–13. [CrossRef]

72. John, H.E.; Price, R.D. Perspectives in the selection of hyaluronic acid fillers for facial wrinkles and aging skin. *Patient Prefer. Adherence* **2009**, *3*, 225–230.

73. Iwasaki, Y.; Ishihara, K. Cell membrane-inspired phospholipid polymers for developing medical devices with excellent biointerfaces. *Sci. Technol. Adv. Mater.* **2012**, *13*, 064101. [CrossRef]

74. Koike, M.; Kurosawa, H.; Amano, Y. A Round-bottom 96-well Polystyrene Plate Coated with 2-methacryloyloxyethyl Phosphorylcholine as an Effective Tool for Embryoid Body Formation. *Cytotechnology* **2005**, *47*, 3–10. [CrossRef] [PubMed]

75. Islam, M.M.; Cėpla, V.; He, C.; Edin, J.; Rakickas, T.; Kobuch, K.; Ruželė, Ž.; Jackson, W.B.; Rafat, M.; Lohmann, C.P.; et al. Functional fabrication of recombinant human collagen–phosphorylcholine hydrogels for regenerative medicine applications. *Acta Biomater.* **2015**, *12*, 70–80. [CrossRef] [PubMed]

76. Pedron, S.; van Lierop, S.; Horstman, P.; Penterman, R.; Broer, D.J.; Peeters, E. Stimuli Responsive Delivery Vehicles for Cardiac Microtissue Transplantation. *Adv. Funct. Mater.* **2011**, *21*, 1624–1630. [CrossRef]

77. Birla, R.K.; Borschel, G.H.; Dennis, R.G.; Brown, D.L. Myocardial Engineeringin Vivo: Formation and Characterization of Contractile, Vascularized Three-Dimensional Cardiac Tissue. *Tissue Eng.* **2005**, *11*, 803–813. [CrossRef]

78. Zustiak, S.P.; Wei, Y.; Leach, J.B. Protein-hydrogel interactions in tissue engineering: Mechanisms and applications. *Tissue Eng. Part B Rev.* **2013**, *19*, 160–171. [CrossRef]

79. DeForest, C.A.; Anseth, K.S. Advances in Bioactive Hydrogels to Probe and Direct Cell Fate. *Annu. Rev. Chem. Biomol. Eng.* **2012**, *3*, 421–444. [CrossRef]

80. Mitra, S.K.; Hanson, D.A.; Schlaepfer, D.D. Focal adhesion kinase: In command and control of cell motility. *Nat. Rev. Mol. Cell Biol.* **2005**, *6*, 56–68. [CrossRef]

81. Lopes, M.M.; Ribeiro, G.C.A.; Tornatore, T.F.; Clemente, C.F.M.Z.; Teixeira, V.P.A.; Franchini, K.G. Increased expression and phosphorylation of focal adhesion kinase correlates with dysfunction in the volume-overloaded human heart. *Clin. Sci.* **2007**, *113*, 195–204. [CrossRef]

82. Clemente, C.F.; Tornatore, T.F.; Theizen, T.H.; Deckmann, A.C.; Pereira, T.C.; Lopes-Cendes, I.; Souza, J.R.M.; Franchini, K.G. Targeting Focal Adhesion Kinase with Small Interfering RNA Prevents and Reverses Load-Induced Cardiac Hypertrophy in Mice. *Circ. Res.* **2007**, *101*, 1339–1348. [CrossRef] [PubMed]

83. Fan, G.-P.; Wang, W.; Zhao, H.; Cai, L.; Zhang, P.-D.; Yang, Z.-H.; Zhang, J.; Wang, X. Pharmacological Inhibition of Focal Adhesion Kinase Attenuates Cardiac Fibrosis in Mice Cardiac Fibroblast and Post-Myocardial-Infarction Models. *Cell. Physiol. Biochem.* **2015**, *37*, 515–526. [CrossRef] [PubMed]

84. Geckil, H.; Xu, F.; Zhang, X.; Moon, S.; Demirci, U. Engineering hydrogels as extracellular matrix mimics. *Nanomedicine* **2010**, *5*, 469–484. [CrossRef]

85. Altomare, L.; Riehle, M.; Gadegaard, N.; Tanzi, M.C.; Farè, S. Microcontact Printing of Fibronectin on a Biodegradable Polymeric Surface for Skeletal Muscle Cell Orientation. *Int. J. Artif. Organs* **2010**, *33*, 535–543. [CrossRef]

86. Bajaj, P.; Reddy, B.; Millet, L.; Wei, C.; Zorlutuna, P.; Bao, G.; Bashir, R. Patterning the differentiation of C2C12 skeletal myoblasts. *Integr. Biol.* **2011**, *3*, 897–909. [CrossRef] [PubMed]

87. Ding, M.; Andersson, H.; Martinsson, S.; Sabirsh, A.; Jonebring, A.; Wang, Q.-D.; Plowright, A.T.; Drowley, L. Aligned nanofiber scaffolds improve functionality of cardiomyocytes differentiated from human induced pluripotent stem cell-derived cardiac progenitor cells. *Sci. Rep.* **2020**, *10*, 13575. [CrossRef]

88. Erratum to: Mesenchymal Progenitors Able to Differentiate into Osteogenic, Chondrogenic, and/or Adipogenic Cells In Vitro Are Present in Most Primary Fibroblast-Like Cell Populations. *Stem Cells* **2014**, *32*, 1995–1996. [CrossRef]
89. Liu, Y.; Gan, L.; Carlsson, D.J.; Fagerholm, P.; Lagali, N.; Watsky, M.; Munger, R.; Hodge, W.G.; Priest, D.; Griffith, M. A Simple, Cross-linked Collagen Tissue Substitute for Corneal Implantation. *Investig. Opthalmol. Vis. Sci.* **2006**, *47*, 1869–1875. [CrossRef]
90. Liu, W.; Deng, C.; McLaughlin, C.R.; Fagerholm, P.; Lagali, N.S.; Heyne, B.; Scaiano, J.C.; Watsky, M.; Kato, Y.; Munger, R.; et al. Collagen–phosphorylcholine interpenetrating network hydrogels as corneal substitutes. *Biomaterials* **2009**, *30*, 1551–1559. [CrossRef]
91. Cėpla, V.; Rakickas, T.; Stankevičienė, G.; Mazėtytė-Godienė, A.; Baradoke, A.; Ruželė, Ž.; Valiokas, R. Photografting and Patterning of Poly(ethylene glycol) Methacrylate Hydrogel on Glass for Biochip Applications. *ACS Appl. Mater. Interfaces* **2020**, *12*, 32233–32246. [CrossRef]
92. Bironaite, D.; Brunk, U.; Venalis, A. Protective induction of Hsp70 in heat-stressed primary myoblasts: Involvement of MAPKs. *J. Cell. Biochem.* **2013**, *114*, 2024–2031. [CrossRef] [PubMed]

International Journal of
Molecular Sciences

MDPI

Article

SH3-Binding Glutamic Acid Rich-Deficiency Augments Apoptosis in Neonatal Rat Cardiomyocytes

Anushka Deshpande [1,2,3], Ankush Borlepawar [1,3], Alexandra Rosskopf [1], Derk Frank [1,3], Norbert Frey [2,4] and Ashraf Yusuf Rangrez [1,2,4,*]

1   Department of Internal Medicine III, Cardiology and Angiology, University Medical Center Schleswig-Holstein, Campus Kiel, 24105 Kiel, Germany; anushka.deshpande@uksh.de (A.D.); ankush.borlepawar@gmail.com (A.B.); alexander.rosskopf@uksh.de (A.R.); derk.frank@uksh.de (D.F.)

2   Department of Cardiology, Angiology and Pneumology, University Hospital Heidelberg, 69120 Heidelberg, Germany; norbert.frey@med.uni-heidelberg.de

3   DZHK (German Centre for Cardiovascular Research), Partner Site Hamburg/Kiel/Lübeck, 24105 Kiel, Germany

4   DZHK (German Centre for Cardiovascular Research), Partner Site Heidelberg/Mannheim, 69120 Heidelberg, Germany

*   Correspondence: ashrafyusuf.rangrez@med.uni-heidelberg.de

Citation: Deshpande, A.; Borlepawar, A.; Rosskopf, A.; Frank, D.; Frey, N.; Rangrez, A.Y. SH3-Binding Glutamic Acid Rich-Deficiency Augments Apoptosis in Neonatal Rat Cardiomyocytes. *Int. J. Mol. Sci.* **2021**, *22*, 11042. https://doi.org/10.3390/ijms222011042

Academic Editors: Nicole Wagner and Kay-Dietrich Wagner

Received: 9 August 2021
Accepted: 8 October 2021
Published: 13 October 2021

Publisher's Note: MDPI stays neutral with regard to jurisdictional claims in published maps and institutional affiliations.

Abstract: Congenital heart disease (CHD) is one of the most common birth defects in humans, present in around 40% of newborns with Down's syndrome (DS). The SH3 domain-binding glutamic acid-rich (SH3BGR) gene, which maps to the DS region, belongs to a gene family encoding a cluster of small thioredoxin-like proteins sharing SH3 domains. Although its expression is confined to the cardiac and skeletal muscle, the physiological role of SH3BGR in the heart is poorly understood. Interestingly, we observed a significant upregulation of SH3BGR in failing hearts of mice and human patients with hypertrophic cardiomyopathy. Along these lines, the overexpression of SH3BGR exhibited a significant increase in the expression of hypertrophic markers (Nppa and Nppb) and increased cell surface area in neonatal rat ventricular cardiomyocytes (NRVCMs), whereas its knockdown attenuated cellular hypertrophy. Mechanistically, using serum response factor (SRF) response element-driven luciferase assays in the presence or the absence of RhoA or its inhibitor, we found that the pro-hypertrophic effects of SH3BGR are mediated via the RhoA–SRF axis. Furthermore, SH3BGR knockdown resulted in the induction of apoptosis and reduced cell viability in NRVCMs via apoptotic Hippo–YAP signaling. Taking these results together, we here show that SH3BGR is vital for maintaining cytoskeletal integrity and cellular viability in NRVCMs through its modulation of the SRF/YAP signaling pathways.

Keywords: SH3BGR; cardiac hypertrophy; SRF signaling; Hippo signaling; apoptosis

1. Introduction

Trisomy 21, the presence of a supernumerary chromosome 21, results in one of the most common chromosomal abnormalities in humans commonly known as Down's syndrome, occurring in about 1 in 1000 babies born each year [1]. DS is among the most genetically complex conditions that are surprisingly compatible with human survival [2]. However, congenital heart disease is frequently described in DS patients as the main cause of death in newborns during the first two years of life. The spectrum of CHD patterns in DS has been subjected to variations due to genetic, social and geographical factors that vary worldwide [3]. Thus, DS is one of the commonest disorders with a huge medical and social cost burden and is associated with several phenotypes, including congenital heart defects, leukemia, Alzheimer's disease, Hirschsprung disease, etc. However, it is still unclear how and why the individuals that are affected by DS display variable phenotypes. Furthermore, the underlying mechanisms of these variable phenotypes are a challenge, making it essential to explore the elementary molecular mechanisms in depth [4].

Gene regulation is one of the mechanisms by which the expression of certain genes is altered, thereby resulting in disease conditions. SRF is one such multifaceted transcription factor that regulates the expression of a variety of genes by binding to the specific promoter sequence CarG box. SRF regulates the expression of a number of heart-specific genes during both embryonic development and pathogenesis [5–7]. A small GTPase RhoA is one of the co-inducers of hypertrophic SRF signaling in the heart [8]. The hypertrophic RhoA–SRF signaling pathway has been closely associated with the apoptotic Hippo–Yap pathway in cancer-associated fibroblasts, revealing mutual dependence. Herein, YAP activity is sensitive to SRF-induced contractility and SRF signaling responds to YAP-dependent TGFβ signaling, establishing an indirect crosstalk to control cytoskeletal dynamics [9]. In the heart, the Hippo–YAP pathway is a kinase cascade that inhibits the Yap transcriptional co-factor and controls organ size during development; epicardial-specific deletion of kinases Lats1/2, for example, is lethal at the embryonic level due to the failure in activating fibroblast differentiation, causing mutant embryos to form/undergo defective coronary vasculature remodeling [10]. This evolutionarily and functionally conserved pathway regulates the size and growth of the heart with crucial roles in cell proliferation, apoptosis and differentiation, thus having great potential for therapeutic manipulation to promote organ regeneration [11,12]. In relation with the highly compartmentalized Hippo pathway in cardiomyocytes, during cardiac stress, Mst1 and Lats2 are activated via a K-Ras–Rassf1A-dependent mechanism in mitochondria or through a NF2-dependent mechanism in the nucleus, respectively, where Mst1 stimulates the mitochondrial mechanism of apoptosis by phosphorylating Bcl-xL and Lats2 induces nuclear exit of Yap [13–15]. The activation of this canonical Hippo pathway leads to the stimulation of cell death and inhibition of compensatory hypertrophy by inhibition of Yap in cardiomyocytes [16,17].

Despite years of molecular biology-based cardiac research and circulatory understanding, several yet uncharacterized genes are expected to be associated with cardiomyopathies. Towards this end, we performed expressed sequence tags (EST)-based bioinformatic screening of genetic databases of heart and skeletal muscle and discovered several novel genes, one of which is SH3 domain-binding glutamic acid-rich (SH3BGR). It belongs to a gene family composed of SH3BGR, SH3BGRL, SH3BGRL2 and SH3BGRL3, which encode a cluster of small thioredoxin-like proteins and shares a Src homology 3 (SH3) domain (Supplementary Figure S1A) [18–22]. SH3BGR, located in the DS chromosomal region, was first reported by Scartezzini et al. over two decades ago [23] and was, interestingly, later found to be expressed in the earliest stages of mouse heart development [24]. Furthermore, transgenic mice with an FVB (friend leukemia virus B) background overexpressing SH3BGR in the heart did not affect cardiac morphogenesis; however, the fate of these mouse hearts at adult stages is not reported [25]. Thus, we believe that the potential role of SH3BGR in cardiomyocytes is still elusive. We observed significant upregulation of SH3BGR in the hearts of human patients suffering cardiac hypertrophy and a mouse model of heart failure due to transverse aortic constriction, consequently pointing towards its potential involvement in cardiac hypertrophy and associated modalities. Thus, in the current manuscript, we aim at characterizing the molecular functions of SH3BGR using gain- and loss-of-function approaches in neonatal rat ventricular cardiomyocytes.

2. Results

2.1. SH3BGR Is Confined to Striated Muscle and Upregulated in Cardiac Hypertrophy

SH3BGR was first reported in association with the critical region for Down's syndrome on chromosome 21 [23,26]. Since then, not much is known about the protein nor its role in cardiac pathophysiology, making it an unusual target to study. In the quest to find a potential role of this protein, we checked its expression in different mouse tissues and found it to be confined to the heart as well as to skeletal muscle, indicating a striated muscle-restricted presence (Figure 1A,B and Supplementary Figure S1B,C). Further, we observed an upregulation of SH3BGR protein levels in the hearts of human patients who suffer from cardiac hypertrophy (as compared to non-failing (NF) human hearts) (Figure 1C,D) and in the mouse hearts suffering cardiac hypertrophy due to biomechanical pressure overload (induced by transverse aortic constriction (TAC)) (Figure 1E,F). Furthermore, endogenous SH3BGR was found to be present at the sarcomere, co-localizing with sarcomeric α-actinin (Figure 1G). Altogether, striated muscle-specific expression, coupled with sarcomeric localization and upregulated protein levels in cardiac hypertrophy, indicates SH3BGR plays an important role in cardiac pathophysiology.

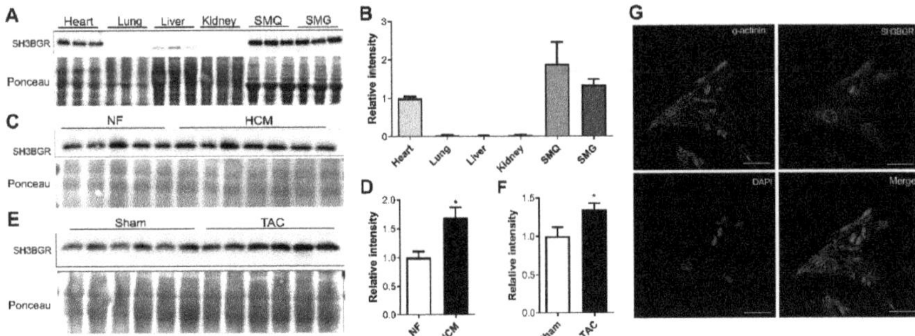

Figure 1. Expression pattern of SH3BGR. SH3BGR expression was observed to be confined to heart and skeletal muscle at protein level from mouse tissue lysates as shown in (**A**); its densitometric analysis is shown in (**B**) (*n* = 3). (**C**) SH3BGR was identified to be upregulated in human patient heart samples suffering from cardiac hypertrophy (HCM, *n* = 7) as compared to non-failing (NF, *n* = 5) samples. Its densitometric analysis is shown in (**D**). SH3BGR upregulation was also observed in mouse hearts subjected to TAC surgery as compared to SHAM as shown in (**E**); its densitometric analysis is shown in (**F**) (*n* = 6). (**G**) Immunofluorescence microscopy suggests sarcomeric and perinuclear localization of SH3BGR in NRVCMs. Statistical calculations were carried out using a two-tailed Student's *t*-test. *, $p < 0.05$; SMQ, skeletal muscle quadriceps; SMG, skeletal muscle gracilis; TAC, transverse aortic constriction.

2.2. SH3BGR Induces Cellular Hypertrophy in NRVCMs

To further assess the impact of elevated levels of SH3BGR, which is also observed in DS, we overexpressed SH3BGR in NRVCMs (Supplementary Figure S2A,C); this resulted in the induction of fetal genes, natriuretic peptides A and B (Nppa and Nppb) (Figure 2A) and increased total cell surface area (Figure 2B,C), thereby suggesting that the overexpression is responsible for the induction of hypertrophy in vitro. Contrastingly, SH3BGR knockdown (Supplementary Figure S2B,C) significantly reduced the levels of the hypertrophic markers NppA and NppB, coupled with further reduction in the cell surface area, compared to the control condition (Figure 2D–F).

2.3. SH3BGR Regulates RhoA–SRF Signaling in NRVCMs

The serum response factor (SRF) is one of the major transcription factors responsible for cardiomyocyte maturation, structural stability and pathological hypertrophy [8,27]. It plays a significant role in the transcriptional activation of natriuretic peptides and cardiac structural genes that form the core structure of the sarcomere, such as myosin heavy chain 6, 7 (myh 6, 7), myosin light chain 2 (myl2), cardiac alpha actin (ACTC1), etc. Interestingly, in terms of mechanistic relevance of our findings, we explored the Harmonizome, a collection of processed datasets gathered to serve and mine knowledge about genes and proteins, which revealed SRF as one of its transcription factors [28]. Thus, we hypothesized that SH3BGR likely induces cardiomyocyte hypertrophy via SRF signaling in vitro. To test this hypothesis, we studied the effect of SH3BGR overexpression and knockdown on SRF signaling using the SRF-response element-driven firefly luciferase assay/activity. In line with cellular hypertrophy data (Figure 2A–F), we observed a strong induction or inhibition of SRF activity upon SH3BGR overexpression or knockdown, respectively (Figure 3A,B). Interestingly, few of the SH3-domain containing proteins, namely, Tuba, SH3BP1, etc., have earlier been shown to mediate Rho-GTPase signaling, where, RhoA is one of the potent modulators of SRF signaling in the heart [29,30]. Thus, to further dissect the mechanistic insights, we performed a series of luciferase assays in several combinations with overexpression and knockdown of SH3BGR in the presence of RhoA or C3-transferase, a RhoA inhibitor. Our luciferase assay data indicate that the modulation of SRF signaling via SH3BGR is RhoA-mediated, since we observed the synergistic effect of RhoA on the activation of SH3BGR-driven SRF-activity, whereas the presence of C3-transferase abrogated this activation (Figure 3C,D). In contrast, SH3BGR knockdown severely hampered RhoA-mediated induction of SRF signaling (Figure 3E). Similar effects were also observed on the expression levels of the hypertrophic markers Nppa and Nppb (Figure 3F,G). As SH3BGR seems to hamper SRF activity, we investigated its effects on SRF downstream signaling. Moreover, we also observed significant downregulation of several downstream targets of SRF, such as Myh6, Myh7, Myl2, Dystrophin, Actc1 and Acta1, upon SH3BGR knockdown (Supplementary Figure S3A). However, the overexpression of SH3BGR, on the other hand, did not have a significant effect on these SRF target genes (Supplementary Figure S3B). Taken together, our data indicate that SH3BGR induces RhoA-mediated SRF signaling in NRVCMs.

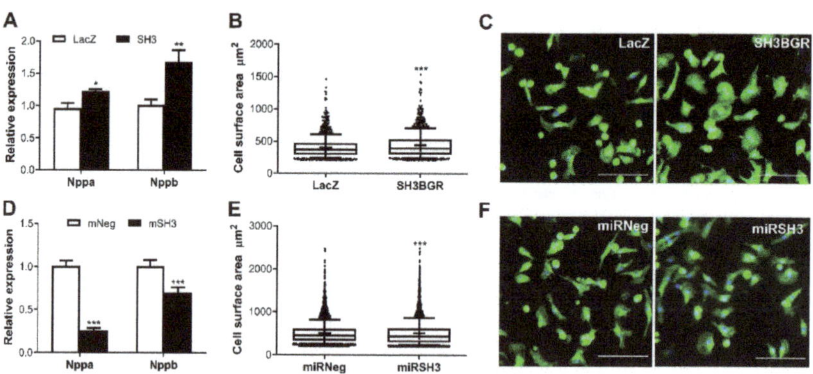

Figure 2. Effect of SH3BGR ablation on hypertrophy in vitro. (**A**) Overexpression of SH3BGR in NRVCMs upregulated fetal genes Nppa and Nppb compared to LacZ control (*n* = 3). (**B**) In line with these results, an increase in cell surface area of NRVCMs was also observed as seen in (**B**); representative images are depicted in (**C**). Contrastingly, on SH3BGR knockdown, this hypertrophic induction was abrogated observed by downregulation of hypertrophic markers (**D**) and reduced cell surface area (**E,F**) in miRSH3 condition as compared to miRNeg. Statistical calculations were carried out using the Student's *t*-test. *, *p* < 0.05; **, *p* < 0.01; ***, *p* < 0.001; SH3, SH3BGR; miRSH3, miRSH3BGR; Nppa, natriuretic peptide A; Nppb, natriuretic peptide B.

2.4. SH3BGR Knockdown Affects NRVCM-Viability and Induces Apoptosis via HIPPO Signaling

As recent literature postulated SH3BGRL2, a homolog of SH3BGR, to affect the Hippo signaling pathway in renal cell carcinoma, we aimed to find whether SH3BGR affects Hippo signaling in neonatal cardiomyocytes [31]. Intriguingly, SH3BGR knockdown significantly upregulated LATS1 (Large tumor suppressor kinase 1), whereas the levels of its phosphorylated form, i.e., pLATS1, were dramatically reduced (Figure 4A,B). In combination, YAP (Yes1-associated transcriptional regulator) protein levels were strongly increased (Figure 4A,B), suggesting the Hippo pathway to be functionally turned off in the cytoplasm, thereby facilitating the translocation of YAP into the nucleus. This translocated nuclear YAP transcriptionally activates a plethora of pro-apoptotic genes that culminate into the cleavage of executionary caspases-3 and -7 (Figure 4C,D) with expectedly reduced cell viability (Figure 4E). Nevertheless, SH3BGR overexpression did not alter Hippo signaling (except for the moderate increase in total YAP levels) and apoptotic markers (Supplementary Figure S4A–E). Overall, our data suggest that SH3BGR deficiency negatively affects RhoA–SRF and Hippo signaling, resulting in cardiomyocyte hypertrophy and apoptosis, respectively, whereas the overexpression of SH3BGR though robustly activated cellular hypertrophy and SRF signaling, did not affect Hippo signaling and associated cell viability (Figure 5).

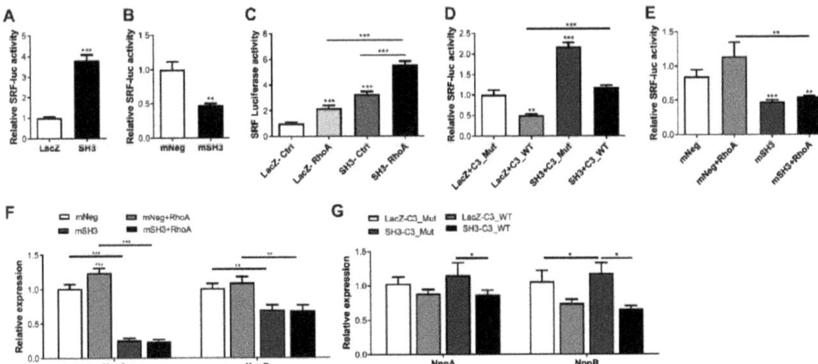

Figure 3. Effect of SH3BGR on SRF signaling-SH3BGR overexpression significantly induced SRF response element-driven luciferase activity (**A**), whereas its knockdown resulted in the inhibition of luciferase activity (**B**). Interestingly, luciferase activity in the presence of RhoA was significantly higher for SH3BGR than LacZ or SH3BGR alone (**C**), whereas the presence of C3 transferase, a RhoA inhibitor, significantly attenuated SH3BGR-mediated luciferase activity (**D**). Similarly, SH3BGR knockdown abrogated the activation of SRF luciferase activity due to RhoA (**E**). These effects were also reflected in the expression of fetal genes, where either knockdown of SH3BGR (**F**) or inhibition of C3 transferase (**G**) reduced the expression levels of Nppa and Nppb. Statistical calculations were carried out using a two-tailed Student's t-test or two-way ANOVA. *, $p < 0.05$; **, $p < 0.01$; ***, $p < 0.001$; mNeg, miRNeg; mSH3, miRSH3BGR; NppA, Natriuretic peptide A; NppB, Natriuretic peptide B; C3Mut, dominant-negative point mutant C3-E174Q; C3WT, Clostridium botulinum–derived exoenzyme C3-transferase; SRF, Serum response factor.

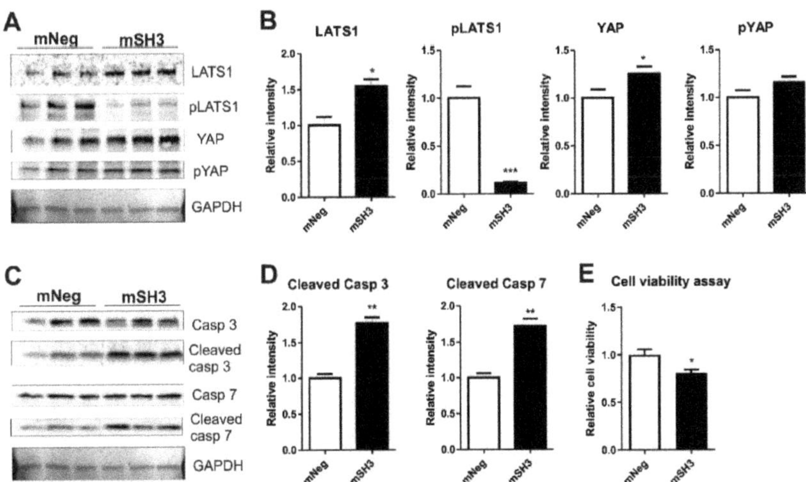

Figure 4. Knockdown of SH3BGR affects cell viability and augments apoptosis via hippo signaling- (**A**) SH3BGR knockdown dramatically reduced the phosphorylation of LATS1 and moderately increased total YAP without altering phosphorylated YAP (pYAP) levels as observed in immunoblots; its densitometric analysis is shown in (**B**). SH3BGR knockdown elevated active, i.e., cleaved, forms of Caspase 3 and 7 (**C,D**), indicating increased apoptosis in NRVCMs, which was also reflected in the reduced cell survival determined by cell viability assay (**E**). Statistical calculations were carried out using a two-tailed Student's t-test. *, $p < 0.05$; **, $p < 0.01$; LATS1, Large tumor suppressor kinase 1; pLATS1, phosphorylated LATS1; YAP, Yes1 Associated Transcriptional Regulator; pYAP, phosphorylated YAP.

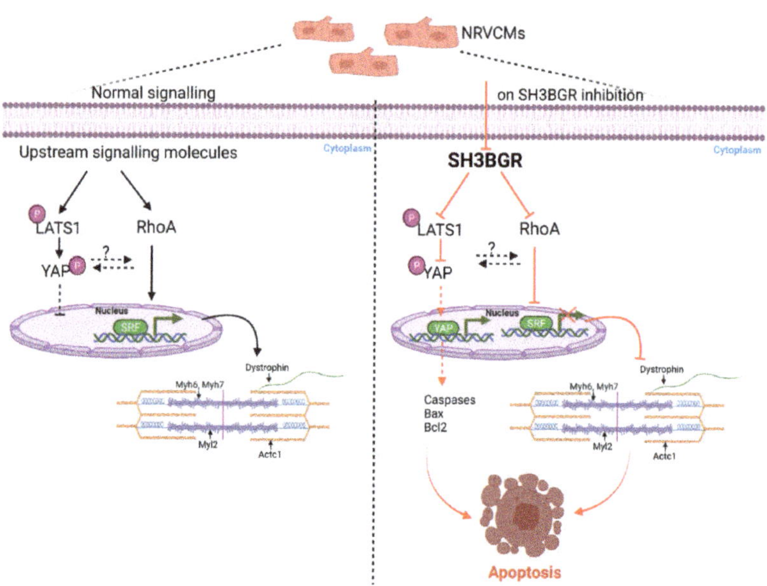

Figure 5. Mechanistic signaling- As observed in the left side panel, LATS1, when phosphorylated, is retained in the cytoplasm and also retains phosphorylated YAP in the cytoplasm, thereby resulting in no transcriptional activation of YAP. Similarly, RhoA activates SRF; the transcriptional activation of SRF is required for genes involved in sarcomere consisting of myosin heavy chain and light chain, actin, etc. On SH3BGR knockdown, as seen in right panel, phosphorylation of YAP is significantly reduced, thereby activating YAP and resulting in probable activation pro-apoptotic genes. Further, the RhoA–SRF axis is inhibited due to SH3BGR inhibition, resulting in hampered SRF activity, thereby resulting in the sarcomeric instability and leading to apoptosis of NRVCMs. SRF, Serum response factor; Myh6, myosin heavy chain 6; Myh7, myosin heavy chain 7; Myl2, Myosin light chain2; Actc1, Actin Alpha Cardiac Muscle 1; Acta1, Actin Alpha 1skeletal muscle; LATS1, Large tumor suppressor kinase 1; pLATS1, phospho Large tumor suppressor kinase 1; YAP, Yes1 Associated Transcriptional Regulator; pYAP, phospho Yes1 Associated Transcriptional Regulator; Bax, BCL2 Associated X; Apoptosis Regulator; Bcl2, B-cell lymphoma 2.

3. Discussion

First discovered in 1997, SH3BGR is associated with the critical region of Down's syndrome on human chromosome 21 [32]. Multiple homologs of SH3BGR were soon identified, making it a small family of proteins [33,34]. Since then, limited studies have speculated its roles in the brain, heart and even in carcinogenesis [25,35,36], but the exact function of SH3BGR and its underlying mechanisms remain elusive. Of note, one of the major outcomes of DS is CHD, that affects nearly 1 in 1000 cases of DS worldwide. Although SH3BGR has recently been shown to play a putative role in cardiogenesis as well as heart physiology [24,25], it is still unknown whether SH3BGR is involved in the acquired cardiac dysfunction. Here, using NRVCM as an in vitro study model, we aimed, on one hand to find if increased levels of SH3BGR, as found in Down's syndrome, affects the cardiomyocyte function. On the other hand, a loss-of-function approach was followed to study the physiological function of SH3BGR.

We observed alterations in SH3BGR levels in the heart of patients with cardiac hypertrophy as well as in the hearts of mouse model of cardiac hypertrophy due to biomechanical stress (TAC). Along these lines, hypertrophic induction and inhibition were observed on SH3BGR via gain- and loss-of-function approaches at in vitro level, respectively, suggesting overexpression, as well as knockdown, has an associated effect against the normal functioning of cells. Digging deeper to understand the mechanistic relevance of these experimental findings, through the Harmonizome database, we identified SH3BGR as one of the transcriptional targets of SRF, a major transcription factor governing the transcription of sarcomeric genes and cardiomyocyte hypertrophy. Interestingly, using a gain-of-function approach, which we believe mimics the Trisomy 21 condition, wherein a gene of interest is overexpressed due to presence of an extra gene copy, we observed that elevated levels of SH3BGR are sufficient for the activation of SRF signaling and cellular hypertrophy in NRVCMs. The loss-of-function approach further strengthened these findings, suggesting a probable mechanism of CHD in DS patients, at least in part. Studies using patient-derived tissue biopsies, induced pluripotent stem cell-derived cardiomyocytes, or in vivo models may be utilized to further validate our findings. Furthermore, we established RhoA to be an intermediary in the activation of SH3BGR-driven SRF signaling, both of which are important in regulating the actin cytoskeleton and sarcomeric homeostasis. The sarcomere is mainly formed of myosin heavy chain and light chain molecules, along with actin. These work in an organized manner for the proper function of contraction and relaxation of beating cells, thereby maintaining cellular homeostasis. Thus, alterations in the levels of any of these proteins could lead to structural and functional instability, which we observed upon SH3BGR deficiency. Thus, we believe that our findings are of direct clinical relevance, particularly in DS patients with CHD, which may be exploited in the future for therapeutic interventions.

Yin et al. [31] have recently provided the first evidence of a homolog of SH3BGR (SH3BGRL2) to regulate the Hippo signaling pathway. In our current study, we found that the knockdown of SH3BGR induces apoptosis via Hippo signaling. Normally, YAP transcriptional activity is shut down on activation of Hippo signaling and vice versa [11]. Being transcriptionally inactive, LATS1 is phosphorylated and, in turn, subsequently phosphorylates YAP, thereby causing YAP retention in the cytoplasm, whereas, on transcriptional activation, YAP is translocated to the nucleus and further triggers downstream genes [11]. Intriguingly, we observed decreased phospho-LATS1 levels and increased YAP levels upon SH3BGR knockdown, indicating YAP to be translocated to the nucleus, which likely resulted in the reduced cell viability and activation of apoptosis via executory caspases 3 and 7. Notwithstanding, though not significant, the overexpression of SH3BGR exerted similar effects on YAP signaling. Although it is not clear from these data why we did not observe contrasting results with SH3BGR overexpression or knockdown on YAP signaling, the plausible reason could be the involvement of yet unidentified player(s) or independent mechanisms, which necessitates further evaluations. Importantly, however, the results from both overexpression and knockdown studies highlight the fact that SH3BGR is essential for normal homeostasis of cardiomyocytes only when present in optimal levels and its either up- or down-regulation is harmful for the cells.

In summary, to the best of our knowledge, this is the first report where the mechanistic insights into how loss- or gain-of SH3BGR differentially affects cardiomyocyte pathophysiology is reported. The overexpression of SH3BGR, which mimics DS condition, significantly activates cardiomyocyte hypertrophy via RhoA/SRF signaling, whereas SH3BGR knockdown abrogates cellular hypertrophy, leading to a combination of sarcomeric dysfunction, activation of apoptosis and reduced cell viability via alterations in the RhoA/SRF and Hippo signaling pathways in cardiomyocytes (Figure 5).

4. Materials and Methods

4.1. Cloning of SH3BGR Vectors

The expression construct for RhoA was generated as described in Rangrez et al. [8]. The construct for overexpression of SH3BGR was cloned from mouse heart cDNA using primers 5′-GCTGGCACCATG-3′ and 3′-GCTGGGTCGCCCTA-5′ in the pDONR221 gateway cloning vector by two sequential ORF and adaptor PCRs. The cleaned product from the adaptor PCR was then cloned into a donor vector pDONR221 following the manufacturer's instructions (Thermo Fisher Scientific, Planegg, Germany). Knockdown of SH3BGR in NRVCMs was achieved by cloning the respective synthetic microRNAs using the BLOCK-iT polymerase II miR RNAi Expression vector kit via a two-step reaction culminating the integration of synthetic microRNAs into the Gateway cloning vector pDONR221 (Thermo Fisher Scientific).

Adenoviruses encoding full-length mouse SH3BGR cDNA and synthetic microRNAs were further generated for use in the NRVCMs system using the ViraPower adenoviral kit (Thermo Fisher Scientific) following the manufacturer's protocol. In brief, previously cloned cDNA or synthetic microRNAs in the pDONR221 vector were transferred into the pAd/CMV/V5-DEST destination vector. The construct was then digested with PacI (10 U/uL; Thermo Fisher Scientific) restriction enzyme and transfected into HEK293A cells to produce the respective adenoviruses. The titration for the adenoviruses was performed by staining virus-infected HEK293A cells with fluorescent anti-Hexon antibody. A β-galactosidase encoding adenovirus (Ad-LacZ; Thermo Fisher Scientific) served as a control for the experimental setup.

4.2. Antibodies

The antibodies used for the various experiments in this study were as follows: α-actinin, mouse monoclonal (1:200; Sigma, Germany); α-actinin, rabbit polyclonal (1:400; Abcam, Germany); α-tubulin, mouse monoclonal (1:8000; Sigma); caspase-3, rabbit polyclonal (1:1000; Cell Signaling Technology, Taufkirchen, Germany); cleaved caspase-3, rabbit monoclonal (1:400; Cell Signaling Technology); caspase-7, rabbit polyclonal (1:1000; Cell Signaling Technology); SH3BGR rabbit polyclonal(1:1000; Proteintech, St. Leon-Rot, Germany); LATS1 (1:1000; Cell Signaling Technology); p-LATS1 (1:1000; Cell Signaling Technology); YAP (1:1000; Cell Signaling Technology); p-YAP (1:1000; Cell Signaling Technology).

4.3. Isolation of NRVCMs

The cell system used for the experiments in this manuscript is the primary neonatal rat ventricular cardiomyocytes, or NRVCMs. These cells were harvested and prepared for experimental use as described previously [37]. In brief, the left ventricles of 1–2-day old Wistar rat babies (Charles River, Lyon, France) were harvested and chopped in ADS buffer containing 120 mmol/liter NaCl, 20 mmol/liter Hepes, 8 mmol/liter NaH2PO$_4$, 6 mmol/liter glucose, 5 mmol/liter KCl and 0.8 mmol/liter MgSO$_4$; pH 7.4. For releasing the individual cardiomyocytes from compound chopped tissue mass, between five and six enzymatic digestion steps were performed with 0.6 mg/mL of pancreatin (Sigma) at 37 °C and 0.5 mg/mL of collagenase type II (Worthington, Columbus, OH, USA) in sterile ADS buffer. Subsequently, the compound cell suspension was passed through a particular cell strainer with the final addition of newborn calf serum to stop enzymatic digestion of cell mass. The cardiomyocytes were separated from cardiac fibroblasts using a Percoll gradient (GE Healthcare, Chicago, IL, USA) centrifugation step and were cultured in DMEM with additives such as 10% FCS, 2 mM penicillin/streptomycin and L-glutamine (PAA Laboratories, Pasching, Austria) to support the growth. Adenovirus infection of NRVCMs in DMEM supplemented with penicillin/streptomycin and L-glutamine, but lacking FCS, was performed 24 h post-harvest. The cells were harvested 72 h post-infection.

4.4. Co-Localization Analysis of SH3BGR with α-Actinin

The co-localization between SH3BGR and α-actinin was observed in NRVCMs using the LSM800 Zeiss laser-scanning microscope with the help of the ZEN-blue software package. The cells were seeded in a 12-well plate that had a collagen-coated coverslip in each well. Following the adenoviral infection and incubation phase, NRVCMs were first fixed with 4% PFA for 5 min and then, in one step, permeabilized and blocked with 0.1% Triton X-100 in 2.5% BSA in saline (PBS) for 1 h. The cells were then incubated for 1 h with primary antibodies using the following dilutions: polyclonal rabbit anti-SH3BGR (1:200) and monoclonal mouse anti- α-actinin (1:200; Sigma) for co-localization observation. The respective secondary antibodies conjugated to either Alexa Fluor-546 (AF546) or Alexa Fluor-488 (AF488) (Thermo Fisher Scientific) were incubated for 1 h with the same dilution of 1:200 in 2.5% BSA in PBS, along with the nuclear stain DAPI (1:500). FluorSave reagent

(Merck Millipore, Burlington, MA, USA) was used as a mounting medium. Fluorescence micrographs were taken using the aforementioned Zeiss LSM800 confocal microscope with a Plan-Apochromat 40/1.4 oil differential interference contrast (UV)-visible IR objective at room temperature. Image pixel size was set to optimal for individual image acquisitions. The pinhole for the acquisition was adjusted to 1 airy unit or less for each laser line. The AF546 and DAPI channels were acquired via GaAsP-Pmt detectors, while the AF488 channel was acquired with a Multialkali-Pmt detector with gain settings between 600 V and 700 V. The laser power for excitation variably ranged from 0.2 to 0.8%.

4.5. Immunofluorescence Microscopy for Cell Size Measurement

The cell size measurement of NRVCMs was studied in NRVCMs by immunofluorescence microscopy. NRVCM preparation and staining were performed as described in two separate sections above. Monoclonal mouse anti-α-actinin (1:200; Sigma) was used as the primary antibody for cell size measurements due to its specificity to sarcomeric α-actinin. The respective secondary antibody conjugated to Alexa Fluor-488 (Thermo Fisher Scientific) was incubated for 1 h at a dilution of 1:200 in 2.5% BSA in PBS along with nuclear stain DAPI (1:500). FluorSave reagent (Merck Millipore) was again used as a mounting medium. Fluorescence micrographs (10 per coverslip) were taken with a Keyence fluorescence microscope of series BZ 9000, at X10 objective (Plan Apochromat, NA: 0.45). Images were acquired using the BZ-II image viewer software (Keyence, Oaska, Japan version 2.1) using a built-in camera at room temperature and were processed and analyzed by a BZ-II Analyzer (Keyence, version 2.1) as detailed below.

Cell size was measured using the HybridCellCount software module from Keyence with the fluorescence harvest kept at single-extraction mode. First, the fluorescence intensity thresholds were set for a moderately intense reference picture. Thereafter, α-actinin whole-cell staining was set as the target area and the blue DAPI-stained nuclei were then extracted from each green target area to determine the number of nuclei per target area, i.e., cell. Then, MacroCellCount was performed, applying the settings from the reference picture to each target picture from one set of experiments. Results were manually filtered in MS excel for the following criteria: (a) 200-μm^2 < target area < 2500 μm^2 (size filter); (b) extraction from target area = 1 (one nucleus filter, as neonatal cardiomyocytes have only one nucleus); (c) area ratio 1 < 30% (cell surface to nucleus ratio filter, to avoid apoptotic cells). The statistical analyses were performed using GraphPad Prism (version 5). An equal distribution of the data in each group was tested by the Shapiro–Wilk test. The samples were then compared according to the Student's t-test. n > 900 (each condition, depending on the cell density).

4.6. MTT Assay for Cell Viability

NRVCMs that were cultured in 24-well plates for viability check were infected with RhoA, SH3 and miRSH3 adenoviruses and incubated for 72 h, where miRNeg and LacZ served as controls as per the mentioned conditions. An MTT labeling reagent (in situ cell proliferation kit, MTT I, Roche Applied Science, Penzberg, Germany) was added at a 10% concentration of total DMEM volume to each well. Plates were then incubated for 4 h in a cell incubators' humidified atmosphere (37 °C, 5% CO_2). After this incubation phase, an MTT solubilization solution (cell proliferation kit MTT, Roche Applied Science) was added to each well in a quantity 10 times higher than the initially added MTT labeling reagent and was further incubated overnight under the same conditions. After complete solubilization of purple formazan crystals, spectrophotometric absorbance was measured using a 96-microplate reader on an Infinite M200 PRO System (Tecan, Life Science). The percentage of viable cells from each condition was plotted as relative to the negative control. Different groups were then compared according to the Student's t-test. All the experiments were performed in hexaplicate or octuplicate and repeated three times.

4.7. RNA Isolation and qRT-PCR

Total RNA was isolated from NRVCMs or human/mouse hearts or other mouse tissue samples using a cell-lysis reagent TriFast (Peqlab, Burladingen, Germany) and a Precellys homogenizer with coarse and fine plastic beads (only for mouse and human samples), following the manufacturer's instructions. A total of 1 µg of total RNA was transcribed into cDNA using the LunaScript RT supermix cDNA synthesis kit (New England Biolabs, Lpswich, MA, USA). For qRT-PCR, the EXPRESS SYBR Green ER reagent (Life Technologies, Inc., Carlsbad, CA, USA) was used in a real-time PCR system CFX96 from Bio-Rad. Cycling conditions used for all the qRT-PCRs were 3 min at 95 °C followed by 40 cycles of 15 s at 95 °C and 45 s at 60 °C, a common step for annealing and extension, at which data were collected. Rpl32 was used as an internal standard for normalization [8]. All experiments with NRVCMs were performed in hexaplicate and repeated three times.

4.8. Protein Preparation and Immunoblotting

For protein isolation, NRVCMs were lysed by two to three freeze-thaw cycles in RIPA lysis buffer containing 50mM Tris, 150mM NaCl, 1% Nonidet P-40, 0.5% sodium deoxycholate and 0.2% SDS, along with phosphatase inhibitor II, phosphatase inhibitor III and protease inhibitor mixture (Roche Applied Science). For protein harvest from mouse tissue or human hearts, a Precellys homogenizer with coarse and fine plastic beads (Peqlab, Germany) was employed. Cell debris in both methods was removed by centrifugation and protein concentration was determined photometrically by the DC assay method (Bio-Rad, Feldkirchen, Germany) against BSA serial dilutions. Protein samples were first resolved by 10% SDS-PAGE, before transferring to a nitrocellulose membrane and subsequently immunoblotted with the target-specific primary antibodies. The overnight application of mono- or poly-clonal primary antibodies was followed by incubation with a suitable HRP-coupled secondary antibody (1:10,000; Santa Cruz Biotechnology, Dallas, TX, USA) or fluorescent antibody Alexa Fluor 546 (for Tubulin only). Finally, protein band visualization was achieved using a chemiluminescence kit (GE Healthcare) and was detected on an imaging system (FluorChem Q; Biozym). A quantitative densitometry analysis was performed using the ImageJ version 1.46 software (National Institutes of Health) and plotted using Graphpad relative to control. All conditions were maintained in triplicates and repeated thrice.

4.9. Human Heart Samples

Left ventricular myocardial samples were taken from the explanted hearts of patients (NF = 5, HCM = 7) with the end-stage heart failure as characterized by the New York Heart Association, heart failure classification IV and thus undergoing heart transplantation. All procedures were performed in accordance with the ethical committee of the medical school of the University of Goettingen in Germany. The explanted hearts were acquired directly in the operation room during surgery and immediately placed in pre-cooled cardioplegic solution (in mmol/l: NaCl 110, KCl 16, MgCl$_2$ 16, NaHCO$_3$ 16, CaCl$_2$ 1.2 and glucose 11). The samples for immunoblots were frozen in liquid nitrogen and stored at -80 degrees immediately after excision.

4.10. SRF Luciferase Assay

The SRF reporter gene assays shown in this study were performed on NRVCMs as described previously [8]. Briefly, cells were infected with several combinations of viruses expressing SH3BGR (50 ifu), miRSH3BGR (100 ifu) and RhoA (50 ifu), where LacZ and miRNeg served as controls or filler viruses to maintain an equal count of viruses, along with adenovirus Ad-SRF-RE-luciferase (20 ifu) carrying a firefly luciferase and Ad-Renilla-luciferase carrying (5 ifu) Renilla luciferase (for normalization of the measurements). SRF reporter gene assays were performed using a dual-luciferase reporter assay kit (Promega, Madison, WI, USA), according to the manufacturer's guidelines. Chemiluminescence was measured photometrically on an Infinite M200 PRO system (Tecan, Life Science). All the experiments were performed in hexaplicate and repeated three times.

4.11. RhoA Inhibitor Usage

Clostridium botulinum-derived exoenzyme C3-transferase and its dominant-negative point mutant C3-E174Q (glutamate to glutamine at aa 174) were used to treat the cells for RhoA inhibition activity. Briefly, C3-E174Q mutant as control and C3-WT as treatment were used. NRVCMs were treated for 12 h at a concentration of 1ug/mL before harvesting.

4.12. Statistical Analysis

All results shown are the means $\pm$ S.E. unless stated otherwise. The statistical analyses of the data were performed using a two-tailed Student's *t*-test for every experimental analysis. If necessary, two-way ANOVA (followed by Student–Newman–Keuls post hoc tests when appropriate) was applied. *p* values of less than 0.05 were considered statistically significant for all experiments.

Supplementary Materials: The following are available online at https://www.mdpi.com/article/10.3390/ijms222011042/s1.

Author Contributions: A.Y.R., D.F. and N.F. conceived and designed the experiments; A.D., A.B., A.R. and A.Y.R. performed the experiments and analyzed the data; D.F. and N.F. contributed reagents/materials/analysis tools; A.D., A.B. and A.Y.R. wrote the manuscript; D.F., N.F. and A.Y.R. revised the manuscript. All authors have read and agreed to the published version of the manuscript.

Funding: This research was funded by the German Centre for Cardiovascular Research (DZHK), with grants awarded to D.F. and N.F.

Conflicts of Interest: The authors declare no conflict of interest. The funders had no role in the design of the study; in the collection, analyses, or interpretation of data; in the writing of the manuscript, or in the decision to publish the results.

Abbreviations

CHD	Congenital heart disease
DS	Down's syndrome
SH3BGR	SH3-binding glutamic acid rich
Nppa	Natriuretic peptide A
Nppb	Natriuretic peptide B
NRVCM	Neonatal rat ventricular cardiomyocytes
RhoA	Ras Homolog family member A
SRF	Serum Response Factor
YAP	Yes associated protein

References

1. Crawford, D.; Dearmun, A. Down's syndrome. *Nurs. Child. Young People* **2016**, *28*, 17. [CrossRef]
2. Antonarakis, S.E.; Skotko, B.G.; Rafii, M.S.; Strydom, A.; Pape, S.E.; Bianchi, D.W.; Sherman, S.L.; Reeves, R.H. Down syndrome. *Nat. Rev. Dis. Primers* **2020**, *6*, 9. [CrossRef] [PubMed]
3. Benhaourech, S.; Drighil, A.; Hammiri, A.E. Congenital heart disease and Down syndrome: Various aspects of a confirmed association. *Cardiovasc. J. Afr.* **2016**, *27*, 287–290. [CrossRef] [PubMed]
4. Asim, A.; Kumar, A.; Muthuswamy, S.; Jain, S.; Agarwal, S. Down syndrome: An insight of the disease. *J. Biomed. Sci.* **2015**, *22*, 41. [CrossRef] [PubMed]
5. Shore, P.; Sharrocks, A.D. The MADS-box family of transcription factors. *Eur. J. Biochem.* **1995**, *229*, 1–13. [CrossRef] [PubMed]
6. Miano, J.M. Role of serum response factor in the pathogenesis of disease. *Lab. Investig. A J. Tech. Methods Pathol.* **2010**, *90*, 1274–1284. [CrossRef] [PubMed]
7. Nishikimi, T.; Kuwahara, K.; Nakao, K. Current biochemistry, molecular biology, and clinical relevance of natriuretic peptides. *J. Cardiol.* **2011**, *57*, 131–140. [CrossRef]
8. Rangrez, A.Y.; Bernt, A.; Poyanmehr, R.; Harazin, V.; Boomgaarden, I.; Kuhn, C.; Rohrbeck, A.; Frank, D.; Frey, N. Dysbindin is a potent inducer of RhoA-SRF-mediated cardiomyocyte hypertrophy. *J. Cell Biol.* **2013**, *203*, 643–656. [CrossRef]

9. Foster, C.T.; Gualdrini, F.; Treisman, R. Mutual dependence of the MRTF-SRF and YAP-TEAD pathways in cancer-associated fibroblasts is indirect and mediated by cytoskeletal dynamics. *Genes Dev.* **2017**, *31*, 2361–2375. [CrossRef]

10. Xiao, Y.; Hill, M.C.; Zhang, M.; Martin, T.J.; Morikawa, Y.; Wang, S.; Moise, A.R.; Wythe, J.D.; Martin, J.F. Hippo Signaling Plays an Essential Role in Cell State Transitions during Cardiac Fibroblast Development. *Dev. Cell* **2018**, *45*, 153–169.e6. [CrossRef]

11. Wang, J.; Liu, S.; Heallen, T.; Martin, J.F. The Hippo pathway in the heart: Pivotal roles in development, disease, and regeneration. *Nat. Rev. Cardiol.* **2018**, *15*, 672–684. [CrossRef]

12. Zhou, Q.; Li, L.; Zhao, B.; Guan, K.L. The hippo pathway in heart development, regeneration, and diseases. *Circ. Res.* **2015**, *116*, 1431–1447. [CrossRef]

13. Del Re, D.P.; Matsuda, T.; Zhai, P.; Maejima, Y.; Jain, M.R.; Liu, T.; Li, H.; Hsu, C.P.; Sadoshima, J. Mst1 promotes cardiac myocyte apoptosis through phosphorylation and inhibition of Bcl-xL. *Mol. Cell* **2014**, *54*, 639–650. [CrossRef]

14. Matsuda, T.; Zhai, P.; Sciarretta, S.; Zhang, Y.; Jeong, J.I.; Ikeda, S.; Park, J.; Hsu, C.P.; Tian, B.; Pan, D.; et al. NF2 Activates Hippo Signaling and Promotes Ischemia/Reperfusion Injury in the Heart. *Circ. Res.* **2016**, *119*, 596–606. [CrossRef] [PubMed]

15. Ikeda, S.; Sadoshima, J. Regulation of Myocardial Cell Growth and Death by the Hippo Pathway. *Circ. J. Off. J. Jpn. Circ. Soc.* **2016**, *80*, 1511–1519. [CrossRef] [PubMed]

16. Del Re, D.P.; Yang, Y.; Nakano, N.; Cho, J.; Zhai, P.; Yamamoto, T.; Zhang, N.; Yabuta, N.; Nojima, H.; Pan, D.; et al. Yes-associated protein isoform 1 (Yap1) promotes cardiomyocyte survival and growth to protect against myocardial ischemic injury. *J. Biol. Chem.* **2013**, *288*, 3977–3988. [CrossRef] [PubMed]

17. Shao, D.; Zhai, P.; Del Re, D.P.; Sciarretta, S.; Yabuta, N.; Nojima, H.; Lim, D.S.; Pan, D.; Sadoshima, J. A functional interaction between Hippo-YAP signalling and FoxO1 mediates the oxidative stress response. *Nat. Commun.* **2014**, *5*, 3315. [CrossRef]

18. Eden, M.; Meder, B.; Volkers, M.; Poomvanicha, M.; Domes, K.; Branchereau, M.; Marck, P.; Will, R.; Bernt, A.; Rangrez, A.; et al. Myoscape controls cardiac calcium cycling and contractility via regulation of L-type calcium channel surface expression. *Nat. Commun.* **2016**, *7*, 11317. [CrossRef]

19. Rangrez, A.Y.; Eden, M.; Poyanmehr, R.; Kuhn, C.; Stiebeling, K.; Dierck, F.; Bernt, A.; Lullmann-Rauch, R.; Weiler, H.; Kirchof, P.; et al. Myozap Deficiency Promotes Adverse Cardiac Remodeling via Differential Regulation of Mitogen-activated Protein Kinase/Serum-response Factor and beta-Catenin/GSK-3beta Protein Signaling. *J. Biol. Chem.* **2016**, *291*, 4128–4143. [CrossRef]

20. Frank, D.; Rangrez, A.Y.; Poyanmehr, R.; Seeger, T.S.; Kuhn, C.; Eden, M.; Stiebeling, K.; Bernt, A.; Grund, C.; Franke, W.W.; et al. Mice with cardiac-restricted overexpression of Myozap are sensitized to biomechanical stress and develop a protein-aggregate-associated cardiomyopathy. *J. Mol. Cell. Cardiol.* **2014**, *72*, 196–207. [CrossRef]

21. Seeger, T.S.; Frank, D.; Rohr, C.; Will, R.; Just, S.; Grund, C.; Lyon, R.; Luedde, M.; Koegl, M.; Sheikh, F.; et al. Myozap, a novel intercalated disc protein, activates serum response factor-dependent signaling and is required to maintain cardiac function in vivo. *Circ. Res.* **2010**, *106*, 880–890. [CrossRef]

22. Dierck, F.; Kuhn, C.; Rohr, C.; Hille, S.; Braune, J.; Sossalla, S.; Molt, S.; van der Ven, P.F.M.; Furst, D.O.; Frey, N. The novel cardiac z-disc protein CEFIP regulates cardiomyocyte hypertrophy by modulating calcineurin signaling. *J. Biol. Chem.* **2017**, *292*, 15180–15191. [CrossRef]

23. Scartezzini, P.; Egeo, A.; Colella, S.; Fumagalli, P.; Arrigo, P.; Nizetic, D.; Taramelli, R.; Rasore-Quartino, A. Cloning a new human gene from chromosome 21q22.3 encoding a glutamic acid-rich protein expressed in heart and skeletal muscle. *Hum. Genet.* **1997**, *99*, 387–392. [CrossRef] [PubMed]

24. Egeo, A.; Di Lisi, R.; Sandri, C.; Mazzocco, M.; Lapide, M.; Schiaffino, S.; Scartezzini, P. Developmental expression of the SH3BGR gene, mapping to the Down syndrome heart critical region. *Mech. Dev.* **2000**, *90*, 313–316. [CrossRef]

25. Sandri, C.; Di Lisi, R.; Picard, A.; Argentini, C.; Calabria, E.; Myklak, K.; Scartezzini, P.; Schiaffino, S. Heart morphogenesis is not affected by overexpression of the Sh3bgr gene mapping to the Down syndrome heart critical region. *Hum. Genet.* **2004**, *114*, 517–519. [CrossRef]

26. Mazzocco, M.; Maffei, M.; Egeo, A.; Vergano, A.; Arrigo, P.; Di Lisi, R.; Ghiotto, F.; Scartezzini, P. The identification of a novel human homologue of the SH3 binding glutamic acid-rich (SH3BGR) gene establishes a new family of highly conserved small proteins related to Thioredoxin Superfamily. *Gene* **2002**, *291*, 233–239. [CrossRef]

27. Guo, Y.; Cao, Y.; Jardin, B.D.; Sethi, I.; Ma, Q.; Moghadaszadeh, B.; Troiano, E.C.; Mazumdar, N.; Trembley, M.A.; Small, E.M.; et al. Sarcomeres regulate murine cardiomyocyte maturation through MRTF-SRF signaling. *Proc. Natl. Acad. Sci. USA* **2021**, *118*, e2008861118. [CrossRef] [PubMed]

28. Rouillard, A.D.; Gundersen, G.W.; Fernandez, N.F.; Wang, Z.; Monteiro, C.D.; McDermott, M.G.; Ma'ayan, A. The harmonizome: A collection of processed datasets gathered to serve and mine knowledge about genes and proteins. *Database* **2016**, *2016*, baw100. [CrossRef] [PubMed]

29. Kilian, L.S.; Voran, J.; Frank, D.; Rangrez, A.Y. RhoA: A dubious molecule in cardiac pathophysiology. *J. Biomed. Sci.* **2021**, *28*, 33. [CrossRef]

30. De Kreuk, B.J.; Hordijk, P.L. Control of Rho GTPase function by BAR-domains. *Small GTPases* **2012**, *3*, 45–52. [CrossRef]

31. Yin, L.; Li, W.; Xu, A.; Shi, H.; Wang, K.; Yang, H.; Wang, R.; Peng, B. SH3BGRL2 inhibits growth and metastasis in clear cell renal cell carcinoma via activating hippo/TEAD1-Twist1 pathway. *EBioMedicine* **2020**, *51*, 102596. [CrossRef]

32. Vidal-Taboada, J.M.; Bergonon, S.; Scartezzini, P.; Egeo, A.; Nizetic, D.; Oliva, R. High-resolution physical map and identification of potentially regulatory sequences of the human SH3BGR located in the Down syndrome chromosomal region. *Biochem. Biophys. Res. Commun.* **1997**, *241*, 321–326. [CrossRef]

33. Mazzocco, M.; Arrigo, P.; Egeo, A.; Maffei, M.; Vergano, A.; Di Lisi, R.; Ghiotto, F.; Ciccone, E.; Cinti, R.; Ravazzolo, R.; et al. A novel human homologue of the SH3BGR gene encodes a small protein similar to Glutaredoxin 1 of Escherichia coli. *Biochem. Biophys. Res. Commun.* **2001**, *285*, 540–545. [CrossRef]

34. Egeo, A.; Mazzocco, M.; Arrigo, P.; Vidal-Taboada, J.M.; Oliva, R.; Pirola, B.; Giglio, S.; Rasore-Quartino, A.; Scartezzini, P. Identification and characterization of a new human gene encoding a small protein with high homology to the proline-rich region of the SH3BGR gene. *Biochem. Biophys. Res. Commun.* **1998**, *247*, 302–306. [CrossRef] [PubMed]

35. Wang, L.; Huang, J.; Jiang, M.; Lin, H. Signal transducer and activator of transcription 2 (STAT2) metabolism coupling postmitotic outgrowth to visual and sound perception network in human left cerebrum by biocomputation. *J. Mol. Neurosci.* **2012**, *47*, 649–658. [CrossRef] [PubMed]

36. Majid, S.M.; Liss, A.S.; You, M.; Bose, H.R. The suppression of SH3BGRL is important for v-Rel-mediated transformation. *Oncogene* **2006**, *25*, 756–768. [CrossRef] [PubMed]

37. Borlepawar, A.; Rangrez, A.Y.; Bernt, A.; Christen, L.; Sossalla, S.; Frank, D.; Frey, N. TRIM24 protein promotes and TRIM32 protein inhibits cardiomyocyte hypertrophy via regulation of dysbindin protein levels. *J. Biol. Chem.* **2017**, *292*, 10180–10196. [CrossRef] [PubMed]

International Journal of
Molecular Sciences

MDPI

Brief Report

Bmp Signaling Regulates Hand1 in a Dose-Dependent Manner during Heart Development

Mingjie Zheng [1], Shannon Erhardt [1,2], Di Ai [3] and Jun Wang [1,2,*]

1   Department of Pediatrics, McGovern Medical School, The University of Texas Health Science Center at Houston, Houston, TX 77030, USA; mingjie.zheng@uth.tmc.edu (M.Z.); shannon.erhardt@uth.tmc.edu (S.E.)
2   The University of Texas MD Anderson Cancer Center UTHealth Graduate School of Biomedical Sciences, The University of Texas Health Science Center at Houston, Houston, TX 77030, USA
3   Department of Pathology and Laboratory Medicine, School of Medicine, Emory University, Atlanta, GA 30322, USA; di.ai@emory.edu
*   Correspondence: jun.wang@uth.tmc.edu

Abstract: The bone morphogenetic protein (Bmp) signaling pathway and the basic helix–loop–helix (bHLH) transcription factor Hand1 are known key regulators of cardiac development. In this study, we investigated the Bmp signaling regulation of *Hand1* during cardiac outflow tract (OFT) development. In *Bmp2* and *Bmp4* loss-of-function embryos with varying levels of *Bmp* in the heart, *Hand1* is sensitively decreased in response to the dose of *Bmp* expression. In contrast, *Hand1* in the heart is dramatically increased in *Bmp4* gain-of-function embryos. We further identified and characterized the Bmp/Smad regulatory elements in *Hand1*. Combined transfection assays and chromatin immunoprecipitation (ChIP) experiments indicated that *Hand1* is directly activated and bound by Smads. In addition, we found that upon the treatment of Bmp2 and Bmp4, P19 cells induced Hand1 expression and favored cardiac differentiation. Together, our data indicated that the Bmp signaling pathway directly regulates *Hand1* expression in a dose-dependent manner during heart development.

Keywords: Bmp signaling; *Hand1*; Smad; transcriptional regulation; heart development; cardiomyocyte differentiation

Citation: Zheng, M.; Erhardt, S.; Ai, D.; Wang, J. Bmp Signaling Regulates Hand1 in a Dose-Dependent Manner during Heart Development. *Int. J. Mol. Sci.* **2021**, *22*, 9835. https://doi.org/10.3390/ijms22189835

Academic Editors: Nicole Wagner and Kay-Dietrich Wagner

Received: 1 August 2021
Accepted: 8 September 2021
Published: 11 September 2021

Publisher's Note: MDPI stays neutral with regard to jurisdictional claims in published maps and institutional affiliations.

1. Introduction

Congenital heart defects (CHDs) are the most common birth defects, with an estimated prevalence of 1% in newborns [1]. Cardiac outflow tract (OFT) defects are the most common CHDs and account for one-third of all reported CHDs in human births. The development of the OFT involves interactions and coordination between two types of progenitor pools: second heart field (SHF) progenitors and cardiac neural crest cells (NCCs), regulated by a complex, fine-tuned molecular regulatory network [2,3].

Bone morphogenetic proteins (Bmps) [4] are a family of growth factors belonging to the transforming growth factor beta (TGF-β) superfamily [5]. In the canonical Bmp pathway, Bmp ligands such as Bmp2 and Bmp4 bind to their dual-specificity kinase and heterodimeric receptor complex, consisting of type I and type II receptors, which phosphorylates downstream receptor-regulated Smads (R-Smads), i.e., Smad1, Smad5, and Smad8 (Smad1/5/8) [6,7]. The phospho-R-Smads then form an oligomeric complex with Smad4 and translocate into the nucleus to regulate the expression of downstream genes. The Smad complex can act as both a transcriptional activator and a repressor to regulate target gene expression. The highly conserved Bmp signaling pathway is essential for heart development [8,9], including OFT formation [10–12]. Mouse models with Bmp signaling disruptions result in embryonic lethality and CHDs [13–17], as evidenced by *Bmp2/4* deletions in the SHF resulting in lethality by embryonic day (E) 12.5 with deficient OFT myocardial differentiation in mice [17].

The hand (the heart- and neural crest derivatives-expressed protein 1) proteins are a subclass of basic helix–loop–helix (bHLH) transcription factors (TFs) that can form homo- and heterodimer combinations with multiple bHLH partners, mediating transcriptional activity in the nucleus [18,19]. Previous studies have shown that Hand1 and Hand2 are core TFs that are expressed in the precardiogenic mesoderm to govern essential gene regulatory networks for cardiovascular growth and morphogenesis [20–24]. *Hand1* plays a vital role in the specification and/or differentiation of extraembryonic structures such as the yolk sac, placenta, and cells of the trophoblast lineages, including cardiac muscle the heart, gut, and sympathetic neuronal development, while also aiding in the proper development of tissues populated by *Hand1*-expressing NCCs [25,26]. *Hand1* deletion in mice results in lethality at E8-8.5, with perturbed heart development at the looping stage [20,27]. Several cardiac studies demonstrated that *Hand1* is an important regulator for cardiac precursor cell fate decision and cardiac morphogenesis [20,21,28,29]. Moreover, mutated *HAND1* has been shown to hinder the effect of *GATA4*, and is associated with congenital heart disease in human patients [30,31]. Mice lacking *Hand2* are embryonically lethal at E10.5, persisting with right ventricular hypoplasia and vascular malformations [22,32]. *HAND2* loss-of-function mutation was found to contribute to human CHDs, and enhanced susceptibility to familial ventricular septal defect (VSD) and double outlet right ventricle (DORV) [33].

In this study, we combined both in vivo mouse genetics and in vitro molecular analyses to investigate the regulation of *Hand1* by Bmp signaling. We found that canonical Bmp-Smad signaling regulates the expression of *Hand1* in a dosage-dependent manner during embryonic heart development, and functions through both cell-autonomous and non-cell-autonomous regulation. Our results suggested that Smads directly bind to the $5'$UTR of *Hand1* and activate its expression. In addition, we found that Bmp treatment can activate Hand1 expression and promote the expression of cardiac TFs such as *Nkx2.5* and *Gata4* in P19 cells. Taken together, our data uncovered a fine-tuned canonical Bmp signaling-*Hand1* regulation during heart development.

2. Results

2.1. Hand1 Expression Decreases in a Dose-Sensitive Manner in Response to Bmp2 and Bmp4 Deficiency during Heart Development

SHF progenitors contribute greatly in the formation of the OFT, inflow tract and right ventricle (RV) [34]. To determine *Hand1* expression changes in response to *Bmp* loss-of-function during embryonic OFT development, we generated compound *Bmp2* and *Bmp4* double-conditional knockout (*Bmp2/4* dCKO) mutants by crossing the SHF-specific *Mef2c^{cre}* driver with the *Bmp2* and *Bmp4* conditional null alleles. Through this cross, we obtained *Bmp2/4* dCKO mutants and *Bmp* compound mutants with varying levels of *Bmp* deficiency, including *Bmp2* homozygous, *Bmp4* heterozygous mutants (*Bmp2−/−; Bmp4+/−*) and *Bmp4* homozygous, *Bmp2* heterozygous mutants (*Bmp2+/−; Bmp4−/−*). Whole-mount in situ hybridization indicated that compared to the control embryos (Figure 1A), both *Bmp2−/−; Bmp4+/−* (Figure 1B) and *Bmp2+/−; Bmp4−/−* (Figure 1C) mutant embryos had a dramatic decrease in *Hand1* expression in the OFT. Strikingly, *Bmp2/4* dCKO mutants (Figure 1D) presented with fully abolished *Hand1* expression in the OFT; however, in all *Bmp* mutant samples (compound and dCKO mutants), *Hand1* was still highly expressed in the non-SHF-derived structures such as the left ventricle (LV). These findings indicate that *Hand1* expression is highly sensitive to a Bmp dose-dependent regulation. Histological section analysis of in situ hybridization further confirmed that *Hand1* expression was fully abolished in the OFT of *Bmp2/4* dCKO mutant hearts (Figure 1F), as compared to the control hearts with a high expression of *Hand1* in the OFT (Figure 1E). The qRT-PCR analysis further indicated that *Hand1* expression in the hearts of *Bmp2/4* dCKO mutants was significantly reduced to around 30% of that of the control hearts at E9.5 (Figure 1G).

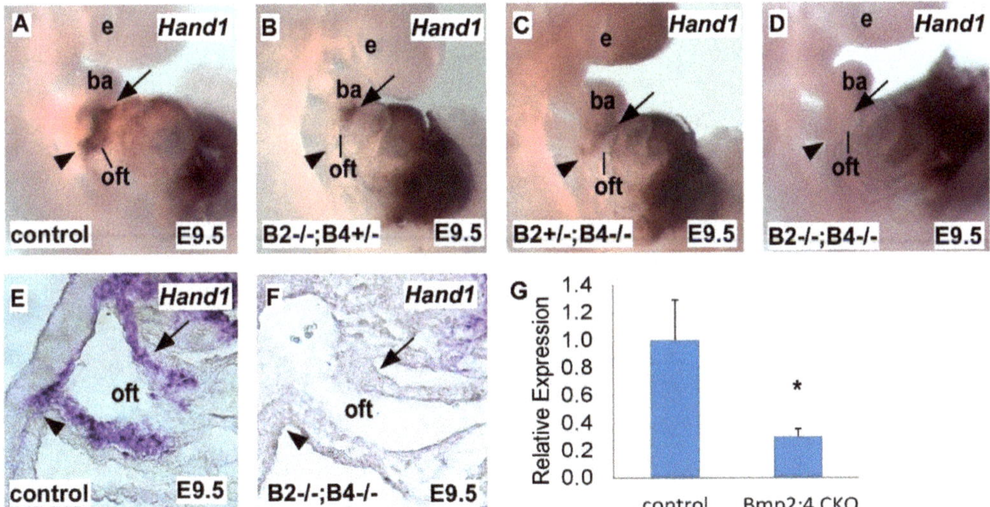

Figure 1. *Hand1* expression is regulated by Bmp signaling in a dose-dependent manner during mouse heart development. (**A–D**) Whole-mount in situ hybridization showing *Hand1* expression in mouse embryos at E9.5. (**E,F**) E9.5 sagittal sections showed *Hand1* in situ hybridization in the branchial arch and OFT. (**G**) qRT-PCR analysis of *Hand1* expression in control and *Bmp2/4* CKO embryos. All genotypes are shown as labeled. *B2−/−; B4+/−: Bmp2* homozygous, *Bmp4* heterozygous mutants; *B2+/−; B4−/−: Bmp4* homozygous, *Bmp2* heterozygous mutants; *B2−/−; B4−/−: Bmp2/4* dCKO. e, eye; ba, branchial arch; oft, outflow tract. Arrows and arrow heads point out in situ hybridization signals in OFT and SHF. Data are presented as means ± s.e.m. * indicates *p*-value < 0.05.

During early cardiac development, proliferating SHF progenitor cells add to the OFT and inflow tract, leading to heart tube elongation and its subsequent asymmetric looping formation by E9.5. The other major progenitor cell population contributing to OFT formation is the cardiac NCCs, a highly migratory, multipotent cell population originating from the cranial/vagal region of the dorsal neural tube that subsequently migrates to the OFT. During cardiac morphogenesis, SHF progenitor cells and cardiac NCCs closely interact with each other and coordinately regulate OFT formation [35]. We evaluated expression of the SHF marker *Hand2* and the NCC marker *Ap2* using whole-mount in situ hybridization, and found that the *Bmp2/4* dCKO mutant heart had *Hand2* (Figure S1B) and *Ap2* (Figure S1D) expression comparable to that of the control embryo (Figure S1A,C). These findings suggest intact SHF and NCC contributions to the OFT, indicating that the abolished *Hand1* expression in the OFT of the *Bmp2/4* dCKO mutant is not caused by reduced cell populations. Importantly, other than in the SHF-derived OFT, *Bmp2/4* deletion in the SHF also caused diminished *Hand1* expression in the NCC-derived components of the OFT, suggesting a non-cell-autonomous regulation by *Bmp2/4*. Together, these data suggested that during development, Bmp signaling regulates *Hand1* expression in the SHF- and NCC-derived OFT through cell-autonomous and non-cell-autonomous regulation, in a dose-dependent manner.

2.2. Hand1 is Upregulated in Bmp4 OE Embryos

Finding that *Hand1* expression is sensitive to *Bmp* loss-of-function, we next detected *Hand1* expression in the heart with elevated Bmp signaling. Using a conditional *Bmp4^{tetO}* gain-of-function allele (tetracycline inducible) crossed with the *Mef2c^{cre}* driver [17,36], we specifically overexpressed *Bmp4* in the SHF-derived heart structures (*Bmp4 OE*). We found that compared with the control heart (Figure 2A), the *Bmp4 OE* mutant heart had robustly expanded *Hand1* expression in the SHF region and SHF-derived structures, including the OFT and RV (Figure 2B), indicated by in situ hybridization staining using a *Hand1* probe.

The qRT-PCR results further indicated that the elevated *Bmp4* expression resulted in a significant increase in *Hand1* in the *Bmp4 OE* mutant heart compared with the control heart at E9.5 (Figure 2C). These results indicated that Bmp signaling activates *Hand1* expression during heart development, further supporting the conclusion that *Hand1* expression sensitively responds to Bmp signaling dosage.

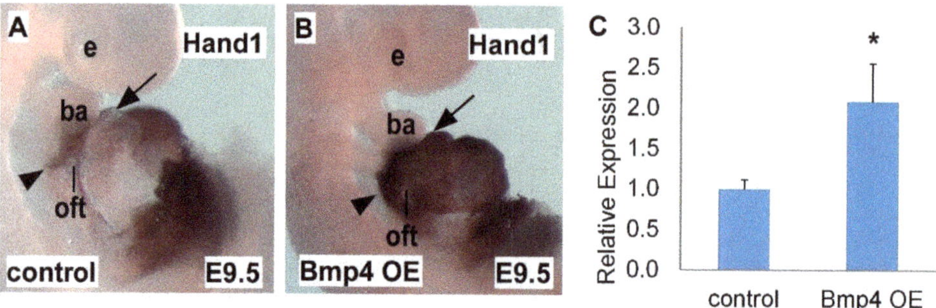

Figure 2. *Hand1* is upregulated upon *Bmp4* overexpression (OE). (**A,B**) Whole-mount in situ hybridization of E9.5 *Bmp4* OE embryo compared with control embryo; *Bmp4* OE embryos expanded *Hand1* expression in the SHF and SHF-derived OFT and RV. (**C**) qRT-PCR indicated increased *Hand1* expression level in *Bmp4* OE embryos compared with control embryos. e, eye; ba, branchial arch; oft, outflow tract; rv, right ventricle. Arrows and arrow heads point out in situ hybridization signals in the OFT and SHF. Data are presented as means ± s.e.m. * indicates *p*-value < 0.05.

2.3. Hand1 Is a Direct Target Activated by the Canonical Bmp/Smad Signaling

Smad TFs function as the major signal transducers for receptors of the Bmp signaling pathway and can interact with specific DNA motifs to regulate gene expression [37–40]. The R-Smads and Smad4 are composed of two evolutionarily conserved domains named Mad Homology 1 and 2 (MH1 and MH2). The MH1 domain is responsible for the Smad binding element's (SBE) DNA-binding activity, while the MH2 domain is important for heterooligomeric Smad complexes formation and transcriptional activation [41,42]. In addition, based on chromatin immunoprecipitation and structural analysis, Smads have been shown to favor recognizing GC-rich elements (also termed BMP response element (BRE) in certain BMP-responsive genes) [43,44] and CAGAC motifs (also termed Smad binding element (SBE)) [45,46]. To determine if the Bmp/Smad signaling directly regulates Hand1, we undertook sequencing analysis and found that several phylogenetically conserved Smad recognition elements, including the GC-rich elements BRE and SBE, were located in the 5′UTR of *Hand1* (Figures 3A and S2).

To determine whether Smads directly bind to *Hand1*, we performed chromatin immunoprecipitation (ChIP) using a Smad1/5/8 antibody in E9.5 wild-type embryonic heart extracts. There was an obvious enrichment in the anti-Smad1/5/8 immunoprecipitated chromatin compared to the controls, indicating that Smad1/5/8 directly bound to the *Hand1* chromatin (Figure 3B). To evaluate whether the potential Bmp/Smad regulatory elements in *Hand1* are functional, we made a *Hand1* 5′UTR (*Hand1* reporter) luciferase (Luc) reporter and performed luciferase assays in P19 cells. We found that Bmp treatment resulted in a dramatic and significant induction of *Hand1* reporter activity (Figure 3C). Overexpression of the constitutively active Bmpr1a (caALK3) [47] also significantly increased *Hand1* reporter activity (Figure 3D). In contrast, overexpression of *Smad6*, an inhibitory Smad, specifically competed with Smad4 for binding to Smad1 [48], and significantly repressed *Hand1* reporter activity (Figure 3E). *Hand1* Luc reporter activity was also dramatically decreased when using a knockdown *Smad1* short hairpin RNA (shRNA) (Figure 3F). Together, these findings supported the idea that *Hand1* is a direct target activated by the canonical Bmp/Smad signaling.

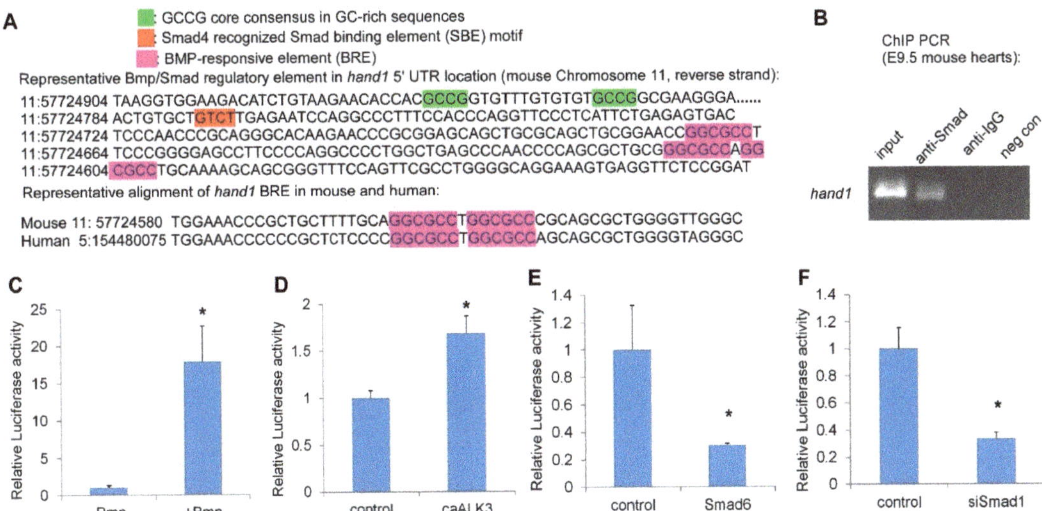

Figure 3. *Hand1* is a direct target activated by the canonical Bmp/Smad signaling. (**A**) Representative Bmp/Smad regulatory element in *Hand1* 5′ UTR location (at upper) and sequence alignment (at lower), showing the conservation among mouse and human (source: Ensembl). (**B**) *In Vivo* ChIP PCR using E9.5 hearts with indicated antibodies to IP chromatin fragment. PCR band (size: 215bp) contains the *Hand1* 5′UTR Bmp/Smad regulatory element. (**C–F**) *Hand1* 5′UTR reporter luciferase assays: treated with Bmp (**C**), co-transfected with constitutively active ALK3 (caALK3) (**D**), pcDNA3.1-*Smad6* (**E**), and pSR si*Smad1* (**F**). Data are presented as means ± s.e.m. * indicates p-value < 0.05.

2.4. Bmp Induces Hand1 Expression during Cardiomyogenesis in P19 Cells

Both in vivo and in vitro studies have established the essential roles of Bmp signals in promoting cardiomyocyte differentiation [49–51]. P19 cells are undifferentiated stem cells derived from murine teratocarcinoma [52], which can differentiate into multiple cell types [53–55]. Previous studies have indicated that P19 cells can undergo cardiomyogenesis after treatment with chemical inducers such as DMSO, cardiac TFs such as *Mef2c*, and various cytokines [56–60]. It has been shown that Bmp treatment can promote cardiomyocyte differentiation in P19 cells by regulating Nkx2.5 activity [60]. To study *Hand1* expression induced by Bmp2 and Bmp4 during cardiomyogenesis, we treated P19 cells with different concentrations of Bmp2 and Bmp4 for 6 days. Our western blot data indicated that both Bmp2 and Bmp4 induced Hand1 protein expression in a dose-dependent manner (Figure 4A,B). The qRT-PCR analysis also indicated that *Hand1* expression was elevated in P19 cells after 6 days of Bmp2 and Bmp4 treatment (Figure 4C). Transcription factor Id1 is a known direct target of the canonical Bmp/Smad signaling pathway [61–63]. Bmp2/4 stimulation also induced *Id1* gene expression, demonstrating that Bmp2 and Bmp4 activate the canonical Bmp/Smad signaling pathway (Figure 4D). Furthermore, we detected an elevated expression of cardiac TFs *Nkx2.5* and *Gata4* with Bmp treatment, indicating undergoing cardiomyogenesis in P19 cells (Figure 4E,F). Taken together, these data showed that the canonical Bmp/Smad signaling pathway induced Hand1 expression during cardiomyogenesis in P19 cells.

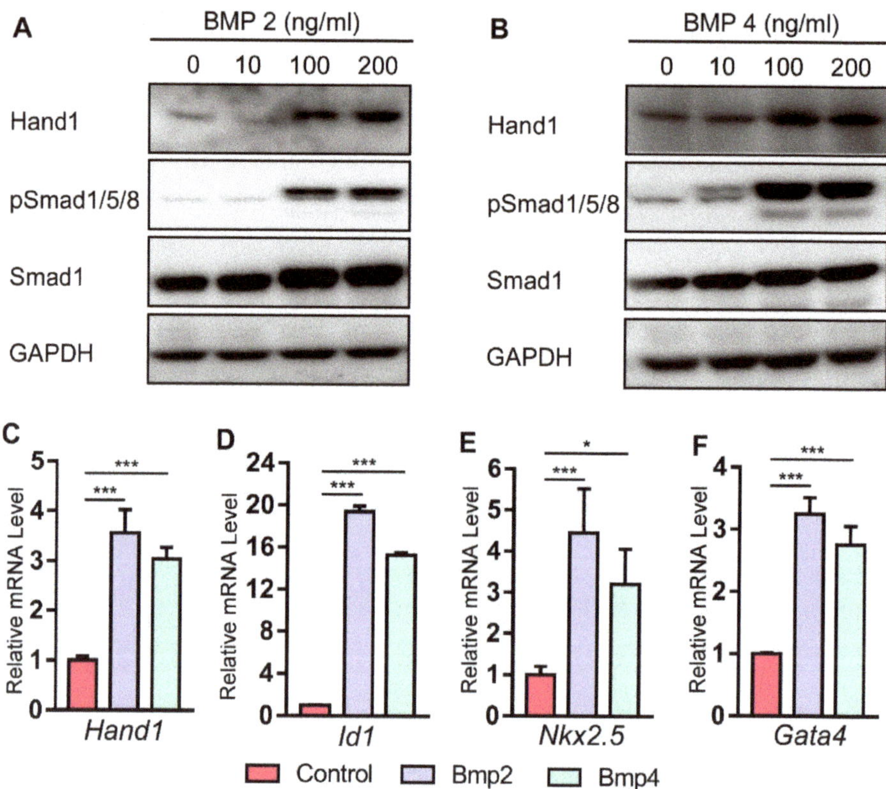

Figure 4. Bmp2 and Bmp4 induce Hand1 expression in P19 cells. (**A**,**B**) P19 cells were stimulated with Bmp2 and 4 for 6 days. Levels of Hand1 were analyzed by western blotting. (**C**–**F**) Total RNA was harvested on day 6 with Bmp treatment for 6 days (100 ng/mL). qRT-PCR was performed for the analysis of *Hand1*, *Id1*, *Nkx2.5*, and *Gata4* mRNA. Data are presented as means ± s.e.m. * indicates *p*-value < 0.05, *** indicates *p*-value < 0.001.

3. Discussion

In this study, we demonstrated that *Hand1* is a direct downstream target of the canonical Bmp/Smad signaling pathway during heart development. Studies have indicated the importance of *Hand1* and *Hand2* during cardiac morphogenesis, including their contribution in NCCs, myocardium, endocardium, and epicardium. *Hand1* and *Hand2* display different restricted expression patterns in the developing heart. In mice, *Hand1* is highly enriched in the OFT, the cardiomyocytes of the LV, and in the myocardial cuff, between E9.5–13.5 [64]. In contrast, *Hand2* is expressed throughout the linear heart tube, including the RV, the atria, and the left ventricular chambers [65]. Here, we found that the *Hand1* expression level is tightly regulated by Bmp signaling in a dose-dependent manner in the OFT, whereas the *Hand2* expression level is not obviously affected by Bmp signaling activity changes. The *Bmp2−/−; Bmp4+/−* and *Bmp2+/−; Bmp4−/−* compound mutant embryos had low levels of *Hand1* expression in the OFT, which indicated a functional redundancy between *Bmp2* and *Bmp4*. The fully abolished *Hand1* expression in the OFT of the *Bmp2/4* dCKO mutant heart indicated that Bmp deletion in the SHF not only regulated *Hand1* expression in the SHF-derived cells, but also *Hand1* expression in neural crest-derived cells, suggesting that Bmp signaling functions in both cell-autonomous and non-cell-autonomous ways. Indeed, Bmp receptors also play essential roles during heart development, such as the Bmp receptor ALK3, that when specifically knocked-out in cardiac myocytes, resulted in cardiac septation and atrioventricular cushion morphogenesis [66]. However, the potential signaling cross talk between SHF progenitors and NCCs in the OFT, mediated by Bmp receptors, will need further investigation. In contrast to *Bmp* loss-of-function, *Bmp* gain-of-function leads to a robust increase in *Hand1* expression, indicating that Bmp signaling is both necessary and sufficient to activate *Hand1* transcription, further supporting the idea that *Hand1* expression sensitively responds to Bmp signaling dosage.

In a facial skeletal development study, Claudio et al. reported that Bmp4 balances self-renewal and differentiation signals in cranial NCCs, and found that compared to the controls, *Hand1* expression was expanded in the developing mandibles of mice with *Bmp4* overexpression in NCCs at E11.5 [36]. In addition, Vincentz et al. found that during mandibular development, Bmp signaling and Hand2 synergistically activate *Hand1* expression, whereas this regulation is inhibited by the homeodomain proteins distal-less homeobox 5 (Dlx5) and Dlx6. However, the Bmp/Hand2 co-regulation and Dlx5/6 antagonism regulation on *Hand1* only occurred in cranial NCCs, not in cardiac NCCs [67]. Here, we found that Bmp signaling in the SHF likely regulates *Hand1* expression in both SHF progenitors and cardiac NCCs during OFT development. However, Hand2 in the SHF likely does not participate in this regulation given that *Hand2* expression was not altered upon Bmp deletion in the SHF.

To further understand the mechanism underlying sensitive expression responses of *Hand1* to Bmp dosages, we analyzed the 5'UTR of *Hand1* and identified conserved Bmp/Smad regulatory elements in the *Hand1* 5'UTR. We made the *Hand1* 5'UTR luciferase reporters and performed a luciferase assay. We found that both the Bmp treatment and overexpression of the constitutively active Bmp receptor (caALK3) induced Hand1 luciferase activity. To further consolidate this result, we also used inhibitory *Smad6* and *Smad1* shRNA to specifically block the Bmp/Smad signaling. We found *Hand1* luciferase reporter activity was decreased when co-transfected with *Smad6* and *Smad1* shRNA. Notably, our ChIP assays' data showed that Smad1/5/8 binds directly to *Hand1* 5'UTRs in the E9.5 wild-type mouse hearts. These data together indicated that the Bmp regulation on *Hand1* functions through the Smads-mediated canonical Bmp signaling pathway.

Both in vivo and in vitro studies of cardiac cardiomyocyte differentiation systems give strong evidence that Bmps can specifically regulate cardiac differentiation and cardiomyogenesis [59,60,68–70]. Our previous work reported that Bmp signaling enhances myocardial differentiation during OFT development [17]. During embryogenesis, *Hand1* is important for the morphogenic patterning and maturation of cardiomyocytes [20,27,29]. The conditional deletion of *Hand1* in cardiomyocytes, using *Nkx2.5*^{Cre} or a-myosin heavy chain Cre (*aMHC*^{Cre}) driver, results in multiple morphological anomalies that include cardiac conduction system defects, survivable interventricular septal defects, and abnormal LV papillary muscles [29]. Monzen et al. reported that Bmps induce P19 cells for cardiomyocyte differentiation through the mitogen-activated protein kinase kinase kinase TAK1 and cardiac TFs Csx/Nkx-2.5 and GATA-4 [59]. In our in vitro experiments examining *Hand1* expression in P19 cells with treatments of varying Bmp2 and 4 concentrations, we found that both Bmp2 and 4 promote *Hand1* expression in a dose-dependent manner. In addition, after Bmp2 and 4 treatment, cardiac TFs *Nkx2.5* and *Gata4* were also induced when *Hand1* expression was increased. These data, together with previously published findings, suggest that Bmps could potentially activate *Hand1* to promote cardiomyocyte differentiation. However, further electro-physiological experiments in P19 cells and in vivo investigations are still needed in the future.

In conclusion, to our knowledge, this study is the first to demonstrate that the canonical Bmp/Smad signaling pathway in the SHF directly activates *Hand1* expression in a dose-dependent manner during OFT development. Our findings also revealed a potential cell-autonomous and non-cell-autonomous function of Bmp signaling in the SHF and provided better insights into the molecular regulation of OFT development.

4. Materials and Methods

4.1. Mouse Alleles and Transgenic Lines

The *Bmp2* and *Bmp4* conditional null, *Bmp4^tetO^* gain-of-function allele and the *Mef2c^cre^* line were previously described [17,36].

4.2. Antibodies and Reagents

Antibodies used in this study include P-Smad1/5/8 (Cell Signaling Technology, #13820, Danvers, MA, USA), Smad1 (Upstate Biotechnology, Lake Placid, NY, USA), Hand1 (R&D systems, AF3168-SP, Minneapolis, MN, USA), and GAPDH antibody (Abcam, #ab9485, Cambridge, UK). Bmp2 (R&D, #355BM, Minneapolis, MN, USA) and Bmp4 (R&D, #314BP, Minneapolis, MN, USA) proteins were purchased from R&D systems.

4.3. Whole-Mount In Situ Hybridization

Whole-mount and section in situ hybridization was performed as previously described [17]. The plasmids for *Hand1* and *Hand2* in situ probes were previously described [71]. For all the experiments, at least three controls and mutant embryos were analyzed for each probe.

4.4. Quantitative Real Time RT-PCR

Total RNA from embryonic hearts was isolated using the RNeasy Micro Kit (QIA-GEN) [17]. Total RNA from P19 cells was extracted using TRIzol reagent (Life technologies, Carlsbad, CA, USA) following the manufacturer's protocol. For qRT-PCR assays, iScript Reverse Transcription Supermix (Bio-Rad, Hercules, CA, USA) was used for RT-PCR, and SYBR Green PCR Master Mix (Applied Biosystems, Waltham, MA, USA) was used for real-time thermal cycling (Applied Biosystems, Waltham, MA, USA). All error bars represent SEM. Primers used for qRT-PCR were Gapdh forward, 5′-TGGCAAAGTGGAGATTGTTGCC-3′. Gapdh reverse, 5′-AAGATGGTGATGGGCTTCCCG-3′. *Hand1* forward, 5′-GCCTACTTGA TGGACGTGCT-3′. *Hand1* reverse, 5′-CAACTCCCTTTTCCGCTTGC-3′. Gata4 forward, 5′-CCCTGGAAGACACCCCAATC-3′. Gata4 reverse, 5′-TTTGAATCCCCTCCTTCCGC-3′. Nkx2.5 forward, 5′-TGCTCTCCTGCTTTCCCAGCC-3′. Nkx2.5 reverse, 5′-CTTTGTCCAGC TCCACTGCCTT-3′. Id1 forward, 5′-TTGGTCTGTCGGAGCAAAGCGT-3′. Id1 reverse, 5′-CGTGAGTAGCAGCCGTTCATGT-3′.

4.5. Chromatin Immunoprecipitation

E9.5 wild-type mice embryonic hearts were dissected and followed by chromatin immunoprecipitation (ChIP) analysis, which was performed using a ChIP assay kit (Up-state) [17]. The two primers for amplifying the Bmp/Smad regulatory element in the 5′ upstream of the *Hand1* genomic sequence were sense, 5′-AACCCGCAGGGCACAAGAA-3′, and antisense, 5′-TGGTTGTGCAAGAGATTGTGA-3′. The PCR product was evaluated for appropriate size on a 2% agarose gel and was confirmed by sequencing. As negative controls, no antibody was used; in addition, normal rabbit immunoglobulin G was used as a replacement for the anti-Smad1/5/8 (sc-6031-R, Santa Cruz) to reveal nonspecific immunoprecipitation of the chromatin.

4.6. Luciferase Reporter Assays

Expression and reporter plasmids were described above. Constitutively active ALK3 (caALK3), pcDNA3.1-*Smad6* expression plasmid, and pSR siSmad1 plasmid were previously described [17]. To generate the *Hand1* luciferase reporter plasmid, 2314bp 5′upstream of *Hand1* genomic sequence was amplified using a high-fidelity PCR system (Roche) with two oligonucleotides, sense, 5′-<u>ACGCGT</u>AGGGTACAAAGGGAAACTGGGTGT-3′ (underlined letters indicate the MluI restriction site introduced for subcloning), and antisense, 5′-<u>CTCGAG</u>TGCTCACTCCCTGTACTGAACCTA-3′ (underlined letters indicate the XhoI restriction site introduced for subcloning), and subcloned into pGL3-Basic vector (Promega).

P19 cells were transfected using Lipofectamine 3000 (Invitrogen). Luciferase activity assays were performed using the Luciferase Assay System (Promega).

4.7. Western Blotting

Western blot was performed as previously described using standard techniques [72]. After 6 days with or without Bmp2/4 treatment, P19 cells were harvested and lysated using 0.5% NP-40 lysis buffer (50 mM Tris-HCl pH 7.5, 150 mM NaCl, 0.5% NP-40, 10% glycerol, phosphatase and protease inhibitors) for 10 min on ice, and centrifuged at 14,000 rpm for 10 min at 4 °C. For Western blot analysis, the proteins were loaded and separated by SDS-PAGE, and transferred onto a PVDF membrane (Millipore, IPVH00010). The membranes were blocked in 5% non-fat milk for 1 h at room temperature and incubated with primary antibodies overnight at 4 °C. The membranes were incubated with HRP-conjugated secondary antibodies for 2 h at room temperature and were imaged by Bio-rad imaging systems. Antibodies used for immunoblotting are mentioned above.

4.8. Cell Culture

Mouse embryonic carcinoma cell line P19 were maintained in Minimum Essential Medium (MEM) supplemented with 10% fetal bovine serum (FBS) at 37 °C in a humidified incubator with 5% CO_2. P19 cells were seeded at a concentration of 0.5×10^6 cells per well in 6-well plates and cultured for 24 h to reach 100% confluence (day 0). To induce differentiation, cells were washed in PBS and cultured in MEM supplemented with 10% fetal bovine serum (FBS), Bmp2 or Bmp4, referred to as differentiation medium.

Supplementary Materials: The following are available online at https://www.mdpi.com/article/10.3390/ijms22189835/s1, Figure S1: *Hand2* is not changed in control and *Bmp2/4* mutant; Figure S2: Diagram of the Bmp/Smad regulatory element in the 5′ upstream of *Hand1* genomic sequence (at upper) and its phylogenetic sequence alignment (at lower).

Author Contributions: J.W. conceived and supervised the project. M.Z., S.E., D.A. and J.W. performed the experiments. M.Z. and J.W. performed the data analyses. M.Z. and S.E. wrote the original draft. D.A. and J.W. reviewed and edited the manuscript. All authors have read and agreed to the published version of the manuscript.

Funding: J.W. was supported by the National Institutes of Health R01HL142704, K01DE026561 and R01DE029014.

Institutional Review Board Statement: Not applicable.

Informed Consent Statement: Not applicable.

Data Availability Statement: Data are contained within the article or Supplementary Materials.

Acknowledgments: We appreciate James F. Martin for providing mouse alleles, Xin-hua Feng and Joan Massagué for providing plasmids and Yan Bai for helping with the luciferase assay, as well as Joshua W. Vincentz and Anthony B. Firulli for discussions.

Conflicts of Interest: The authors declare no conflict of interest.

References

1. Warnes, C.A.; Liberthson, R.; Danielson, G.K.; Dore, A.; Harris, L.; Hoffman, J.I.; Somerville, J.; Williams, R.G.; Webb, G.D. Task force 1: The changing profile of congenital heart disease in adult life. *J. Am. Coll. Cardiol.* **2001**, *37*, 1170–1175. [CrossRef]
2. High, F.A.; Jain, R.; Stoller, J.Z.; Antonucci, N.B.; Lu, M.M.; Loomes, K.M.; Kaestner, K.H.; Pear, W.S.; Epstein, J.A. Murine Jagged1/Notch signaling in the second heart field orchestrates Fgf8 expression and tissue-tissue interactions during outflow tract development. *J. Clin. Investig.* **2009**, *119*, 1986–1996. [CrossRef]
3. Erhardt, S.; Zheng, M.J.; Zhao, X.L.; Le, T.P.; Findley, T.O.; Wang, J. The Cardiac Neural Crest Cells in Heart Development and Congenital Heart Defects. *J. Cardiovasc. Dev. Dis.* **2021**, *8*, 89. [CrossRef] [PubMed]
4. Urist, M.R. Bone: Formation by autoinduction. *Science* **1965**, *150*, 893–899. [CrossRef] [PubMed]
5. Feng, X.H.; Derynck, R. Specificity and versatility in tgf-beta signaling through Smads. *Annu. Rev. Cell Dev. Biol.* **2005**, *21*, 659–693. [CrossRef] [PubMed]
6. ten Dijke, P.; Hill, C.S. New insights into TGF-beta-Smad signalling. *Trends Biochem. Sci.* **2004**, *29*, 265–273. [CrossRef] [PubMed]

7.   Derynck, R.; Zhang, Y.E. Smad-dependent and Smad-independent pathways in TGF-beta family signalling. *Nature* **2003**, *425*, 577–584. [CrossRef] [PubMed]
8.   Bai, Y.; Wang, J.; Morikawa, Y.; Bonilla-Claudio, M.; Klysik, E.; Martin, J.F. Bmp signaling represses Vegfa to promote outflow tract cushion development. *Development* **2013**, *140*, 3395–3402. [CrossRef]
9.   Wang, J.; Greene, S.B.; Martin, J.F. BMP signaling in congenital heart disease: New developments and future directions. *Birth Defects Res. Part A Clin. Mol. Teratol.* **2011**, *91*, 441–448. [CrossRef] [PubMed]
10.   Liu, W.; Selever, J.; Wang, D.; Lu, M.F.; Moses, K.A.; Schwartz, R.J.; Martin, J.F. Bmp4 signaling is required for outflow-tract septation and branchial-arch artery remodeling. *Proc. Natl. Acad. Sci. USA* **2004**, *101*, 4489–4494. [CrossRef] [PubMed]
11.   Prall, O.W.; Menon, M.K.; Solloway, M.J.; Watanabe, Y.; Zaffran, S.; Bajolle, F.; Biben, C.; McBride, J.J.; Robertson, B.R.; Chaulet, H.; et al. An Nkx2-5/Bmp2/Smad1 negative feedback loop controls heart progenitor specification and proliferation. *Cell* **2007**, *128*, 947–959. [CrossRef] [PubMed]
12.   Ma, L.; Lu, M.F.; Schwartz, R.J.; Martin, J.F. Bmp2 is essential for cardiac cushion epithelial-mesenchymal transition and myocardial patterning. *Development* **2005**, *132*, 5601–5611. [CrossRef] [PubMed]
13.   Zhang, H.; Bradley, A. Mice deficient for BMP2 are nonviable and have defects in amnion/chorion and cardiac development. *Development* **1996**, *122*, 2977–2986. [CrossRef] [PubMed]
14.   Rivera-Feliciano, J.; Tabin, C.J. Bmp2 instructs cardiac progenitors to form the heart-valve-inducing field. *Dev. Biol.* **2006**, *295*, 580–588. [CrossRef]
15.   Jiao, K.; Kulessa, H.; Tompkins, K.; Zhou, Y.; Batts, L.; Baldwin, H.S.; Hogan, B.L. An essential role of Bmp4 in the atrioventricular septation of the mouse heart. *Genes Dev.* **2003**, *17*, 2362–2367. [CrossRef]
16.   Chen, H.; Shi, S.; Acosta, L.; Li, W.; Lu, J.; Bao, S.; Chen, Z.; Yang, Z.; Schneider, M.D.; Chien, K.R.; et al. BMP10 is essential for maintaining cardiac growth during murine cardiogenesis. *Development* **2004**, *131*, 2219–2231. [CrossRef] [PubMed]
17.   Wang, J.; Greene, S.B.; Bonilla-Claudio, M.; Tao, Y.; Zhang, J.; Bai, Y.; Huang, Z.; Black, B.L.; Wang, F.; Martin, J.F. Bmp signaling regulates myocardial differentiation from cardiac progenitors through a MicroRNA-mediated mechanism. *Dev. Cell* **2010**, *19*, 903–912. [CrossRef]
18.   Hollenberg, S.M.; Sternglanz, R.; Cheng, P.F.; Weintraub, H. Identification of a new family of tissue-specific basic helix-loop-helix proteins with a two-hybrid system. *Mol. Cell. Biol.* **1995**, *15*, 3813–3822. [CrossRef] [PubMed]
19.   Scott, I.C.; Anson-Cartwright, L.; Riley, P.; Reda, D.; Cross, J.C. The HAND1 basic helix-loop-helix transcription factor regulates trophoblast differentiation via multiple mechanisms. *Mol. Cell. Biol.* **2000**, *20*, 530–541. [CrossRef]
20.   Firulli, A.B.; McFadden, D.G.; Lin, Q.; Srivastava, D.; Olson, E.N. Heart and extra-embryonic mesodermal defects in mouse embryos lacking the bHLH transcription factor Hand1. *Nat. Genet.* **1998**, *18*, 266–270. [CrossRef]
21.   Srivastava, D.; Cserjesi, P.; Olson, E.N. A subclass of bHLH proteins required for cardiac morphogenesis. *Science* **1995**, *270*, 1995–1999. [CrossRef] [PubMed]
22.   Srivastava, D.; Thomas, T.; Lin, Q.; Kirby, M.L.; Brown, D.; Olson, E.N. Regulation of cardiac mesodermal and neural crest development by the bHLH transcription factor, dHAND. *Nat. Genet.* **1997**, *16*, 154–160. [CrossRef] [PubMed]
23.   Cserjesi, P.; Brown, D.; Lyons, G.E.; Olson, E.N. Expression of the novel basic helix-loop-helix gene eHAND in neural crest derivatives and extraembryonic membranes during mouse development. *Dev. Biol.* **1995**, *170*, 664–678. [CrossRef] [PubMed]
24.   Vincentz, J.W.; Toolan, K.P.; Zhang, W.; Firulli, A.B. Hand factor ablation causes defective left ventricular chamber development and compromised adult cardiac function. *PLoS Genet.* **2017**, *13*, e1006922. [CrossRef]
25.   Smart, N.; Dube, K.N.; Riley, P.R. Identification of Thymosin beta4 as an effector of Hand1-mediated vascular development. *Nat. Commun.* **2010**, *1*, 46. [CrossRef] [PubMed]
26.   Firulli, A.B. A HANDful of questions: The molecular biology of the heart and neural crest derivatives (HAND)-subclass of basic helix-loop-helix transcription factors. *Gene* **2003**, *312*, 27–40. [CrossRef]
27.   Riley, P.; Anson-Cartwright, L.; Cross, J.C. The Hand1 bHLH transcription factor is essential for placentation and cardiac morphogenesis. *Nat. Genet.* **1998**, *18*, 271–275. [CrossRef]
28.   Risebro, C.A.; Smart, N.; Dupays, L.; Breckenridge, R.; Mohun, T.J.; Riley, P.R. Hand1 regulates cardiomyocyte proliferation versus differentiation in the developing heart. *Development* **2006**, *133*, 4595–4606. [CrossRef]
29.   Firulli, B.A.; George, R.M.; Harkin, J.; Toolan, K.P.; Gao, H.; Liu, Y.; Zhang, W.; Field, L.J.; Liu, Y.; Shou, W.; et al. HAND1 loss-of-function within the embryonic myocardium reveals survivable congenital cardiac defects and adult heart failure. *Cardiovasc. Res.* **2020**, *116*, 605–618. [CrossRef]
30.   Reamon-Buettner, S.M.; Ciribilli, Y.; Traverso, I.; Kuhls, B.; Inga, A.; Borlak, J. A functional genetic study identifies HAND1 mutations in septation defects of the human heart. *Hum. Mol. Genet.* **2009**, *18*, 3567–3578. [CrossRef]
31.   Zhou, Y.M.; Dai, X.Y.; Qiu, X.B.; Yuan, F.; Li, R.G.; Xu, Y.J.; Qu, X.K.; Huang, R.T.; Xue, S.; Yang, Y.Q. HAND1 loss-of-function mutation associated with familial dilated cardiomyopathy. *Clin. Chem. Lab. Med.* **2016**, *54*, 1161–1167. [CrossRef]
32.   Yamagishi, H.; Olson, E.N.; Srivastava, D. The basic helix-loop-helix transcription factor, dHAND, is required for vascular development. *J. Clin. Investig.* **2000**, *105*, 261–270. [CrossRef]
33.   Sun, Y.M.; Wang, J.; Qiu, X.B.; Yuan, F.; Li, R.G.; Xu, Y.J.; Qu, X.K.; Shi, H.Y.; Hou, X.M.; Huang, R.T.; et al. A HAND2 Loss-of-Function Mutation Causes Familial Ventricular Septal Defect and Pulmonary Stenosis. *G3* **2016**, *6*, 987–992. [CrossRef] [PubMed]

34. Buckingham, M.; Meilhac, S.; Zaffran, S. Building the mammalian heart from two sources of myocardial cells. *Nat. Rev. Genet.* **2005**, *6*, 826–835. [CrossRef] [PubMed]

35. Brade, T.; Pane, L.S.; Moretti, A.; Chien, K.R.; Laugwitz, K.L. Embryonic heart progenitors and cardiogenesis. *Cold Spring Harb. Perspect. Med.* **2013**, *3*, a013847. [CrossRef] [PubMed]

36. Bonilla-Claudio, M.; Wang, J.; Bai, Y.; Klysik, E.; Selever, J.; Martin, J.F. Bmp signaling regulates a dose-dependent transcriptional program to control facial skeletal development. *Development* **2012**, *139*, 709–719. [CrossRef] [PubMed]

37. Macias, M.J.; Martin-Malpartida, P.; Massague, J. Structural determinants of Smad function in TGF-beta signaling. *Trends Biochem. Sci.* **2015**, *40*, 296–308. [CrossRef]

38. Martin-Malpartida, P.; Batet, M.; Kaczmarska, Z.; Freier, R.; Gomes, T.; Aragon, E.; Zou, Y.; Wang, Q.; Xi, Q.; Ruiz, L.; et al. Structural basis for genome wide recognition of 5-bp GC motifs by SMAD transcription factors. *Nat. Commun.* **2017**, *8*, 2070. [CrossRef]

39. Derynck, R.; Zhang, Y.; Feng, X.H. Smads: Transcriptional activators of TGF-beta responses. *Cell* **1998**, *95*, 737–740. [CrossRef]

40. Massague, J.; Seoane, J.; Wotton, D. Smad transcription factors. *Genes Dev.* **2005**, *19*, 2783–2810. [CrossRef] [PubMed]

41. Morikawa, M.; Koinuma, D.; Miyazono, K.; Heldin, C.H. Genome-wide mechanisms of Smad binding. *Oncogene* **2013**, *32*, 1609–1615. [CrossRef] [PubMed]

42. Shi, Y.; Massague, J. Mechanisms of TGF-beta signaling from cell membrane to the nucleus. *Cell* **2003**, *113*, 685–700. [CrossRef]

43. Morikawa, M.; Koinuma, D.; Tsutsumi, S.; Vasilaki, E.; Kanki, Y.; Heldin, C.H.; Aburatani, H.; Miyazono, K. ChIP-seq reveals cell type-specific binding patterns of BMP-specific Smads and a novel binding motif. *Nucleic Acids Res.* **2011**, *39*, 8712–8727. [CrossRef]

44. Kim, J.; Johnson, K.; Chen, H.J.; Carroll, S.; Laughon, A. Drosophila Mad binds to DNA and directly mediates activation of vestigial by Decapentaplegic. *Nature* **1997**, *388*, 304–308. [CrossRef]

45. BabuRajendran, N.; Palasingam, P.; Narasimhan, K.; Sun, W.; Prabhakar, S.; Jauch, R.; Kolatkar, P.R. Structure of Smad1 MH1/DNA complex reveals distinctive rearrangements of BMP and TGF-beta effectors. *Nucleic Acids Res.* **2010**, *38*, 3477–3488. [CrossRef] [PubMed]

46. Chai, N.; Li, W.X.; Wang, J.; Wang, Z.X.; Yang, S.M.; Wu, J.W. Structural basis for the Smad5 MH1 domain to recognize different DNA sequences. *Nucleic Acids Res.* **2017**, *45*, 6255–6257. [CrossRef] [PubMed]

47. Fujii, M.; Takeda, K.; Imamura, T.; Aoki, H.; Sampath, T.K.; Enomoto, S.; Kawabata, M.; Kato, M.; Ichijo, H.; Miyazono, K. Roles of bone morphogenetic protein type I receptors and Smad proteins in osteoblast and chondroblast differentiation. *Mol. Biol. Cell* **1999**, *10*, 3801–3813. [CrossRef]

48. Hata, A.; Lagna, G.; Massague, J.; Hemmati-Brivanlou, A. Smad6 inhibits BMP/Smad1 signaling by specifically competing with the Smad4 tumor suppressor. *Genes Dev.* **1998**, *12*, 186–197. [CrossRef]

49. Callis, T.E.; Cao, D.; Wang, D.Z. Bone morphogenetic protein signaling modulates myocardin transactivation of cardiac genes. *Circ. Res.* **2005**, *97*, 992–1000. [CrossRef]

50. Brown, C.O., 3rd; Chi, X.; Garcia-Gras, E.; Shirai, M.; Feng, X.H.; Schwartz, R.J. The cardiac determination factor, Nkx2-5, is activated by mutual cofactors GATA-4 and Smad1/4 via a novel upstream enhancer. *J. Biol. Chem.* **2004**, *279*, 10659–10669. [CrossRef]

51. Shi, Y.; Katsev, S.; Cai, C.; Evans, S. BMP signaling is required for heart formation in vertebrates. *Dev. Biol.* **2000**, *224*, 226–237. [CrossRef]

52. McBurney, M.W.; Jones-Villeneuve, E.M.; Edwards, M.K.; Anderson, P.J. Control of muscle and neuronal differentiation in a cultured embryonal carcinoma cell line. *Nature* **1982**, *299*, 165–167. [CrossRef] [PubMed]

53. McBurney, M.W. P19 embryonal carcinoma cells. *Int. J. Dev. Biol.* **1993**, *37*, 135–140. [PubMed]

54. Wilton, S.; Skerjanc, I. Factors in serum regulate muscle development in P19 cells. *In Vitro Cell. Dev. Biol. Anim.* **1999**, *35*, 175–177. [CrossRef]

55. Skerjanc, I.S. Cardiac and skeletal muscle development in P19 embryonal carcinoma cells. *Trends Cardiovasc. Med.* **1999**, *9*, 139–143. [CrossRef]

56. Grepin, C.; Robitaille, L.; Antakly, T.; Nemer, M. Inhibition of transcription factor GATA-4 expression blocks in vitro cardiac muscle differentiation. *Mol. Cell. Biol.* **1995**, *15*, 4095–4102. [CrossRef]

57. Jamali, M.; Rogerson, P.J.; Wilton, S.; Skerjanc, I.S. Nkx2-5 activity is essential for cardiomyogenesis. *J. Biol. Chem.* **2001**, *276*, 42252–42258. [CrossRef]

58. Skerjanc, I.S.; Petropoulos, H.; Ridgeway, A.G.; Wilton, S. Myocyte enhancer factor 2C and Nkx2-5 up-regulate each other's expression and initiate cardiomyogenesis in P19 cells. *J. Biol. Chem.* **1998**, *273*, 34904–34910. [CrossRef]

59. Monzen, K.; Shiojima, I.; Hiroi, Y.; Kudoh, S.; Oka, T.; Takimoto, E.; Hayashi, D.; Hosoda, T.; Habara-Ohkubo, A.; Nakaoka, T.; et al. Bone morphogenetic proteins induce cardiomyocyte differentiation through the mitogen-activated protein kinase kinase kinase TAK1 and cardiac transcription factors Csx/Nkx-2.5 and GATA-4. *Mol. Cell. Biol.* **1999**, *19*, 7096–7105. [CrossRef]

60. Jamali, M.; Karamboulas, C.; Rogerson, P.J.; Skerjanc, I.S. BMP signaling regulates Nkx2-5 activity during cardiomyogenesis. *FEBS Lett.* **2001**, *509*, 126–130. [CrossRef]

61. Benezra, R.; Davis, R.L.; Lockshon, D.; Turner, D.L.; Weintraub, H. The protein Id: A negative regulator of helix-loop-helix DNA binding proteins. *Cell* **1990**, *61*, 49–59. [CrossRef]

62. Norton, J.D.; Deed, R.W.; Craggs, G.; Sablitzky, F. Id helix-loop-helix proteins in cell growth and differentiation. *Trends Cell Biol.* **1998**, *8*, 58–65.
63. Hollnagel, A.; Oehlmann, V.; Heymer, J.; Ruther, U.; Nordheim, A. Id genes are direct targets of bone morphogenetic protein induction in embryonic stem cells. *J. Biol. Chem.* **1999**, *274*, 19838–19845. [CrossRef]
64. Firulli, B.A.; McConville, D.P.; Byers, J.S., 3rd; Vincentz, J.W.; Barnes, R.M.; Firulli, A.B. Analysis of a Hand1 hypomorphic allele reveals a critical threshold for embryonic viability. *Dev. Dyn.* **2010**, *239*, 2748–2760. [CrossRef]
65. Thomas, T.; Yamagishi, H.; Overbeek, P.A.; Olson, E.N.; Srivastava, D. The bHLH factors, dHAND and eHAND, specify pulmonary and systemic cardiac ventricles independent of left-right sidedness. *Dev. Biol.* **1998**, *196*, 228–236. [CrossRef]
66. Gaussin, V.; Van de Putte, T.; Mishina, Y.; Hanks, M.C.; Zwijsen, A.; Huylebroeck, D.; Behringer, R.R.; Schneider, M.D. Endocardial cushion and myocardial defects after cardiac myocyte-specific conditional deletion of the bone morphogenetic protein receptor ALK3. *Proc. Natl. Acad. Sci. USA* **2002**, *99*, 2878–2883. [CrossRef]
67. Vincentz, J.W.; Casasnovas, J.J.; Barnes, R.M.; Que, J.; Clouthier, D.E.; Wang, J.; Firulli, A.B. Exclusion of Dlx5/6 expression from the distal-most mandibular arches enables BMP-mediated specification of the distal cap. *Proc. Natl. Acad. Sci. USA* **2016**, *113*, 7563–7568. [CrossRef] [PubMed]
68. Winnier, G.; Blessing, M.; Labosky, P.A.; Hogan, B.L. Bone morphogenetic protein-4 is required for mesoderm formation and patterning in the mouse. *Genes Dev.* **1995**, *9*, 2105–2116. [CrossRef] [PubMed]
69. Schultheiss, T.M.; Burch, J.B.; Lassar, A.B. A role for bone morphogenetic proteins in the induction of cardiac myogenesis. *Genes Dev.* **1997**, *11*, 451–462. [CrossRef]
70. Yuasa, S.; Itabashi, Y.; Koshimizu, U.; Tanaka, T.; Sugimura, K.; Kinoshita, M.; Hattori, F.; Fukami, S.; Shimazaki, T.; Ogawa, S.; et al. Transient inhibition of BMP signaling by Noggin induces cardiomyocyte differentiation of mouse embryonic stem cells. *Nat. Biotechnol.* **2005**, *23*, 607–611. [CrossRef] [PubMed]
71. McFadden, D.G.; Barbosa, A.C.; Richardson, J.A.; Schneider, M.D.; Srivastava, D.; Olson, E.N. The Hand1 and Hand2 transcription factors regulate expansion of the embryonic cardiac ventricles in a gene dosage-dependent manner. *Development* **2005**, *132*, 189–201. [CrossRef] [PubMed]
72. Wang, J.; Klysik, E.; Sood, S.; Johnson, R.L.; Wehrens, X.H.; Martin, J.F. Pitx2 prevents susceptibility to atrial arrhythmias by inhibiting left-sided pacemaker specification. *Proc. Natl. Acad. Sci. USA* **2010**, *107*, 9753–9758. [CrossRef] [PubMed]

International Journal of
Molecular Sciences

Article

Compromised Biomechanical Properties, Cell–Cell Adhesion and Nanotubes Communication in Cardiac Fibroblasts Carrying the Lamin A/C D192G Mutation

Veronique Lachaize [1], Brisa Peña [2,3,4], Catalin Ciubotaru [5], Dan Cojoc [5], Suet Nee Chen [2], Matthew R. G. Taylor [2], Luisa Mestroni [2] and Orfeo Sbaizero [1,2,*]

[1] Department of Engineering and Architecture, University of Trieste, Via Valerio 10, 34127 Trieste, Italy; lachaize.v@gmail.com
[2] CU-Cardiovascular Institute, University of Colorado Anschutz Medical Campus, 12700 E. 19th Ave., Aurora, CO 80045, USA; brisa.penacastellanos@cuanschutz.edu (B.P.); suet.chen@cuanschutz.edu (S.N.C.); matthew.taylor@cuanschutz.edu (M.R.G.T.); luisa.mestroni@cuanschutz.edu (L.M.)
[3] Consortium for Fibrosis Research & Translation, Anschutz Medical Campus, University of Colorado, 12700 E. 19th Ave., Aurora, CO 80045, USA
[4] Bioengineering Department, University of Colorado Denver Anschutz Medical Campus, Bioscience 2 1270 E. Montview Ave., Suite 100, Aurora, CO 80045, USA
[5] Institute of Materials, National Research Council of Italy (CNR_IOM), Area Science Park Basovizza, 34149 Trieste, Italy; ciubotaru@iom.cnr.it (C.C.); cojoc@iom.cnr.it (D.C.)
* Correspondence: sbaizero@units.it

Citation: Lachaize, V.; Peña, B.; Ciubotaru, C.; Cojoc, D.; Chen, S.N.; Taylor, M.R.G.; Mestroni, L.; Sbaizero, O. Compromised Biomechanical Properties, Cell–Cell Adhesion and Nanotubes Communication in Cardiac Fibroblasts Carrying the Lamin A/C D192G Mutation. *Int. J. Mol. Sci.* **2021**, *22*, 9193. https:// doi.org/10.3390/ijms22179193

Academic Editors: Nicole Wagner and Kay-Dietrich Wagner

Received: 25 June 2021
Accepted: 5 August 2021
Published: 25 August 2021

Publisher's Note: MDPI stays neutral with regard to jurisdictional claims in published maps and institutional affiliations.

Abstract: Clinical effects induced by arrhythmogenic cardiomyopathy (ACM) originate from a large spectrum of genetic variations, including the missense mutation of the lamin A/C gene (*LMNA*), *LMNA* D192G. The aim of our study was to investigate the biophysical and biomechanical impact of the *LMNA* D192G mutation on neonatal rat ventricular fibroblasts (NRVF). The main findings in mutated NRVFs were: (i) cytoskeleton disorganization (actin and intermediate filaments); (ii) decreased elasticity of NRVFs; (iii) altered cell–cell adhesion properties, that highlighted a strong effect on cellular communication, in particular on tunneling nanotubes (TNTs). In mutant-expressing fibroblasts, these nanotubes were weakened with altered mechanical properties as shown by atomic force microscopy (AFM) and optical tweezers. These outcomes complement prior investigations on *LMNA* mutant cardiomyocytes and suggest that the *LMNA* D192G mutation impacts the biomechanical properties of both cardiomyocytes and cardiac fibroblasts. These observations could explain how this mutation influences cardiac biomechanical pathology and the severity of ACM in *LMNA*-cardiomyopathy.

Keywords: arrhythmogenic cardiomyopathy (ACM); lamin A/C; atomic force microscopy (AFM); cell–cell adhesion; neonatal rat ventricular fibroblasts (NRVF); tunneling nanotubes (TNT)

1. Introduction

Arrhythmogenic cardiomyopathy (ACM) is a myocardial disease characterized by a high risk of life-threatening ventricular arrhythmias, sudden cardiac death, and progression towards heart failure [1,2]. This disease, which can predominantly affect the right ventricle (arrhythmogenic right ventricular cardiomyopathy or ARVC), the left ventricle (arrhythmogenic left ventricular cardiomyopathy (ALVC), or arrhythmogenic dilated cardiomyopathy (DCM)), or both (biventricular ACM) is generated by a large spectrum of genes including the lamin A/C gene (*LMNA*) [1]. Many ACM genes, such as *LMNA* [3], induce important alterations of the signal transduction and the intercellular communication mechanism in cardiac cells [4–6]. In 2015, Lanzicher et al. investigated the biomechanical impact of *LMNA* missense mutations in neonatal rat ventricular myocytes (NRVM), and found damages on the actin cytoskeleton, suggesting loss of actin filament anchorage by dysfunction of the laminar proteins at the nuclear membrane as the initiating alteration [7,8]. Their results were further confirmed by different studies, in which it was demonstrated that lamina network mutations induce an alteration of actin filaments, microtubule, and intermediate filament anchorage in the nuclear membrane, a region critical for cytoskeleton organization [9,10]. Moreover, the link between the lamina network and the cytoskeleton filaments is ensured by a protein complex: LInker of Nucleoskeleton and Cytoskeleton (LINC). A functional nuclear lamin cortex is therefore necessary for the cytoskeleton anchorage, at the nuclear membrane LINC complex, and for maintaining the cytoskeleton integrity [11,12]. As previously reported, atomic force microscopy (AFM) studies highlighted that cardiomyocytes expressing a LMNA D192G mutation have alterations of membrane biomechanical properties (stiffness and fragility), due to cytoskeletal structural modifications, and decreased adhesion between the cell membrane and the AFM probe [7,13]. Furthermore, LMNA mutations appear to disrupt cell adhesion, potentially hampering the communication between cardiomyocytes [7]. However, the link between the defective lamina network, defective cell adhesion, and the ACM phenotype has not been demonstrated.

In 2004, A. Rustom et al. reported a cell–cell communication system which offers a new perspective: cellular communication by tunneling nanotubes (TNT). In their work, this communication works as a "highway" connecting two cells and it is composed by nanoscale membrane tubes with diameters between 50 and 200 nm, containing actin filaments and allowing the exchange of biological material (such as organelles, plasma membrane components, and cytoplasmic small particles [14] like mitochondria [15], HLA proteins [16], viruses [17], lysosomes [18], and calcium [16]) (Figure 1A).

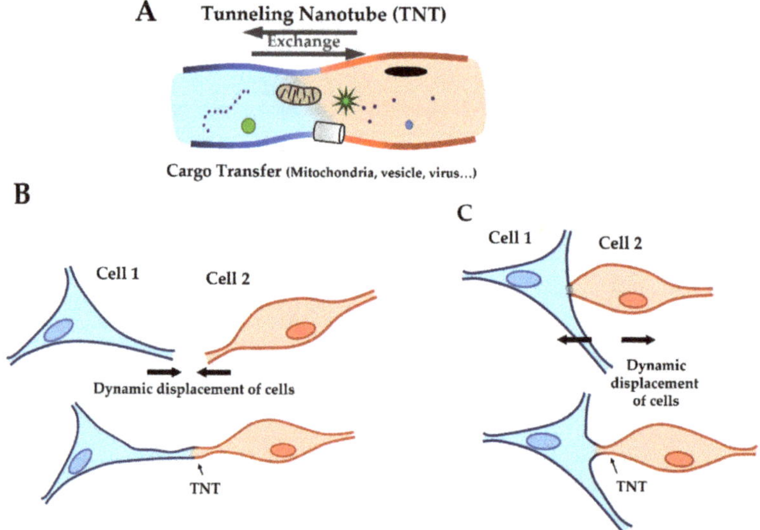

Figure 1. Cartoon showing tunneling nanotubes (TNT). (**A**) Cargo transfer through TNT with exchange of particles (such as mitochondria, ions, virus …) between two cells. (**B**) TNT construction by extension of filopodia. (**C**) TNT construction by contact, formation of TNT after physical contact between two cells and distancing of cells.

In particular conditions (such as cell rescue phenomena), microtubules can be found in TNTs, but in typical conditions, TNTs are composed mainly of actin filaments [19–21]. In the literature, different studies assessed the impact of this communication mechanism in physio/pathology condition, like the infection of HIV [17] and cancer cell proliferation [18,22,23]. In in vivo conditions, TNT formation uses two different mechanisms: the filopodia production, followed with lamellipodia and, finally, the TNT between two cells

(Figure 1B), or by direct contact between two cells (Figure 1C). Some studies suggest that the interaction of two connexin 43 (Cx43) proteins can initiate TNT formation [14,24,25]. Cx43 interactions stabilize the connection between two cells, allowing the cytoskeleton filament of each cell to organize and form TNTs. Recently, K. He et al. not only demonstrated the presence of TNT in the cardiac tissue between two cardiomyocytes, but also between cardiac fibroblasts and cardiomyocytes [26]. These data suggest that the presence of TNT between fibroblasts and cardiomyocytes may be important in cellular crosstalk. Indeed, cardiac fibroblasts have an essential role: they are the largest cell populations in the heart, followed by cardiomyocytes, and cardiac fibroblasts contribute to the structural, biochemical, mechanical, and electrical properties of the myocardium tissue [27,28], modulation of the myocardial response [29], and production of the extra cellular matrix (ECM) [30]. If chronically stimulated they may induce fibrosis [31]. Fibroblasts could therefore have an important role in ACM. Indeed, the multiple functions of fibroblasts could be altered by TNT damages. Decrease of communication by TNT dysfunction could impact the modulation of the myocardial response to stress and trigger the production of the extracellular matrix. Therefore, the aim of our study was to investigate the impact of an *LMNA* mutation on cardiac fibroblasts. To this purpose, biophysical and biomechanical measurements were carried out on living neonatal rat ventricular fibroblasts (NRVF) expressing the LMNA D192G mutation. This mutation was selected because it represents a mutation associated with disruption of nuclear envelope morphology [32]. It has been described in one family with a severe phenotype where symptom onset and death/transplantation happened at 30.5 ($\pm$6.4) and 32.0 ($\pm$7.1) years, respectively: both patients died before the age of 40 years [33]. Although the small number of observations of this rare mutation must be taken into consideration, clinical data suggest that LMNA D192G have a very severe outcome.

We used a multidisciplinary approach including cell–cell adhesion measurements by cell spectroscopy using atomic force microscopy [34]. The cell–cell adhesion tests identified differences in adhesion patterns such as energy, maximum adhesion, rupture, and nanotube characteristics between two living NRVFs [35,36]. Moreover, we analyzed adhesion parameters in wild type (WT) NRVFs after blocking Cx43 to understand the role of this protein in the TNT. Furthermore, we studied the nanotubes' mechanical properties by optical tweezers (OT) and their structure by fluorescence staining experiments. OT measurements allowed us to quantify forces involved in the extrusion of only one nanotube under different conditions (mutated or WT). Here, we report how complementary and multidisciplinary biophysical and biological tools generated a new mechanistic insight into the impact of the LMNA D192G mutation on the fibroblast structure and function.

2. Results

Since adenovirus causes a transient transfection, to control if our cells were still expressing the episome, we checked them, during and after the experiment, using GFP fluorescence.

2.1. Immunofluorescence Showed Actin Disorganization and Brittle TNTs in Fibro-MT

Neonatal rat ventricular fibroblasts (NRVFs) were prepared as (i) non-treated, transfection control NRVFs (Fibro-CT), as (ii) wild type NRVFs (Fibro-WT), and as (iii) NRVFs with LMNA D192G mutation (Fibro-MT). For each condition, actin filaments and intermediate filaments were labeled by immunofluorescence, in order to observe the effect of LMNA D192G mutation on their cytoskeleton. Fibro-MT showed actin network damage which stands out as disorganization of the cytoskeleton when compared with both Fibro-WT and Fibro-CT (Figure 2A).

As far as the TNTs are concerned, when they were produced by Fibro-MT they displayed a "brittle" morphology with a thickness of 0.66 $\pm$ 0.36 µm (Figure 2B), while TNTs produced by Fibro-WT and Fibro-CT were thicker: 2.55 $\pm$ 1.29 µm and 2.63 $\pm$ 1.48 µm respectively.

In all three conditions actin and intermediate filaments were present inside the TNTs (Figure 2). When TNTs linked two cells, they could be as long as 278 $\pm$ 23 nm, but this length was greatly reduced in Fibro-MT (70 ± 11 nm). Our TNT observations showed therefore that the morphological damages in both the actin network and the intermediate filament in the LMNA D192G mutant fibroblasts also influence the TNTs structure making them thinner, shorter, and more brittle.

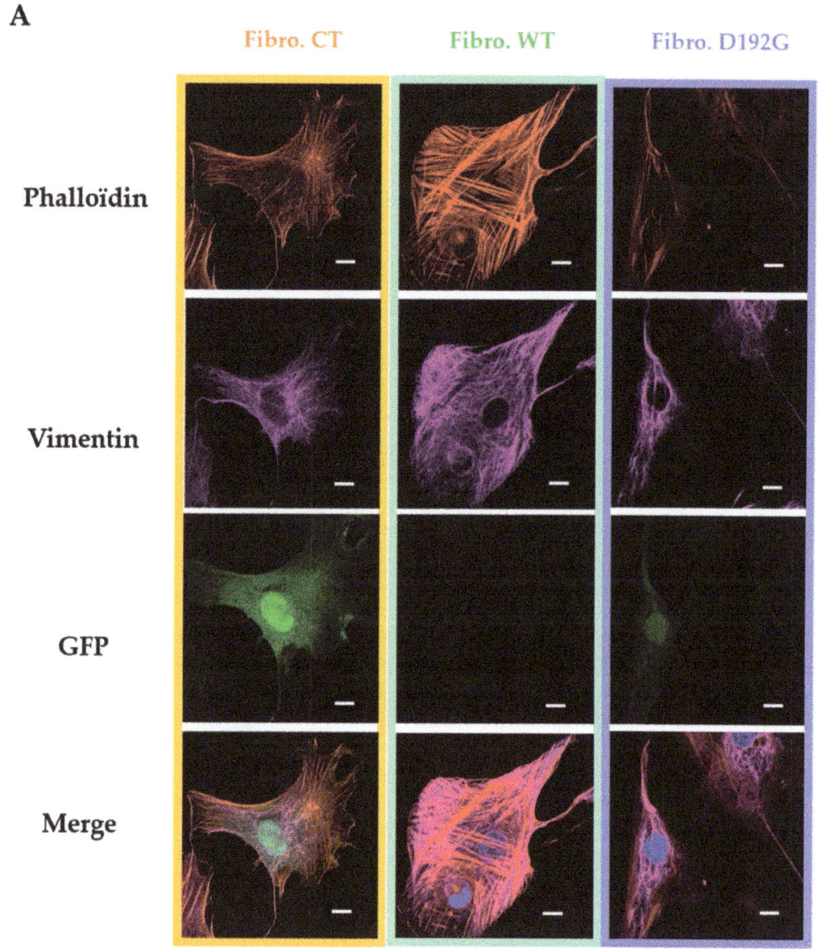

Figure 2. *Cont.*

B

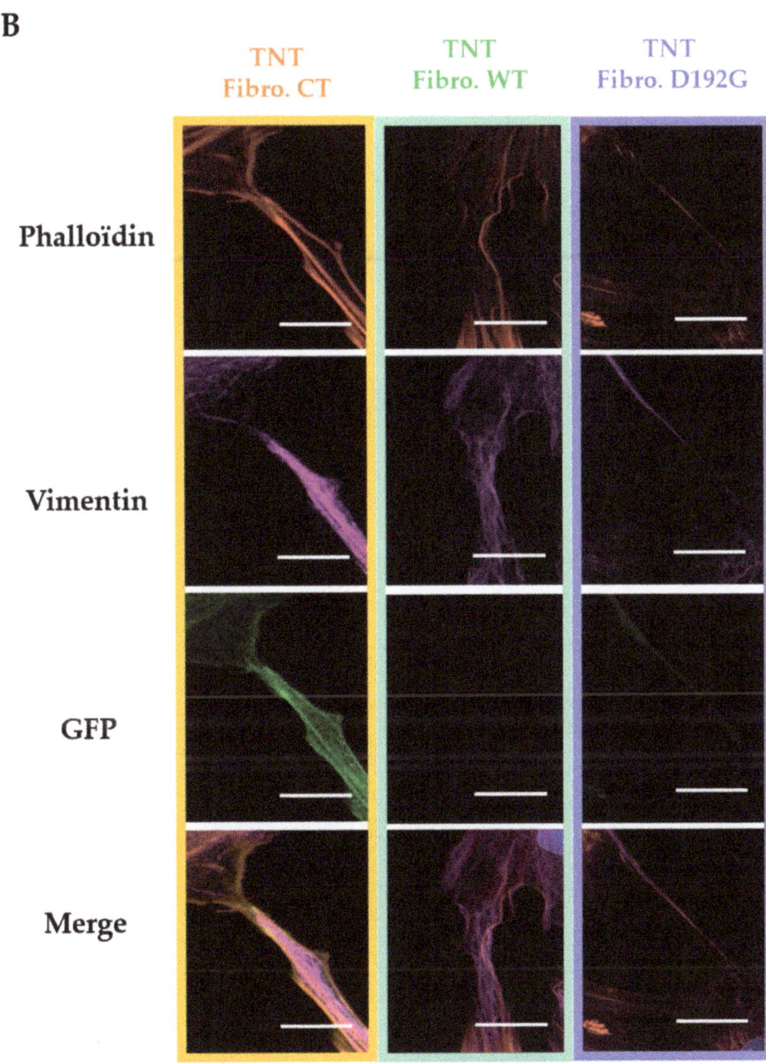

Figure 2. (**A**) Fluorescence staining of NRVF cytoskeleton. NRVF fluorescence staining in CT, WT, and lamin A/C D192G mutated condition with (i) Phalloidin to target both actin filaments and vimentin intermediate filaments, (ii) DAPI to target the nucleus, and (iii) green fluorescent protein (GFP) used as the transfection control after infection of adenoviral with, or without, the mutation lamin A/C D192 G. Scale bar (white), 10 µm. (**B**) Fluorescence staining of NRVF nanotubes. NRVF nanotubes' fluorescence staining in CT, WT, and lamin A/C D192G mutated condition with (i) Phalloidin to target both actin filaments and vimentin intermediate filaments, (ii) DAPI to target the nucleus, and (iii) green fluorescent protein (GFP) used as the transfection control after infection of adenoviral with, or without, the mutation lamin A/C D192 G. Scale bar (white), 10 µm.

The cylindrical shape of membrane nanotubes results from a compromise between surface tension and bending rigidity (1). Taking the analogy between tethers and filopodia as a guide, the force that the cell membrane exerts on the cytoskeleton at the ends of the TNT can be calculated if we idealize the TNT as an actin bundle cylinder of radius *r*. In this case, the force *Fmem* exerted on this cylinder by the membrane is equal to the product of the surface tension N of the membrane and the circumference of the cylinder:

$$F_{mem} = 2\pi r\, N \tag{1}$$

This force does not consider membrane bending and/or micro-membrane structures such as invaginations and protrusion deformation, both of which could provide further resistance to TNT extraction, but it is a good approximation that allows us to compare the optical tweezer (OT) results.

As far as the TNT mechanical properties are concerned (2), we can consider it as a beam of length L with one fixed end, and with an axial force F applied at the free end. The beam will buckle if F exceeds the buckling force

$$F_{buckle} = \pi^2 E\, I/4\, L^2 \tag{2}$$

where E is the Youngs modulus (elasticity) of the beam, and I is its moment of inertia.

The maximum TNT length before buckling occurs can be estimated assuming two contrasting hypotheses: (i) actin without cross-linkers inside the TNT (3) and (ii) highly cross-linked actin. In the latter hypothesis, the filaments can be seen as a single unit of effective radius *rbundle* (4).

The first hypothesis, actin without cross-linkers, produces a maximum length before buckling ($L_{N\text{-}CL}$):

$$L_{N-CL} = \sqrt{\frac{\pi^2\, \dfrac{\pi\, r_{Ac}^4}{4}\, E_{Ac}}{4\, \dfrac{F_{mem}}{n}}} \tag{3}$$

while highly cross-linked actin will have a maximum length (L_{CL}):

$$L_{CL} = \sqrt{\frac{\pi^2\, \dfrac{\pi\, \left(\sqrt{\pi}\, r_{Ac}\right)^4}{4}\, E_{Ac}}{4\, F_{mem}}} \tag{4}$$

where E_{AC} is the actin Young's modulus and r_{AC} is the actin filament radius, respectively. Therefore, the ratio of $(L_{CL})/(L_{N\text{-}CL})$ is $\sqrt{n}$. This latter result leads to the conclusion that the highest the number of filaments inside each TNT, the more distinct the dissimilarity in the maximum lengths between the cross-linked and uncross-linked TNT. Basically, if the actin inside the TNT is well organized and highly cross-linked, the TNT is more stable mechanically, and can develop much longer. The reduction of the slope of the tether force in Fibro-MT is consistent with two concomitant factors, (i) in order to form tethers, the membrane-cytoskeleton links must be ruptured, (ii) the much smaller contribution from the cytoskeleton remnants left inside the tether itself.

2.2. AFM Tests Highlighted Biomechanical Changes in Fibro-MT

In order to support and quantify these initial microscopic observations, measurements of the fibroblasts' biomechanical properties were also carried out using AFM. Cell elasticity (Young Modulus) was estimated using the Hertz–Sneddon model after collecting force curves obtained by force spectroscopy. Preliminary assessment of these two conditions, morphological and biomechanical comparison between Fibro-CT and Fibro-WT, namely NRVFs infected by the same protocol used for the mutated condition but with an empty plasmid, were also performed to check the adenovirus infection's impact on NRVFs. Fibro-CT and Fibro-WT shows no significant differences, regarding mechanical properties (using AFM), cell morphology and dimensions (using fluorescence labeling) (Supplementary Figure S1). Figure 3A shows the Young's modulus for Fibro-CT, Fibro-WT, and Fibro-MT.

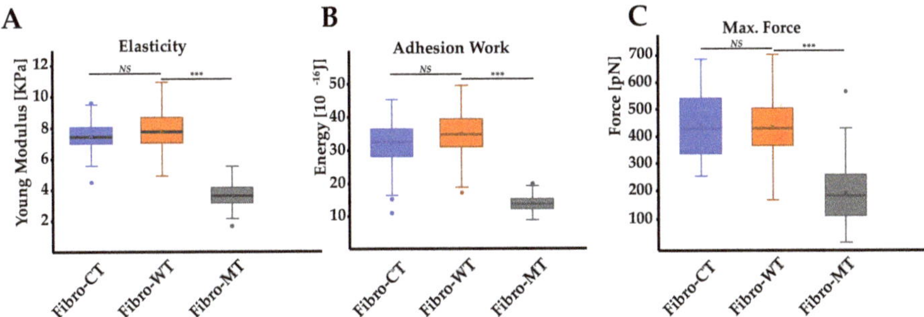

Figure 3. Biophysical properties of living NRVFs. (**A**) Comparison of elasticity (expressed as Young's modulus) between NRVFs with the WT adenoviral infection (Fibro-CT), NRVFs wild type (Fibro-WT), and NRVFs with *LMNA* D192G mutation (Fibro-MT) ($n^{Fibro-CT}$ = 61, $n^{Fibro-WT}$ = 63, $n^{Fibro-MT}$ = 56 analyzed curves, respectively). (**B**) Cell–cell adhesion energy defined as the energy to separate two NRVFs. ($n^{Fibro-CT}$ = 152, $n^{Fibro-WT}$ = 221, $n^{Fibro-MT}$ = 177 analyzed curves, respectively). (**C**) Comparison of maximum force in cell–cell adhesion/de-adhesion expresses as the maximum force applied to separate two NRVFs, ($n^{Fibro-CT}$ = 152, $n^{Fibro-WT}$ = 221, $n^{Fibro-MT}$ = 177 analyzed curves, respectively), (*** $p < 0.0001$, NS = non-significant). Numerical data is divided into quartiles, and a box is drawn between the first and third quartiles, with an additional line drawn along the second quartile to mark the median. The minimums and maximums outside the first and third quartiles are depicted with lines.

The Young's modulus values were obtained from the AFM force–deformation data with a cell deformation of 10%, since this range is considered typical of a linear elastic deformation. The elasticity mean value for Fibro-CT is 7.33 ± 0.88 kPa, for Fibro-WT is 7.77 ± 1.21 kPa, while for Fibro-MT the elasticity mean value is reduced to 3.63 ± 0.65 kPa. The elasticity difference between these different conditions proves that the resistance needed to compress the cell with the AFM tip is two times lower for the mutated condition. These differences in elasticity could be explained by the presence of cytoskeleton damages as previously observed in the mutated condition by immunofluorescent labeling (Figure 2). Indeed, Lanzicher et al. already detected this type of damage on the cardiomyocyte actin network expressing LMNA D192G mutation with a lower actin fibers density in the cytoskeleton and the presence of blebs on nuclear membrane [7].

2.3. AFM on Cell–Cell Adhesion Confirmed Changes in Fibro-MT

In addition to these results, an innovative approach has been explored: the cell–cell adhesion measurements conducted by AFM. This method allows us to generate nanotubes between two cells and measure the cell walls' adhesive properties (Figure 4A). As previously mentioned, in the AFM description (Materials and Methods), one cell has been immobilized on a tipless cantilever and put in contact (for 25 s) with a second cell, immobilized in a fibronectin-coated support. During a retraction step, described as the separation between these two cells, one force curve was acquired. By using these retract curves, several adhesion parameters could be quantified, as previously mentioned (Figure 4B): (i) the energy defined as the total energy applied to separate two cells and given by the area under each retract curves; (ii) the maximum force i.e., the maximum value of the force needed to separate the cells; (iii) the rupture pattern which provides information about separation phenomena at short distances; (iv) the nanotube patterns generated during the separation of two cells. In order to analyze the mutation's overall incidence on adhesive parameters between two NRVFs, energy, and maximum force parameters between Fibro-WT and Fibro-MT were compared. Figure 3B shows boxplots which highlight differences between these cell lines. In the Fibro-CT and Fibro-MT, the separation of the two fibroblasts required less energy compared to Fibro-WT, in particular: for Fibro-CT 32.26 ± 6.38 × 10^{-16} J, for Fibro-WT 34.64 ± 6.47 × 10^{-16} J, and for Fibro-WT 13.76 ± 2.35 × 10^{-16} J, respectively. Furthermore, the resistance to separate the cells in the Fibro-CT and Fibro-WT is stronger:

this could be explained if the cell surface generates more links/connections, and it has more affinity with the second cell. The maximum force was indeed 407.97 ± 173.14 pN for Fibro-CT and 431.12 ± 107.1 pN for Fibro-WT, but only 332.54 ± 196.43 pN for Fibro-MT. A force ≈29% lower was therefore enough to separate two mutated NRVFs. In summary, Figure 5 shows the decrease of the compression resistance (as elasticity, Figure 3A) and the decrease of the interaction between cells (by adhesion, Figure 3B,C) for the Fibro-MT.

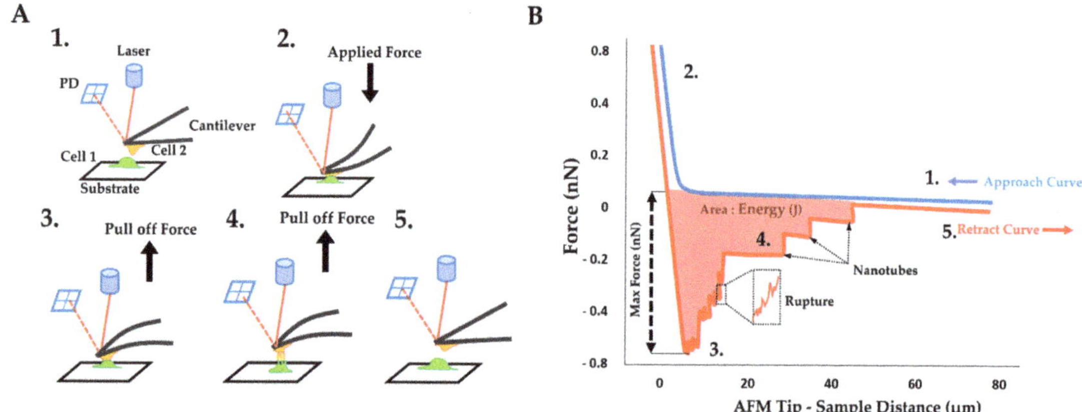

Figure 4. Cell-cell adhesion experiments by AFM. (**A**) Cell-cell adhesion setup: 1: Initial position with a fibroblast (cell 1) immobilized by fibronectin on support and a second cell (cell 2, yellow) attached to a tipless AFM cantilever coated with fibronectin driven by a wide-range (100 µm) piezo element. 2: Contact between cells by vertical force applied on the cantilever. The force on the cantilever is monitored by the deflection of a laser beam focused on the cantilever end and recorded by the photodiode (PD). 3: Cantilever retract step, producing a cells separation by a short distance. 4: Cantilever retract step, separation of cells over a longer distance and TNT formation. 5: Return to the initial step and total separation between cell 1 and cell 2. (**B**) Force–curve principle. Characteristic force curve recorded by PD, representing the force on the cantilever and the displacement during a cell–cell detachment and separation experiment: during the approach step (blue), the cell on the cantilever is brought into contact at a constant velocity, then the cell is retracted (red curve), until it goes back to the initial position going through a different pattern.

2.4. Force Rupture Distribution and Nanotube Patterns Are Different in Fibro-MT

For a better understanding, rupture and nanotube patterns were analyzed to obtain more information on the phenomena that occur during the two cells' separation at both short and long distance. Rupture patterns provide information on protein interaction during the first range of separation, defined as withdrawal from 0 to 5 µm. Nanotube patterns characterize the nanotubes formed between two cells and their following break in the second separation range, defined as withdrawal from 5 to 80 µm. The breaking force showed a slight decrease in the mutated condition, but even though there is this trend, this decrease was not statistically significant (Figure 5A). The mean of the rupture pattern for the breaking force is 32.23 ± 19.40 pN for Fibro-CT, 28.98 ± 12.60 pN for Fibro-WT, and 26.7 ± 13.74 pN for Fibro-MT, respectively. Even if the number of rupture pattern analysis is important (cf. Supplementary Table S1), the standard deviation is too heterogeneous to highlight a trend.

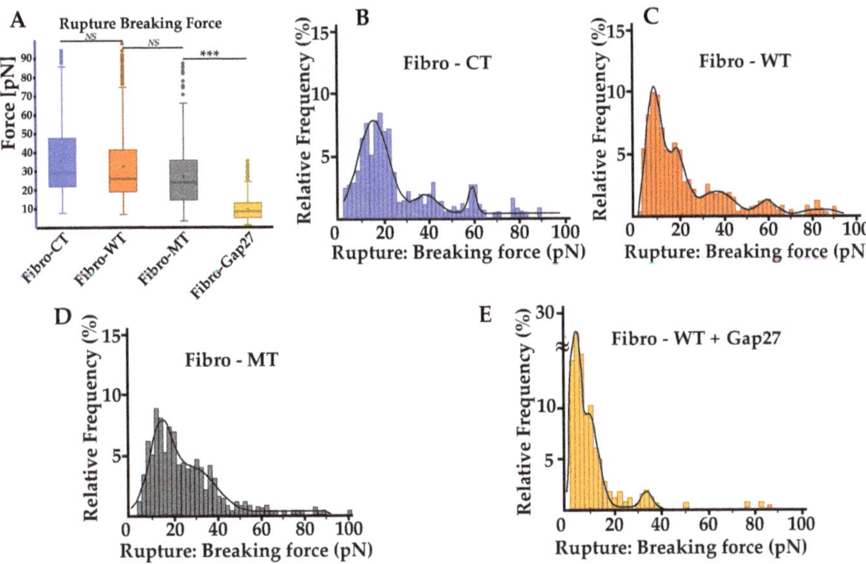

Figure 5. Rupture pattern obtain using the AFM. (**A**) Comparison of rupture pattern between control, wild type, *LMNA* D192G mutation, and NRVFs WT incubate with Gap27 (Fibro-WT + Gap 27). Rupture breaking force is the force applied to break the link between the adhesion proteins of cells in contact during the "short distance (0 to 5 μm)" de-adhesion step. ($n^{Fibro-CT}$ = 152, $n^{Fibro-WT}$ = 221, $n^{Fibro-MT}$ = 177, $n^{Fibro-WT + Gap27}$ = 122 analyzed curves, respectively). (**B–E**) Rupture breaking force distribution histogram for the different NRVFs conditions. The black curves represent the overall Gaussian fit distributions, (*** $p < 0.0001$; NS = non-significant).

However, both the energy and maximum force, previously discussed, showed that the adhesion decreases significantly in the mutated conditions and therefore a detailed rupture breaking force distribution was investigated to understand if there were differences in the adhesion proteins involved or in their density when the mutated cells are tested (Figure 3). Therefore, the analysis of breaking force distribution might allow us to get better information on the standard deviation of the mean breaking force. The distribution of the breaking force is shown in (Figure 5B–E) and it displays a multimodal distribution. Peaks are highlighted by fitting with Gaussian function where each peak of the curve represent the interaction break of a specific adhesion protein population [36]. Even though some sub-peaks may be hidden, this assumption allows us to compare the behavior of the different cell lines. Fibro-CT and Fibro-WT show four and five populations of breaking force after Gaussian deconvolutions, respectively. For Fibro-WT the first at 7.8 pN, the second at 17.45, then 36.1 pN, 59.1, and 83.5 pN (Figure 5C). The breaking force distribution for Fibro-MT identified only 3 peaks, the first at 13.1 pN, the second at 28.1 pN, and the last one at 85.2 pN. A fourth peak around 20 pN is present, but the signal is perturbed by the adjacent peaks' spread (Figure 5D). In both conditions, the same three populations of adhesion protein were identified. The first adhesion protein population presents a breaking force value around 10 pN, the second around 30 pN, and the third around 80 pN. However, both the Fibro-CT and the Fibro-WT present another population around 60 pN, which is not observed in the mutated condition. As previously reported, Cx43 is an adhesion protein that makes the TNTs' formation possible [14,24,25], we postulate that in Fibro-MT the function of this protein is altered, and could be responsible for the missing population around 60 pN.

2.5. Cx43 Is Correlated to the Biomechanical Changes in Fibro-MT

To verify the aforementioned hypothesis, a mimetic protein of Cx43, Gap 27, was used to block the link between two Cx43 and subsequently, the TNTs' formation. In previous work [10], we examined the levels and localization of Cx43 in cells with LMNA D192G mutation, since Cx43 protein is important for ensuring coordinated contractile action and its amount and localization might be one of the leading mechanisms for arrhythmias. We found that in the presence of this mutation, there was an altered Cx43 localization even though its overall cellular levels were unchanged. Prior to the cell–cell adhesion measurement, Fibro-WT were incubated with Gap27 solution for 1 h. In Figure 5A, the mean of rupture pattern for the breaking force is 7.35 $\pm$ 4.68 pN for NRVFs WT incubated with Gap27. Furthermore, Figure 5E shows that the distribution of the breaking forces, and the deconvolution by Gaussian, presents three populations: the first and the highest at 3.4 pN, the second peak at 9.3 pN, and the third at 33.5 pN. Moreover, in all the studied conditions, the adhesion population has a part of breaking forces lower than 10 pN, that could be associated with unspecific interactions [35,37]. All conditions present a peak around 30 pN, which could be one specific adhesion protein population such as integrin. Integrins are cell surface receptors that are instrumental in mediating cell–matrix interactions in all cells, and specifically in fibroblast with its matrix remodeling functions [38]. Few articles present some adhesion measurements of specific protein such as integrin $\alpha_5\beta_1$ on fibronectin support. In particular, Z. Sun et al. report that the breaking force between integrin $\alpha_5\beta_1$ and fibronectin is around 40 pN [39]. The value of the breaking force between two integrins in in vivo condition has not been measured yet, but the 30 pN peak might probably be inferred as the integrin population. The M. Horton's team studied, using the AFM, the breaking force of ligand–receptor interaction [40], and found that the interaction between integrin and RGD-peptide is around 30 to 120 pN. However, they highlighted that the force interaction between integrins is difficult to assess with accuracy. As explained in the article, the affinity of the receptor for a given ligand depends not only on the integrin type but also on conformational changes in the receptor and hence its "activation" status. Moreover, working with living cells leads to a reduced accuracy in measurements due to membrane deformability, ECM, etc.

All conditions present a range of breaking forces around 80 pN, which could be another type of adhesion protein population with a strong interaction force and low frequency on the membrane surface. What is remarkable is that in the Fibro-CT and Fibro-WT conditions, one peak is always present at around 60 pN and this population is lacking after the Gap 27 blocking process. This breaking force likely represents the Cx43 population as previously suggested.

2.6. In Fibro-MT, a Decrease of the Highest Value of the TNT's Breaking Force Was Found

The average of nanotube breaking force presented no significant difference between the CT and WT and the difference between the WT and the MT is small, but in any case, statistically significant: Fibro-CT showed an average force value of 36.71 $\pm$ 4.54 pN, Fibro-WT showed a force value of 39.53 $\pm$ 4.90 pN, and Fibro-MT a force of 33.61 $\pm$ 5.59 pN (Figure 6A).

Additionally, the distribution of nanotube breaking forces was also analyzed. As previously mentioned in the energy and maximum force parameters study, a significant decrease of the cell adhesion in the mutant condition was observed (Figure 3). However, the nanotube breaking-force distribution (Figure 6B–E) presents different multimodal force distribution for the three conditions studied. Each multimodal distribution peak represents a specific nanotube population generated during the cell–cell separation. In the Fibro-WT condition, three populations of nanotubes were recognized: the first population presents a breaking force at 11.1 pN, the second at 31.7, and the third at 58.0 pN. In the Fibro-MT, a similar trend was observed: three populations were underscored, the first at 12.4 pN, the second at 28.2 pN, and the third at 43.3 pN. Even though both conditions present two similar populations of nanotubes, the first is around 10 pN and the second is around 30 pN, in the Fibro-WT, the third population presents a breaking force at 58.0 pN, 34% higher compared to the third population in the Fibro-MT. This decrease of the highest value was also found in the Fibro-WT condition after Gap 27 blocking. Two hypotheses could be suggested to explain this observation: (i) a model where periodicity could suggest that the peak around 10 pN corresponds to the value of the breaking force for a single nanotube, the second peak around 30 pN corresponds to the simultaneous breaking force of two nanotubes, where the third corresponds to three nanotubes, etc.; (ii) a model where each peak corresponds to a specific type of nanotube. The 10 pN peak could be a nanotube composed only by the lipid bilayer of the plasma membrane. The population with a high nanotube breaking force around 60 pN could be a TNT with Cx43 anchor function. The intermediate value could be a hybrid TNT, with the presence of Cx43. In the mutated condition, the lower values of adhesion energy and maximum force due to decrease on Cx43 on the cell membrane might represent their difficulty to initiate the TNT formation.

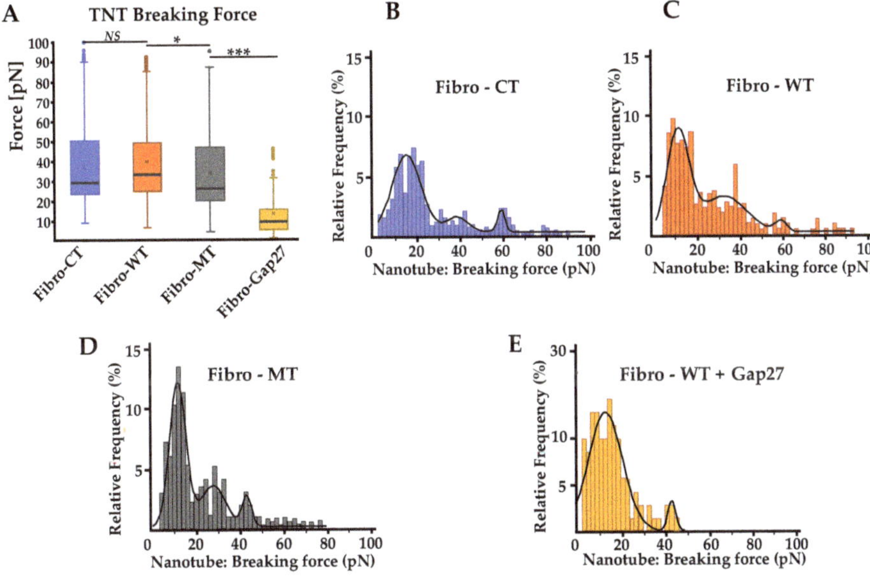

Figure 6. Nanotube pattern obtain using the AFM. (**A**) Comparison of nanotube pattern between NRVF control and wild type, NRVFs with *LMNA* D192G mutation and NRVFs WT incubate with Gap 27. Nanotube breaking force represents the force applied to break the link between both cells during the "long distance (5 to 80 μm) de-adhesion steps". ($n^{Fibro-CT}$ = 152, $n^{Fibro-WT}$ = 221, $n^{Fibro-MT}$ = 177, $n^{Fibro-WT + Gap27}$ = 122 analyzed curves, respectively). (**B–E**) Breaking force distribution histogram of nanotube pattern in different NRVF conditions. The black curves represent the overall Gaussian fit distributions. (*** $p < 0.0001$; * $p < 0.05$, NS = non-significant).

2.7. Optical Tweezers

Optical tweezer (OT) observations were carried out in parallel to the AFM experiments to collect a new range of information about the nanotubes' pattern (Figure 7A). OT allows us to measure the mechanical properties of a single nanotube and complements the AFM results. Direct tether extraction is also an important source of information about the elastic properties of the cell membrane. During the nanotube extrusion, force curves have been acquired and nanotube elasticity has been calculated. A typical force-deformation curve in tether extraction is shown in Figure 8B.

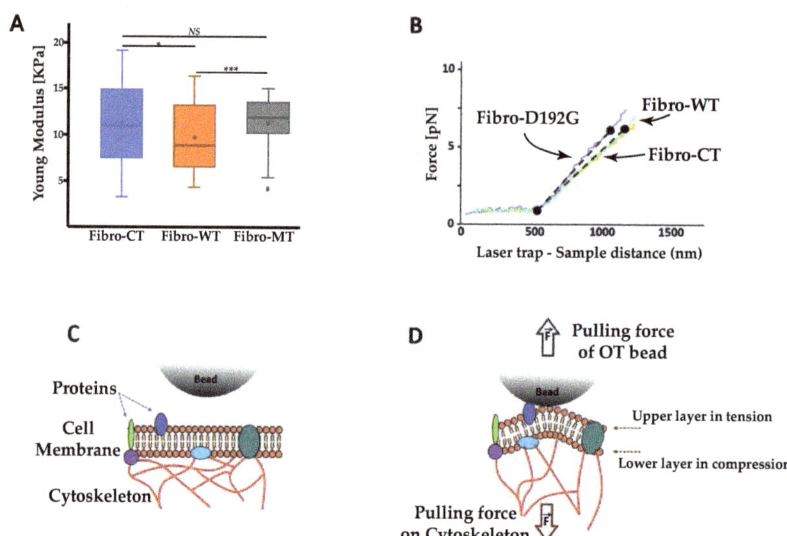

Figure 7. Nanotube extrusion measured using the optical tweezers. (**A**) Nanotube stiffness (expresses as Young's modulus) between different conditions NRVFs. ($n^{Fibro-CT}$ = 60, $n^{Fibro-WT}$ = 58, $n^{Fibro-MT}$ = 70, analyzed curves, respectively). (**B**) Force curve between NRVFs CT, WT, and NRVFs with *LMNA* D192G mutation. (**C**) Cartoon showing the resistance force impacting the nanotube extrusion. (**D**) The pulling force of the OT bead after the interaction with the plasma membrane is counteracted by bending rigidity (with stretching and compression of membrane layer) and surface tension (with pulling force of cytoskeleton), (*** $p < 0.0001$; * $p < 0.05$; NS = non-significant).

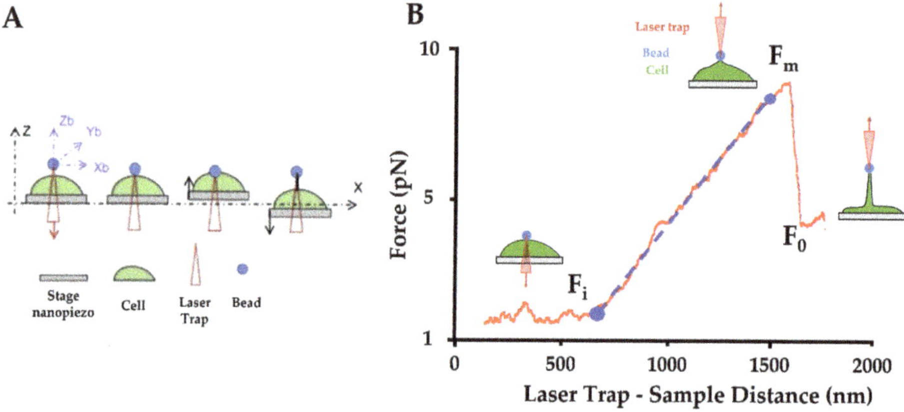

Figure 8. Optical tweezers (OT) setup. (**A**) Schematic procedure of an optically trapped bead applying a force to form a nanotube from the cell membrane. The trapped bead can be moved in three directions ($Z_b/Y_b/X_b$), in this way the bead can be put in contact with the cell. Then the trap remains in the same position, and the stage moves in Z and X direction to extract a nanotube (black line). (**B**) Force curve obtained by OT: The initial force (F_i) is the force, between bead and cell, measured during the contact step. The maximum force (F_m), is the maximum force applied to extrude the nanotube from the cell surface. The plateau force (F_0), is the constant force applied during the nanotube extension. The slope between the two points F_i and F_0 is used to calculate the nanotube's stiffness (blue line: slope used to stiffness; red curve: force applied on bead during the laser displacement).

Figure 7B shows the comparison between the three conditions. There is an initial situation with the trapped microsphere attached to the cell and then a clear change in the slope when cell membrane deformation and initial tether formation begins. This part displays a monotonically rising portion and it slope provides the nanotube extrusion resistance. Fibro-CT and WT had a stretch resistance mean value of 9.7 $\pm$ 3.31 pN/μm, while in the mutated condition, nanotubes present a higher resistance mean value of 11.4 $\pm$ 4.71 pN/μm.

3. Discussion

3.1. Lamin A/C D192G Gene Mutation Effect on the LINC Complex

Previous studies described the role of the LINC complex on anchorage of cytoskeleton in nuclear membrane allowing its stability [32,41–43]. The LINC complex is composed of outer and inner nuclear membrane Klarsicht, ANC-1, and Syne homology (KASH), and Sad1 and UNC-84 (SUN) proteins. LINC connects the nucleus to cytoskeletal filaments and performs several functions such as nuclear positioning, mechanotransduction, meiotic chromosome movements, and cell stabilization [44–47]. In our study, immunofluorescence studies showed actin network damage and a disorganized cytoskeleton (Figure 2) and a significant decrease in the Young's modulus in the mutated condition. As already described in the literature, the LMNA D192G mutation induces a loss of lamina cortex integrity and impact the LINC function [13,48]. In NRVFs expressing LMNA D192G mutation, the defective lamin cortex caused a decrease of the actin filament anchorage by LINC complex protein on the nuclear membrane. The cytoskeleton is anchored on the nuclear lamina cortex by a succession of LINC proteins. From inside the nuclear membrane to outside, SUN domain, KASH domain, and Nesprin proteins attach actin filaments on the lamina cortex, the plectin protein bind intermediate filament and the Kif5E protein bind microtubule on this same lamina cortex [49]. Thus, we hypothesized that in LMNA D192G NRVFs, a damaged lamina cortex could induce an alteration of the whole cell structure, and in particular the defective cytoskeleton could impact cell functions including the extracellular communication with the TNTs' adjacent cells.

3.2. Impact of LMNA D192G Mutation on TNT Mechanical Properties

Actin and intermediate filaments showed disorganization of the cytoskeleton not only in the cell but also inside the TNTs in mutant-expressing fibroblasts. The literature confirms the impact on actin network disorganization in cells expressing LMNA D192G mutation [7,13], but these studies were carried out on cardiomyocytes. Our results demonstrate that the mutation D192G is also expressed in other cardiac cells like cardiac fibroblast and influences their overall cytoskeleton organization including the TNT structure. TNTs were characterized by a length between 50 to 270 nm, with cytoskeleton filament inside [19,23] (Figure 2). In the mutated NRVFs, actin filaments inside the TNT (in red) were reduced in comparison with intermediate filaments. This decrease could weaken the TNT structure. Furthermore, the information exchange by TNT between two cells could be altered and impact the fibroblasts functions in the cardiac tissue. Indeed, AFM tests confirmed that in Fibro-MT, both elasticity (cells are more deformable) and adhesion properties (Figure 3) were reduced when compared to Fibro-WT. Furthermore, AFM data proved that the separation between two Fibro-MT requires less force than the force required to separate two Fibro-WT. The analysis of the AFM retracting curves [35,39,50] also provided the distribution on the nanotube breaking force, highlighting the presence of several force populations (Figure 6). A model from Dérényi et al. [50] proposes that each force population represents TNTs with similar breaking points but in different numbers (the first population represents the breaking force of a single nanotube, the second population the breaking force of two nanotubes, etc.). A second model by Sun et al. [51] proposes that each population is considered a breaking force of specific nanotube family. These breaking forces are impacted by the sum of all macromolecular contributions to nanotube formation. In this model, the type of nanotube is dependent on the cytoskeleton properties and interaction of adhesion proteins during the contact step. Our results offered a better fit with the second model, in which one nanotube population extruded from the mutated condition has been impacted by disorganization in the cytoskeleton and by altering the function of one adhesion protein such as Cx43, which it is necessary for priming TNT formation. As previously mentioned, in the WT condition, one nanotube population has been detected with a mean breaking force around 60 pN. This population has no longer been detected after Cx43 blocking by Gap 27 (Figure 6E). In the mutated condition, this population with a breaking force around 60 pN decreases significantly. Therefore, on the membrane surface of NRVFs expressing the LMNA mutation, the nanotubes' formation could be impacted by alteration on the Cx43 protein, as observed in WT condition after blocking Gap 27. AFM experiments allowed us to quantify the adhesion force during the cell-cell separation and the nanotube formation. In the LMNA mutant condition, these parameters were altered: adhesion properties were weaker, and the generated TNTs presented lower breaking force values.

3.3. Extrusion of Single Nanotube by OT

As shown in Figure 7, the elasticity of nanotube extrusion by OT bead in the NRVF-MT is 17.5% higher than in the NRVF-WT condition. To understand this trend, a comparison between the AFM and OT test parameters must be considered: in particular, the size of contact surface between bead and cell, interaction types in the contact zone and the force applied during the contact time. During OT experiments, TNTs have been extruded after a contact between the bead in laser trap and the membrane surface. The size of this contact zone is lower that the bead diameter, therefore lower than 1 μm^2, while during AFM acquisition, the size of contact surface was equal to the NRVF size immobilized on the cantilever surface (around 100 μm^2). Thus, cell mechanics (resistance to mechanical stress, change in cell shape, membrane mechanical parameters) involved during OT and AFM tests are quite different. Moreover, the force applied during OT test is about 40 pN vs. 2 nN in the AFM cell–cell adhesion measurements. Therefore, the force applied in the OT experiment involves a quite different cell mechanical response than during the AFM measurements [52,53]. As described by H. Moysés Nussenzveig [54], the nanotube extruded using OT consists of a bilayer plasma membrane and is different from nanotubes studied in AFM tests. In particular, three membrane phenomena influence the extrusion of the bilayer membrane nanotube: cell membrane bending rigidity [55], surface tension [50,56,57], and membrane reservoir [58]. As shown in Figure 7C,D, the bending rigidity of the bilayer membrane arises from the fact that during TNT extraction the lower membrane layer is compressed and the upper one is stretched. This stress affects the bilayer equilibrium spacing, and it requires energy input to extrude the membrane nanotube. A second parameter is the surface tension, this resistance of nanotube extrusion results from the difference in pressure across the bilayer. The pressure differences arise from the pulling force by the OT bead and from cytoskeleton filaments bounded to the membrane (Figure 7C). The last phenomenon represents a fundamental mechanism of cell shape change: reservoir membrane. As described in the literature [58], cells have a plasma membrane unfolding to keep their capacity to use membrane extension for different biological mechanisms such as growth [59] and cytokinesis [60]. However, even considering these parameters, the nanotube extrusion in the mutated condition has higher elasticity than in the WT condition. Indeed, to extrude a nanotube in the WT condition, the force applied is higher to the bending rigidity and surface tension in order to unfold the reservoir membrane. In the mutated condition, the cytoskeleton damaged induces a decrease of surface tension. Therefore, the force applied to have access to the reservoir membrane and extrude a nanotube should be lower in the mutated condition than in the WT condition. However, the study of L. Figard et al. [58] characterized the reservoir membrane mechanism and demonstrated that the stabilization of the reservoir membrane is performed by the actin filament. In the NRVF expressing LMNA D192G mutation, the actin network is damaged and induces an alteration on the reservoir membrane function. In the mutated condition, this network disorganization decreases the reservoir membrane access during the nanotube extrusion experiments performed by OT. In Figure 7B, the absence of significant differences between mutated and WT conditions could be explained by the equilibrium between the reservoir membrane alteration and the higher cell deformability induced by cytoskeleton damage. Moreover, cell size has been measured (Supplementary Figure S1) to determine if the reservoir membrane alteration has an impact on the cell size. A significant difference in measurements of cell size have been observed. In the mutated condition, NRVFs present both less length and width than CT and WT condition. These results could explain a decrease on the deformability capacity in mutated condition when compared with the NRFVs in WT condition.

3.4. Impact on Adhesion Protein in Mutated Condition

LMNA mutations alter the lamina cortex and induce cytoskeleton disorganization by loss of LINC complex anchorage on the nuclear membrane [48]. Without this anchor, the cytoskeleton is damaged and the capacity to generate a TNT is compromised. Girao's works [25] showed that Cx43 is an important actor of intercellular communication, mediated by extracellular vesicles, gap junctions, and TNTs. The analysis of adhesive pattern by cell–cell adhesion shows the impacts of Cx43 (Figures 5 and 6). When Cx43 proteins are blocked by Gap 27 mimetic peptide in the WT NRVFs, a decrease in the adhesion protein and nanotube population values takes place. The same population of adhesion protein decreases in the NRVFs expressing the LMNA 192G mutation (Figure 5D). This decrease suggests an alteration on the Cx43 function. The Cx43 migration to the membrane surface requires an organized cytoskeleton as suggest by Thomas et al. [61]. The transport of Cx43 to the membrane surface is dependent upon intact cytoskeletal filaments. In LMNA mutant NRVFs, the expression of Cx43 could be decreased on the membrane surface and thus, could induce altered adhesion properties during the separation of two cells. As seen previously, Cx43 proteins initiate TNT formation [25]: the nanotubes' breaking force distribution (Figure 6D) suggests that the nanotubes population representing TNT were decreased in the mutated condition. Moreover, the number of retract curves analyzed and the number of events recognized from cell–cell adhesion acquisitions suggest that the mutated NRVFs capacity to generate points of interaction with another cell is reduced (Supplementary Table S1). Adhesive areas are lower on the cell wall of NRVFs expressing a laminar mutation in comparison with the WT NRVFs. The analysis of force curves events by cell–cell adhesion highlights that the LMNA D192G mutation could have an important impact on the TNTs' formation and on the extracellular TNT communication with adjacent cells.

3.5. Impacts of the LMNA D192G Mutation in NRVFs on the Heart in ACM

The cardiac muscle is composed of several cell types. Cardiomyocytes account for roughly 75% of normal heart tissue volume, but they account for only 30–40% of cell numbers [62]. Among the remaining cells, fibroblasts have an important role because a component of the adult human cardiac muscle is also collagen type I and collagen type III, and for both, synthesis and turnover are mainly regulated by cardiac fibroblasts. Cardiac fibroblasts are also responsible for (physiological and pathological) ECM synthesis in the heart muscle and play an important role for the structural, mechanical, and electrical cardiac functions. Every cardiomyocyte is closely connected to a fibroblast in normal cardiac tissue. However, pathological states like ACM are frequently associated with myocardial remodeling involving fibrosis. Fibroblasts and myocytes are also interconnected in their response to mechanical stresses. The preservation of physiological levels of cardiac stiffness not only determines overall ventricular diastolic function but also ensures correct cardiomyocyte functionality. Cardiomyocytes are embedded within an ECM whom protein composition and levels of cross-linking can affect the physiological stiffness of the heart. In this regard, collagens are important players due to their ability to form fibrils that supply tissue stiffness. Therefore, it is evident that fibroblasts are a multipurpose dynamic cell that controls collagen and other ECM components affording the neighboring cells to properly function. Furthermore, the decrease of fibroblast elasticity could reduce the heart capacity to contract. Therefore, the presence of LMNA mutant fibroblasts necessarily impacts the overall heart function. Furthermore, ACM is a pathological condition generated by a large spectrum of genetic variations that can target the *LMNA* gene. According to our results, the mutation of the lamina cortex could induce the decrease of extracellular communication by TNT during ACM pathology. Lack of communication from fibroblasts in cardiac tissue could bring fibrofatty infiltration in cardiomyocytes with tissue fibrosis [63–65]. Several reports illustrated the function of fibroblasts in the ECM production and their role in fibrosis and adipogenesis [63–66]. Cells in the cardiac tissue secrete fibrogenic mediators and matricellular proteins that bind to the fibroblast surface receptors. Then, the fibroblast receptors transduce intracellular signaling cascades that regulate genes involved in synthesis, processing, and metabolism of the extracellular matrix. Fibroblast endogenous pathways are involved in the negative regulation of fibrosis and are critical for cardiac tissue as they may protect the myocardium from excessive fibrogenic responses. Therefore, cellular communication is necessary to receive and exchange the biological information in order to have an adapted regulation response. In ACM cells, LMNA mutations can modify this regulation by a communication of altered TNTs. The restoration of TNT formation in cardiac cells of ACM patients could be an interesting research perspective to decrease cardiac damages.

3.6. Study Limitations

A limitation of our study is that an important assumption has been made, specifically that the cohesive force between cells were only due to the nanotubes, while the junctions formed by desmosomal proteins, adherens junctions, or CAMs have not been taken into account. In our study, the cells confluence was relatively low and therefore TNTs were the principal players of the adhesion between cells.

Furthermore, although the study of pro-fibrotic and pro-adipogenic changes in LMNA mutant cardiac fibroblasts was beyond the scope of our study, it should be noted that Chen et al. have previously reported altered mechanotransduction/mechanosignalling in LMNA mutant cardiomyocytes, leading to the activation of the Hippo pathway, that triggers fibro-adipogenic signals in fibroblasts [67]. Finally, other mechanisms may be implicated in the defective mechanotransduction of LMNA mutant cardiac cells. In cardiomyocytes with three different LMNA missense mutations, we previously reported defective adhesion, altered actin, and microtubule networks and dislocated connexin 43 [10] leading to higher frequency of beating but with a reduced beating force. However, future studies should investigate the complex molecular changes in LMNA mutant cardiac fibroblasts.

4. Materials and Methods

4.1. Isolation and Culture of Ventricular Fibroblasts from Neonatal Rat (NRVFs)

The University of Colorado Denver institutional review board gave approval for these experiments (protocol number 00235-30 July 2019). All animal studies have been carried out according to the guidelines Animal Care and Use Committee. All animal experiments were performed using all possible methods to alleviate or minimize potential pain, suffering, or distress, and enhance animal welfare. Animals were provided with housing in an enriched environment, with at least some freedom of movement, food, water, and daily care and cleaning. The well-being and state of health of experimental animals were observed by competent persons, dedicated to the management of the Animal House, able to prevent pain or avoidable suffering, distress, or lasting harm. Experiments were performed solely by competent authorized persons. Primary cultures of neonatal rat ventricular fibroblasts (NRVFs) were isolated and cultured from 1 to 3-day-old Sprague Dawley rat pups (Charles River), following decapitation and enzymatic digestion as previously described with minor modifications [68,69]. Briefly, ventricles were separated from the atria using scissors and then dissociated in CBFHH buffer (calcium and bicarbonate-free Hanks with Hepes) containing 0.5 mg/mL of Collagenase type 2 (Worthington Biochemical Corporation, Lakewood, NJ, USA), and 1 mg/mL of Pancreatine (Sigma-Aldrich, St. Louis, MO, USA). To separate Myocytes and Fibroblasts, mix cells were incubated on tissue culture plates at 37 °C for 1 h. Then, the cells were washed with minimum essential media (MEM), supplemented with 5% bovine calf serum to collect the unattached myocytes, and the adhered NRVFs were further used for our experiments. After one day of culture, the NRVFs were re-plated at a density of 2×10^5 cells/mL in primary Petri dishes (Falcon) [70] for further experiments.

4.2. Isolation Adenoviral Constructs and Infection

Methods for adenoviral infection have been previously reported [71,72]. In brief, shuttle constructs were generated in Dual CCM plasmid DNA containing GFP gene and human *LMNA* cDNA. Constructs were bicistronicity with the two inserts (*LMNA* and GFP) driven by two different CMV promoters to identify cells expressing LMNA protein using GFP as a marker of cellular infection. NRVFs were infected by adenoviruses at a multiplicity of infection (MOI) of 30 in serum free medium; 6 h post-infection, complete medium was replaced, and the cells were incubated at 37 °C and 5% CO_2. Previous control experiments showed that GFP transfection and expression did not affect endpoints of interest in NRVF in our experimental conditions.

4.3. Immunofluorescence

NRVFs were fixed in PBS containing 4% PFA for 15 min at room temperature. Cells were permeabilized at room temperature, with 1% Triton X-100 for 90 min, blocked with 2% BSA in PBS for 45 min, and incubated overnight with vimentin 1:1000 (Sigma-Aldrich, St. Louis, MO, USA). Goat anti-mouse antibody conjugated to Cy5 (Abcam, Cambridge, United Kingdom) was used as secondary antibody 1:300 (Invitrogen, Waltham, MA, USA). Each sample was stained with Dapi 1:2000 (Thermo Fisher Scientific, Waltham, MA, USA) to counter-stain the nuclei and with phalloidin 1:1000 (Sigma-Aldrich, St. Louis, MO, USA) prepared in 1% BSA + 0.3% Triton 1X to staining the cytoskeleton. Representative immunofluorescence images were acquired using a Zeiss LSM780 confocal.

4.4. Single Cell Force Spectroscopy by AFM

Atomic force microscopy is widely used for measuring the elasticity (Young's modulus) of living cells. In this case, AFM experiments were carried out by single cell force spectroscopy on NRVFs with complete medium, at 37 °C and on immobilized NRVFs using a human fibronectin coating (20 μg/mL). During acquisition, MLCT AFM probes were used. These probes have a pyramidal tip made of Si_3N_4, with a curvature radius of 35° and are manufactured by Bruker. The cantilever spring constants were systematically checked using the thermal tune method and were found to range from 10 to 30 pN/nm [73]. The maximal force applied to the cell was limited to 2 nN to preserve the cell membrane integrity. The acquired force (F) versus displacement curves were converted into indentation (δ) curves and fitted with the Hertz–Sneddon model [74]. The AFM experiments were carried out at a velocity of 5 μm/s [75,76]. All experiments were performed at the same velocity since AFM tests are rate-dependent.

4.5. Cell–Cell Adhesion Experiments by AFM

4.5.1. Cell Preparation

For AFM cell–cell adhesion assay [36], two areas were demarcated in the TPP Petri dish: the first one was aimed at obtaining individual adherent cells on fibronectin coating after 24 h, and the second one to prepare non-adherent cells on the dish surface without coating. Briefly, the previous day, the first area was coated for 1 h at 37 °C with 20 μg/mL human fibronectin in phosphate buffered saline (PBS, Thermo Fisher Scientific, Waltham, MA, USA) and then washed twice with PBS. Then 15,000 cells were seeded on the coating side and cultured in complete DMEM medium (Thermo Fisher Scientific, Waltham, MA, USA) at 37 °C in 5% CO_2. The day of AFM measurements, a second Petri dish with the same seeded cells were trypsinized and washed to obtain a tube of non-adherent cells. These cells were stocked at 37 °C in 5% CO_2. Just before the AFM experiments, the non-adherent cells were added in the first Petri dish on the non-coating side at 37 °C [35].

4.5.2. Data Acquisition

The cell biomechanical assessments were performed with the CellHesion 200 apparatus (JPK Instruments Bruker Nano GmbH, Berlin, Germany), which provides a large vertical piezoelectric range of 100 µm, mounted on a Zeiss inverted optical microscope with a controlled temperature of 37 °C. Tipless cantilevers from Bruker (Bruker, Billerica, MA, USA) (NP-O10) were used with the nominal spring constant of 30 mN/m, the exact value for each cantilever was inferred from the thermal tuning method [72]. For each experimental condition, the cantilever was replaced after every 3 cell contacts. The cantilevers were coated with fibronectin (10 µg/mL for 30 min before acquisition) after being activated by ozone (0.1 mBar, 3 min) in order to promote cell attachment. The cell-adherent cantilever was brought into contact with one non-adherent cell localized in the zone without fibronectin coating on the Petri dish (providing low adhesion to the substrate) (Figure 4A). This "fishing" step was carried out with a contact duration of 30 s applying 1.5 nN. After 30 s the cantilever was retracted and stopped in this position for 1 min to allow the stabilization of cell caught at the end of the cantilever. The approach–retraction steps were performed with a ramp size of 80 µm, cantilever speed of 5 µm/s, contact time of 25 s, and force setpoint of 2 nN.

4.5.3. Data Analysis

The acquired force curves were analyzed using the JPK Data processing software. The first numerical treatment was the reduction of curve noise, followed by the baseline definition and the tilt correction. After these corrections, several adhesion parameters, identifiable on the retract force curve, were measured, exported, and analyzed (Figure 4B): in particular (i) the total energy applied during the separation of 2 cells, represented by the area under the retracting force distance curve; (ii) the maximum force is the maximum value of the force applied during the separation step; (iii) the rupture pattern which illustrates interaction during short distance separation (0 to 5 µm of piezo withdraw). This pattern is characterized by jumps in the force retracting curve everyone preceded by a slope >20%. These ruptures events were induced by broken adhesion protein link between two cells [36,50]; (iv) nanotube pattern i.e., nanotubes generated during long-distance separation (5 to 80 µm of piezo withdraw). In this latter case, this pattern of jumps in the retract force curve was preceded by a plateau with <15% derivation and persisting for > 1 µm, representing the force needed to break these plasma membranes, pulled out to form a nanotube between two cells after their contact [35,50,51,77]. The distribution of breaking forces was analyzed using OriginPro 9.7 2020 software, and Gaussian function was fitted to highlight a specific population.

4.5.4. Synthetic Connexin 43 Mimetic Peptide (Gap 27) for Cx43 Blocking

In this study, Gap 27 was used to block the function of Cx43 gap junction. Gap 27 is mimicking amino acids 204–214 on extracellular loop 2 of Cx43 (SRPTEKTIFII). Before the AFM tests of cell–cell adhesion, selected cell cultures were incubated with Gap 27 (Sigma-Aldrich) in PBS at 150 µM for 1h, at 37 °C [78–80].

4.6. Optical Tweezers

In order to extract the nanotubes' membrane and measure their stiffness, a custom optical tweezers was used with the technique previously described [81] and summarized in Figure 8A.

An infrared trapping laser from an Ytterbium fiber laser (YLM-5–1064-LP, IPG Photonics, Oxford, MA, USA) with a wavelength of 1064 nm and maximum power of 5 W at a nominal power of 250 mW (60 mW at the sample) was used. The vertical stiffness of the optical trap was $k_z = 0.04$ pN/nm. A silica microbead of 1 µm diameter (SS04000 Bangs Lab

Inc., Fishers, IN, USA) was used to extract the tether membrane. This step was achieved by manipulating the bead to get in contact with the cell, then moving the cell/stage vertically against the bead by 0.5 μm and keeping the contact for 30 s to create the cell adhesion. Finally, a nanotube was extracted by moving the cell away from the bead by 2 μm. Precise displacement of the bead was obtained by moving the optical trap with a Focus Tunable Lens FTL (EL-10-30-C, Optotune AG, Dietikon, Switzerland), as previously described [82]. The displacement of the cell, which was adhered on the coverslip, was controlled in x-y-z by a 3-Axis NanoMax Flexure Stage (MAX311D/M Thorlabs Inc., Newton, NJ, USA). The displacement of the bead from the center of the trap, which allows us to measure the force, was measured by a Quadrant Detector Sensor (PDQ080A, Thorlabs Inc., Newton, NJ, USA). The vertical force exerted by the tether during extraction was calculated as $F = k_z B$ and the length of the tether L was given by the displacement of the nanopiezo. The nanotube's stiffness was deduced as the slope of the force–tether length (F–L) curve (Figure 8B).

4.7. Statistics

For all experiments, data was collected from at least three independent experiments. All data were first subjected to the Shapiro–Wilk normality test. The unpaired one-way ANOVA with Dunnet's or Tukey's correction for normal distributions or the Kruskal–Wallis with Dunn's correction test was employed. A confidence interval of 95% (p value of <0.05) was used to identify significant differences among the samples. Data in the text are reported as mean of values ± standard deviation and graphically presented as scattered dot plots with mean (and median for those deviating from normality test) and standard deviation.

Statistical analysis was performed using GraphPad Prism 7 software (San Diego, CA, USA).

5. Conclusions

The goal of this study was to define the multiple impacts of the mutation LMNA D192G on cardiac fibroblast cellular mechanisms, such as cytoskeleton anchorage on nuclear membrane, cell deformability, cell shape evolution, adhesive properties, and TNT formation. Using a multidisciplinary approach including immunofluorescence, AFM, and OT, we investigated the changes in biomechanical properties and cell–cell adhesion of cardiac fibroblasts with the LMNA D192G mutation, which can cause an inherited arrhythmogenic cardiomyopathy. We found that mutant NRVFs have a damaged cytoskeleton, as previously shown in cardiomyocytes [7], leading to a decrease of cell elasticity and a lower force of adhesion between two NRVFs. Furthermore, we found altered mechanical properties of nanotubes generated by contact with another cell (by AFM) or by contact with a silicate bead (using OT). Our results complement our previous investigations on LMNA mutant cardiomyocytes [7,67] and suggest that this mutation also impacts biomechanical properties of cardiac fibroblasts. Future studies should investigate the impact of the LMNA mutations on TNT communication in each type of cardiac cells and also the TNT communication at a higher scale, such as in the cardiac tissue.

Supplementary Materials: The following are available online at https://www.mdpi.com/article/10.3390/ijms22179193/s1.

Author Contributions: Conceptualization, D.C., M.R.G.T., L.M. and O.S.; data curation, B.P. and S.N.C.; formal analysis, D.C. and S.N.C.; funding acquisition, M.R.G.T., L.M. and O.S.; investigation, V.L., B.P., C.C. and S.N.C.; methodology, V.L., B.P. and D.C.; supervision, D.C., L.M. and O.S.; writing—original draft, V.L.; writing—review and editing, B.P., M.R.G.T., L.M. and O.S. All authors have read and agreed to the published version of the manuscript.

Funding: This research was funded by the National Institutes of Health, grants (K25HL148386 to (B.P.)), (R01 HL147064 to (M.R.G.T., L.M., B.P.)), it was supported in part by a Trans-Atlantic Network of Excellence grant from the Leducq Foundation (14 CVD 03 to (M.R.G.T., O.S., L.M., D.B., B.P.)), the project PRIN 20173ZW ACS awarded by the MIUR-Italy and by generous grants of the John Patrick Albright (to (M.R.G.T., L.M., B.P.)).

Conflicts of Interest: The authors declare no competing interests.

References

1. Towbin, J.A.; McKenna, W.J.; Abrams, D.; Ackerman, M.J.; Calkins, H.; Darrieux, F.; Daubert, J.P.; de Chillou, C.; DePasquale, E.C.; Desai, M.Y.; et al. 2019 HRS expert consensus statement on evaluation, risk stratification, and management of arrhythmogenic cardiomyopathy. *Heart Rhythm.* **2019**, *16*, e301–e372. [CrossRef]
2. Pilichou, K.; Thiene, G.; Bauce, B.; Rigato, I.; Lazzarini, E.; Migliore, F.; Marra, M.P.; Rizzo, S.; Zorzi, A.; Daliento, L.; et al. Arrhythmogenic cardiomyopathy. *Orphanet J. Rare Dis.* **2016**, *11*, 1–17. [CrossRef] [PubMed]
3. Chen, S.N.; Sbaizero, O.; Taylor, M.R.G.; Mestroni, L. Lamin A/C Cardiomyopathy: Implications for Treatment. *Curr. Cardiol. Rep.* **2019**, *21*, 160. [CrossRef] [PubMed]
4. Saffitz, J.E. Molecular mechanisms in the pathogenesis of arrhythmogenic cardiomyopathy. *Cardiovasc. Pathol.* **2017**, *28*, 51–58. [CrossRef] [PubMed]
5. Vimalanathan, A.K.; Ehler, E.; Gehmlich, K. Genetics of and pathogenic mechanisms in arrhythmogenic right ventricular cardiomyopathy. *Biophys. Rev.* **2018**, *10*, 973–982. [CrossRef] [PubMed]
6. Zhao, G.; Qiu, Y.; Zhang, H.M.; Yang, D. Intercalated discs: Cellular adhesion and signaling in heart health and diseases. *Heart Fail. Rev.* **2019**, *24*, 115–132. [CrossRef] [PubMed]
7. Lanzicher, T.; Martinelli, V.; Puzzi, L.; Del Favero, G.; Codan, B.; Long, C.S.; Mestroni, L.; Taylor, M.R.G.; Sbaizero, O. The Cardiomyopathy Lamin A/C D192G Mutation Disrupts Whole-Cell Biomechanics in Cardiomyocytes as Measured by Atomic Force Microscopy Loading-Unloading Curve Analysis. *Sci. Rep.* **2015**, *5*, 13388. [CrossRef]
8. Lanzicher, T.; Martinelli, V.; Long, C.; Del Favero, G.; Puzzi, L.; Borelli, M.; Mestroni, L.; Taylor, M.R.G.; Sbaizero, O. AFM single-cell force spectroscopy links altered nuclear and cytoskeletal mechanics to defective cell adhesion in cardiac myocytes with a nuclear lamin mutation. *Nucleus* **2015**, *6*, 394–407. [CrossRef] [PubMed]
9. Van Loosdregt, I.A.E.W.; Kamps, M.A.F.M.; Oomens, C.W.J.; Loerakker, S.; Broers, J.L.V.; Bouten, C.C. Lmna knockout mouse embryonic fibroblasts are less contractile than their wild-type counterparts. *Integr. Biol.* **2017**, *9*, 709–721. [CrossRef]
10. Borin, D.; Peña, B.; Chen, S.N.; Long, C.S.; Taylor, M.R.; Mestroni, L.; Sbaizero, O. Altered microtubule structure, hemichannel localization and beating activity in cardiomyocytes expressing pathologic nuclear lamin A/C. *Heliyon* **2020**, *6*, e03175. [CrossRef]
11. Khatau, S.B.; Bloom, R.J.; Bajpai, S.; Razafsky, D.; Zhang, S.; Giri, A.; Wu, P.-H.; Marchand, J.; Celedon, A.; Hale, C.; et al. The distinct roles of the nucleus and nucleus-cytoskeleton connections in three-dimensional cell migration. *Sci. Rep.* **2012**, *2*, 488. [CrossRef]
12. Houben, F.; Ramaekers, F.; Snoeckx, L.; Broers, J. Role of nuclear lamina-cytoskeleton interactions in the maintenance of cellular strength. *Biochim. Biophys. Acta Mol. Cell Res.* **2007**, *1773*, 675–686. [CrossRef]
13. Sbaizero, O.; Lanzicher, T.; Martinelli, V.; Long, C.; Slavov, D.; Del Favero, G.; Taylor, M.; Mestroni, L. Altered Nuclear and Cytoskeletal Mechanics and Defective Cell Adhesion in Cardiac Myocytes Carrying the Cardiomyopathy LMNA D192G Mutation. *Circulation* **2014**, *130* (Suppl. S2), A17200. [CrossRef]
14. Gerdes, H.-H.; Bukoreshtliev, N.V.; Barroso, J.F. Tunneling nanotubes: A new route for the exchange of components between animal cells. *FEBS Lett.* **2007**, *581*, 2194–2201. [CrossRef]
15. Koyanagi, M.; Brandes, R.; Haendeler, J.; Zeiher, A.M.; Dimmeler, S. Cell-to-Cell Connection of Endothelial Progenitor Cells with Cardiac Myocytes by Nanotubes. *Circ. Res.* **2005**, *96*, 1039–1041. [CrossRef]
16. Watkins, S.; Salter, R.D. Functional Connectivity between Immune Cells Mediated by Tunneling Nanotubules. *Immunity* **2005**, *23*, 309–318. [CrossRef]
17. Eugenin, E.; Gaskill, P.; Berman, J. Tunneling nanotubes (TNT) are induced by HIV-infection of macrophages: A potential mechanism for intercellular HIV trafficking. *Cell. Immunol.* **2009**, *254*, 142–148. [CrossRef]
18. Vidulescu, C.; Clejan, S.; O'Connor, K.C. Vesicle traffic through intercellular bridges in DU 145 human prostate cancer cells. *J. Cell. Mol. Med.* **2004**, *8*, 388–396. [CrossRef]
19. Gurke, S.; Barroso, J.F.V.; Gerdes, H.-H. The art of cellular communication: Tunneling nanotubes bridge the divide. *Histochem. Cell Biol.* **2008**, *129*, 539–550. [CrossRef] [PubMed]
20. Sherer, N.M.; Mothes, W. Cytonemes and tunneling nanotubules in cell–cell communication and viral pathogenesis. *Trends Cell Biol.* **2008**, *18*, 414–420. [CrossRef] [PubMed]
21. Wang, X.; Gerdes, H.-H. Transfer of mitochondria via tunneling nanotubes rescues apoptotic PC12 cells. *Cell Death Differ.* **2015**, *22*, 1181–1191. [CrossRef] [PubMed]
22. Pasquier, J.; Guerrouahen, B.S.; Al Thawadi, H.; Ghiabi, P.; Maleki, M.; Abu-Kaoud, N.; Jacob, A.; Mirshahi, M.; Galas, L.; Rafii, S.; et al. Preferential transfer of mitochondria from endothelial to cancer cells through tunneling nanotubes modulates chemoresistance. *J. Transl. Med.* **2013**, *11*, 94. [CrossRef] [PubMed]
23. Lou, E.; Fujisawa, S.; Barlas, A.; Romin, Y.; Manova-Todorova, K.; Moore, M.A.; Subramanian, S. Tunneling Nanotubes. *Commun. Integr. Biol.* **2012**, *5*, 399–403. [CrossRef]
24. Li, X. Gap junction protein connexin43 and tunneling nanotubes in human trabecular meshwork cells. *Int. J. Physiol. Pathophysiol. Pharmacol.* **2019**, *11*, 212–219. [PubMed]

25. Ribeiro-Rodrigues, T.M.; Martins-Marques, T.; Morel, S.; Kwak, B.R.; Girão, H. Role of connexin 43 in different forms of intercellular communication—Gap junctions, extracellular vesicles and tunnelling nanotubes. *J. Cell Sci.* **2017**, *130*, 3619–3630. [CrossRef]
26. He, K.; Shi, X.; Zhang, X.; Dang, S.; Ma, X.; Liu, F.; Xu, M.; Lv, Z.; Han, D.; Fang, X.; et al. Long-distance intercellular connectivity between cardiomyocytes and cardiofibroblasts mediated by membrane nanotubes. *Cardiovasc. Res.* **2011**, *92*, 39–47. [CrossRef] [PubMed]
27. Furtado, M.B.; Nim, H.T.; Boyd, S.; Rosenthal, N. View from the heart: Cardiac fibroblasts in development, scarring and regeneration. *Development* **2016**, *143*, 387–397. [CrossRef]
28. Ivey, M.; Tallquist, M.D. Defining the Cardiac Fibroblast. *Circ. J.* **2016**, *80*, 2269–2276. [CrossRef]
29. Pellman, J.; Zhang, J.; Sheikh, F. Myocyte-fibroblast communication in cardiac fibrosis and arrhythmias: Mechanisms and model systems. *J. Mol. Cell. Cardiol.* **2016**, *94*, 22–31. [CrossRef]
30. Souders, C.A.; Bowers, S.L.; Baudino, T.A. Cardiac Fibroblast. *Circ. Res.* **2009**, *105*, 1164–1176. [CrossRef]
31. Fan, D.; Takawale, A.; Lee, J.; Kassiri, Z. Cardiac fibroblasts, fibrosis and extracellular matrix remodeling in heart disease. *Fibrogenes. Tissue Repair* **2012**, *5*, 15. [CrossRef]
32. Sylvius, N.; Bilinska, Z.; Veinot, J.P.; Fidzianska, A.; Bolongo, P.M.; Poon, S.; McKeown, P.; Davies, R.A.; Chan, K.-L.; Tang, A.S.L.; et al. In vivo and in vitro examination of the functional significances of novel lamin gene mutations in heart failure patients. *J. Med. Genet.* **2005**, *42*, 639–647. [CrossRef] [PubMed]
33. Bilińska, Z.T.; Sylvius, N.; Grzybowski, J.; Fidziańska, A.; Michalak, E.; Walczak, E.; Walski, M.; Bieganowska, W.; Szymaniak, E.; Kusmierczyk, B.; et al. Original article Dilated cardiomyopathy caused by LMNA mutations. Clinical and morphological studies. *Kardiol. Pol. Pol. Heart J.* **2006**, *64*, 8.
34. Binnig, G.G.; Quate, C.F. atomic force microscope. *At. Force Microsc.* **1986**, *56*, 930–933. [CrossRef] [PubMed]
35. Smolyakov, G.; Thiébot, B.; Campillo, C.; Labdi, S.; Severac, C.; Pelta, J.; Dague, É. Elasticity, Adhesion, and Tether Extrusion on Breast Cancer Cells Provide a Signature of Their Invasive Potential. *ACS Appl. Mater. Interfaces* **2016**, *8*, 27426–27431. [CrossRef]
36. Puech, P.-H.; Poole, K.; Knebel, D.; Muller, D.J. A new technical approach to quantify cell–cell adhesion forces by AFM. *Ultramicroscopy* **2006**, *106*, 637–644. [CrossRef]
37. Formosa, C.; Lachaize, V.; Gales, C.; Rols, M.P.; Martin-Yken, H.; François, J.M.; Duval, R.; Dague, E. Mapping HA-tagged protein at the surface of living cells by atomic force microscopy. *J. Mol. Recognit.* **2014**, *28*, 1–9. [CrossRef]
38. Manso, A.M.; Kang, S.-M.; Ross, R.S. Integrins, Focal Adhesions and Cardiac Fibroblasts. *J. Investig. Med.* **2009**, *57*, 856–860. [CrossRef]
39. Sun, Z.; Martinez-Lemus, L.A.; Trache, A.; Trzeciakowski, J.P.; Davis, G.E.; Pohl, U.; Meininger, G.A. Mechanical properties of the interaction between fibronectin and $\alpha5\beta1$-integrin on vascular smooth muscle cells studied using atomic force microscopy. *Am. J. Physiol. Circ. Physiol.* **2005**, *289*, H2526–H2535. [CrossRef]
40. Horton, M.; Charras, G.; Lehenkari, P. Analysis of ligand–receptor interactions in cells by atomic force microscopy. *J. Recept. Signal Transduct.* **2002**, *22*, 169–190. [CrossRef]
41. Chang, E.; Heo, K.-S.; Woo, C.-H.; Lee, H.; Le, N.-T.; Thomas, T.N.; Fujiwara, K.; Abe, J.-I. MK2 SUMOylation regulates actin filament remodeling and subsequent migration in endothelial cells by inhibiting MK2 kinase and HSP27 phosphorylation. *Blood* **2011**, *117*, 2527–2537. [CrossRef]
42. De Leeuw, R.; Gruenbaum, Y.; Medalia, O. Nuclear Lamins: Thin Filaments with Major Functions. *Trends Cell Biol.* **2018**, *28*, 34–45. [CrossRef]
43. Hah, J.; Kim, D.-H. Deciphering Nuclear Mechanobiology in Laminopathy. *Cells* **2019**, *8*, 231. [CrossRef] [PubMed]
44. Crisp, M.; Liu, Q.; Roux, K.; Rattner, J.B.; Shanahan, C.; Burke, B.; Stahl, P.D.; Hodzic, D. Coupling of the nucleus and cytoplasm: Role of the LINC complex. *J. Cell Biol.* **2005**, *172*, 41–53. [CrossRef] [PubMed]
45. Östlund, C.; Folker, E.S.; Choi, J.C.; Gomes, E.; Gundersen, G.G.; Worman, H.J. Dynamics and molecular interactions of linker of nucleoskeleton and cytoskeleton (LINC) complex proteins. *J. Cell Sci.* **2009**, *122*, 4099–4108. [CrossRef] [PubMed]
46. Sosa, B.A.; Kutay, U.; Schwartz, T.U. Structural insights into LINC complexes. *Curr. Opin. Struct. Biol.* **2013**, *23*, 285–291. [CrossRef]
47. Sullivan, T.; Escalante-Alcalde, D.; Bhatt, H.; Anver, M.; Bhat, N.; Nagashima, K.; Stewart, C.L.; Burke, B. Loss of a-Type Lamin Expression Compromises Nuclear Envelope Integrity Leading to Muscular Dystrophy. *J. Cell Biol.* **1999**, *147*, 913–920. [CrossRef]
48. Yang, L.; Munck, M.; Swaminathan, K.; Kapinos, L.E.; Noegel, A.A.; Neumann, S. Mutations in LMNA Modulate the Lamin A–Nesprin-2 Interaction and Cause LINC Complex Alterations. *PLoS ONE* **2013**, *8*, e71850. [CrossRef]
49. Bouzid, T.; Kim, E.; Riehl, B.D.; Esfahani, A.M.; Rosenbohm, J.; Yang, R.; Duan, B.; Lim, J.Y. The LINC complex, mechanotransduction, and mesenchymal stem cell function and fate. *J. Biol. Eng.* **2019**, *13*, 1–12. [CrossRef]
50. Derényi, I.; Julicher, F.; Prost, J. Formation and Interaction of Membrane Tubes. *Phys. Rev. Lett.* **2002**, *88*, 238101. [CrossRef]
51. Sun, M.; Graham, J.S.; Hegedus, B.; Marga, F.; Zhang, Y.; Forgacs, G.; Grandbois, M. Multiple Membrane Tethers Probed by Atomic Force Microscopy. *Biophys. J.* **2005**, *89*, 4320–4329. [CrossRef] [PubMed]
52. Saif, T. Mechanical response of single living cells under controlled stretch and indentation using functionalized micro force sensors. In Proceedings of the 2006 International Conference on Microtechnologies in Medicine and Biology, Okinawa, Japan, 9–12 May 2006; p. 3.

53. Basoli, F.; Giannitelli, S.M.; Gori, M.; Mozetic, P.; Bonfanti, A.; Trombetta, M.; Rainer, A. Biomechanical Characterization at the Cell Scale: Present and Prospects. *Front. Physiol.* **2018**, *9*, 1449. [CrossRef]

54. Nussenzveig, H.M. Cell membrane biophysics with optical tweezers. *Eur. Biophys. J.* **2018**, *47*, 499–514. [CrossRef] [PubMed]

55. Ramakrishnan, N.; Kumar, P.S.; Radhakrishnan, R. Mesoscale computational studies of membrane bilayer remodeling by curvature-inducing proteins. *Phys. Rep.* **2014**, *543*, 1–60. [CrossRef]

56. Derényi, I.; Koster, G.; van Duijn, M.; Czövek, A.; Dogterom, M.; Prost, J.; van Duijn, M. Membrane Nanotubes. *Control. Nanoscale Motion* **2007**, *711*, 141–159. [CrossRef]

57. Tian, F.; Yue, T.; Dong, W.; Yi, X.; Zhang, X. Size-dependent formation of membrane nanotubes: Continuum modeling and molecular dynamics simulations. *Phys. Chem. Chem. Phys.* **2018**, *20*, 3474–3483. [CrossRef] [PubMed]

58. Figard, L.; Sokac, A.M. A membrane reservoir at the cell surface. *BioArchitecture* **2014**, *4*, 39–46. [CrossRef]

59. Kapustina, M.; Elston, T.C.; Jacobson, K. Compression and dilation of the membrane-cortex layer generates rapid changes in cell shape. *J. Cell Biol.* **2013**, *200*, 95–108. [CrossRef]

60. Deschamps, C.; Echard, A.; Niedergang, F. Phagocytosis and Cytokinesis: Do Cells Use Common Tools to Cut and to Eat? Highlights on Common Themes and Differences. *Traffic* **2013**, *14*, 355–364. [CrossRef] [PubMed]

61. Thomas, T.; Jordan, K.; Simek, J.; Shao, Q.; Jedeszko, C.; Walton, P.; Laird, D.W. Mechanisms of Cx43 and Cx26 transport to the plasma membrane and gap junction regeneration. *J. Cell Sci.* **2005**, *118*, 4451–4462. [CrossRef]

62. Camelliti, P.; Borg, T.K.; Kohl, P. Structural and functional characterisation of cardiac fibroblasts. *Cardiovasc. Res.* **2005**, *65*, 40–51. [CrossRef]

63. Lombardi, R.; Chen, S.N.; Ruggiero, A.; Gurha, P.; Czernuszewicz, G.Z.; Willerson, J.T.; Marian, A.J. Cardiac Fibro-Adipocyte Progenitors Express Desmosome Proteins and Preferentially Differentiate to Adipocytes Upon Deletion of the Desmoplakin Gene. *Circ. Res.* **2016**, *119*, 41–54. [CrossRef]

64. Maione, A.S.; Pilato, C.A.; Casella, M.; Gasperetti, A.; Stadiotti, I.; Pompilio, G.; Sommariva, E. Fibrosis in Arrhythmogenic Cardiomyopathy: The Phantom Thread in the Fibro-Adipose Tissue. *Front. Physiol.* **2020**, *11*, 279. [CrossRef] [PubMed]

65. Maione, A.; Stadiotti, I.; Pilato, C.; Perrucci, G.; Saverio, V.; Catto, V.; Vettor, G.; Casella, M.; Guarino, A.; Polvani, G.; et al. Excess TGF-β1 Drives Cardiac Mesenchymal Stromal Cells to a Pro-Fibrotic Commitment in Arrhythmogenic Cardiomyopathy. *Int. J. Mol. Sci.* **2021**, *22*, 2673. [CrossRef] [PubMed]

66. Brown, R.D.; Ambler, S.K.; Mitchell, M.D.; Long, C. THE CARDIAC FIBROBLAST: Therapeutic Target in Myocardial Remodeling and Failure. *Annu. Rev. Pharmacol. Toxicol.* **2005**, *45*, 657–687. [CrossRef] [PubMed]

67. Laurini, E.; Martinelli, V.; Lanzicher, T.; Puzzi, L.; Borin, D.; Chen, S.N.; Long, C.S.; Lee, P.; Mestroni, L.; Taylor, M.R.G.; et al. Biomechanical defects and rescue of cardiomyocytes expressing pathologic nuclear lamins. *Cardiovasc. Res.* **2018**, *114*, 846–857. [CrossRef]

68. Martinelli, V.; Cellot, G.; Fabbro, A.; Bosi, S.; Mestroni, L.; Ballerini, L. Improving cardiac myocytes performance by carbon nanotubes platforms. *Front. Physiol.* **2013**, *4*, 239. [CrossRef]

69. Martinelli, V.; Cellot, G.; Toma, F.M.; Long, C.; Caldwell, J.H.; Zentilin, L.; Giacca, M.; Turco, A.; Prato, M.; Ballerini, L.; et al. Carbon Nanotubes Promote Growth and Spontaneous Electrical Activity in Cultured Cardiac Myocytes. *Nano Lett.* **2012**, *12*, 1831–1838. [CrossRef]

70. Golden, H.B.; Gollapudi, D.; Gerilechaogetu, F.; Li, J.; Cristales, R.J.; Peng, X.; Dostal, D.E. Isolation of Cardiac Myocytes and Fibroblasts from Neonatal Rat Pups. In *Advanced Structural Safety Studies*; Peng, X., Entonyak, M., Eds.; Humana Press: Totowa, NJ, USA, 2012; pp. 205–214. [CrossRef]

71. Kass-Eisler, A.; Falck-Pedersen, E.; Alvira, M.; Rivera, J.; Buttrick, P.M.; Wittenberg, B.A.; Cipriani, L.; Leinwand, L. Quantitative determination of adenovirus-mediated gene delivery to rat cardiac myocytes in vitro and in vivo. *Proc. Natl. Acad. Sci. USA* **1993**, *90*, 11498–11502. [CrossRef]

72. Hajjar, R.J.; Kang, J.X.; Gwathmey, J.K.; Rosenzweig, A. Physiological Effects of Adenoviral Gene Transfer of Sarcoplasmic Reticulum Calcium ATPase in Isolated Rat Myocytes. *Circulation* **1997**, *95*, 423–429. [CrossRef]

73. Emerson, R.; Camesano, T. On the importance of precise calibration techniques for an atomic force microscope. *Ultramicroscopy* **2006**, *106*, 413–422. [CrossRef] [PubMed]

74. Hertz, H. Ueber die Berührung fester elastischer Körper. *J. Reine Angew. Math* **1882**, *92*, 156–171. [CrossRef]

75. Dague, E.; Genet, G.; Lachaize, V.; Frugier, C.; Fauconnier, J.; Mias, C.; Payré, B.; Chopinet, L.; Alsteens, D.; Kasas, S.; et al. Atomic force and electron microscopic-based study of sarcolemmal surface of living cardiomyocytes unveils unexpected mitochondrial shift in heart failure. *J. Mol. Cell. Cardiol.* **2014**, *74*, 162–172. [CrossRef] [PubMed]

76. Lachaize, V.; Formosa-Dague, C.; Smolyakov, G.; Guilbeau-Frugier, C.; Galés, C.; Dague, E. Atomic Force Microscopy: An innovative technology to explore cardiomyocyte cell surface in cardiac physio/pathophysiology. *Lett. Appl. NanoBioScience* **2015**, *4*, 321–324.

77. Marcus, W.D.; Hochmuth, R.M. Experimental Studies of Membrane Tethers Formed from Human Neutrophils. *Ann. Biomed. Eng.* **2002**, *30*, 1273–1280. [CrossRef] [PubMed]

78. Pollok, S.; Pfeiffer, A.-C.; Lobmann, R.; Wright, C.S.; Moll, I.; Martin, P.E.M.; Brandner, J.M. Connexin 43 mimetic peptide Gap27 reveals potential differences in the role of Cx43 in wound repair between diabetic and non-diabetic cells. *J. Cell. Mol. Med.* **2011**, *15*, 861–873. [CrossRef] [PubMed]

79. Wright, C.S.; van Steensel, M.A.M.; Hodgins, M.B.; Martin, P.E.M. Connexin mimetic peptides improve cell migration rates of human epidermal keratinocytes and dermal fibroblasts in vitro. *Wound Repair Regen.* **2009**, *17*, 240–249. [CrossRef]
80. Evans, W.H.; Boitano, S. Connexin mimetic peptides: Specific inhibitors of gap-junctional intercellular communication. *Biochem. Soc. Trans.* **2001**, *29*, 606–612. [CrossRef]
81. Tavano, F.; Bonin, S.; Pinato, G.; Stanta, G.; Cojoc, D. Custom-Built Optical Tweezers for Locally Probing the Viscoelastic Properties of Cancer Cells. *Int. J. Optomechatron.* **2011**, *5*, 234–248. [CrossRef]
82. Falleroni, F.; Torre, V.; Cojoc, D. Cell Mechanotransduction with Piconewton Forces Applied by Optical Tweezers. *Front. Cell. Neurosci.* **2018**, *12*, 130. [CrossRef]

International Journal of
Molecular Sciences

Article

YAP/TEAD1 Complex Is a Default Repressor of Cardiac Toll-Like Receptor Genes

Yunan Gao [1,2,†], Yan Sun [1,†], Adife Gulhan Ercan-Sencicek [1,3], Justin S. King [4], Brynn N. Akerberg [4], Qing Ma [4], Maria I. Kontaridis [1], William T. Pu [4] and Zhiqiang Lin [1,*]

[1] Masonic Medical Research Institute, 2150 Bleecker St, Utica, NY 13501, USA; gaoyunan2014@163.com (Y.G.); ysun@mmri.edu (Y.S.); gercansencicek@mmri.edu (A.G.E.-S.); mkontaridis@mmri.edu (M.I.K.)

[2] Department of Cardiology, The Fourth Affiliated Hospital of Harbin Medical University, Harbin 150001, China

[3] Department of Neurosurgery, Program on Neurogenetics, Yale School of Medicine, Yale University, New Haven, CT 06510, USA

[4] Department of Cardiology, Boston Children's Hospital, 300 Longwood Ave, Boston, MA 02115, USA; kingjustinscott@gmail.com (J.S.K.); brynn.akerberg@childrens.harvard.edu (B.N.A.); Qing.Ma@childrens.harvard.edu (Q.M.); william.pu@enders.tch.harvard.edu (W.T.P.)

* Correspondence: zlin@mmri.edu

† These authors contributed equally.

Citation: Gao, Y.; Sun, Y.; Ercan-Sencicek, A.G.; King, J.S.; Akerberg, B.N.; Ma, Q.; Kontaridis, M.I.; Pu, W.T.; Lin, Z. YAP/TEAD1 Complex Is a Default Repressor of Cardiac Toll-Like Receptor Genes. *Int. J. Mol. Sci.* **2021**, 22, 6649. https://doi.org/10.3390/ijms22136649

Academic Editor: Nicole Wagner

Received: 30 May 2021
Accepted: 18 June 2021
Published: 22 June 2021

Abstract: Toll-like receptors (TLRs) are a family of pattern recognition receptors (PRRs) that modulate innate immune responses and play essential roles in the pathogenesis of heart diseases. Although important, the molecular mechanisms controlling cardiac TLR genes expression have not been clearly addressed. This study examined the expression pattern of *Tlr1, Tlr2, Tlr3, Tlr4, Tlr5, Tlr6, Tlr7, Tlr8*, and *Tlr9* in normal and disease-stressed mouse hearts. Our results demonstrated that the expression levels of cardiac *Tlr3, Tlr7, Tlr8*, and *Tlr9* increased with age between neonatal and adult developmental stages, whereas the expression of *Tlr5* decreased with age. Furthermore, pathological stress increased the expression levels of *Tlr2, Tlr4, Tlr5, Tlr7, Tlr8*, and *Tlr9*. Hippo-YAP signaling is essential for heart development and homeostasis maintenance, and YAP/TEAD1 complex is the terminal effector of this pathway. Here we found that TEAD1 directly bound genomic regions adjacent to *Tlr1, Tlr2, Tlr3, Tlr4, Tlr5, Tlr6, Tlr7*, and *Tlr9*. In vitro, luciferase reporter data suggest that YAP/TEAD1 repression of *Tlr4* depends on a conserved TEAD1 binding motif near *Tlr4* transcription start site. In vivo, cardiomyocyte-specific YAP depletion increased the expression of most examined TLR genes, activated the synthesis of pro-inflammatory cytokines, and predisposed the heart to lipopolysaccharide stress. In conclusion, our data indicate that the expression of cardiac TLR genes is associated with age and activated by pathological stress and suggest that YAP/TEAD1 complex is a default repressor of cardiac TLR genes.

Keywords: YAP; TEAD1; Toll-like receptor; heart; TLR4; cardiomyocyte; innate immune responses

1. Introduction

Heart failure is one of the leading causes of mortality and morbidity in the developed countries [1]. Involved in the pathogenesis of heart failure, Toll-like receptors (TLR) are a family of pattern recognition receptors that sense pathogenic stimuli and signal the cardiac residential cells to cope with harsh conditions [2]. Human and mouse genomes contain 10 (*TLR1–10*) and 13 (*Tlr1–13*) TLR genes, respectively. *Tlr11, Tlr12*, and *Tlr13* exist in mouse but not in the human genome, and murine *Tlr10* is a pseudogene [3]. In the heart, TLR genes are expressed in both cardiomyocytes (CMs) and non-CMs [4], and disturbance of TLR genes expression has been implicated in a range of heart diseases, including pathogen and non-pathogen related heart failure [5,6]. Despite the importance of TLR genes in the pathogenesis of heart failure, the molecular mechanisms that regulate these genes' expression are largely unknown.

Activation of CM innate immune signaling is one of the underlying mechanisms of ischemic and non-ischemic heart failure [7]. As crucial pattern recognition receptors, TLRs sense extracellular or intracellular danger signals and activate innate immune responses that lead to the synthesis and release of pro-inflammatory cytokine peptides [8]. The roles of several TLRs have been studied in the CMs, laying out a consensus direction that activation of TLRs is detrimental. For example, depletion of TLR4 attenuates myocardial infarction injury [9], whereas activation of TLR2, or TLR4, or TLR5 impairs CMs contractility [4]. Among the TLR pathways, TLR4/NF-κB signaling axis has been best established. TLR4 signals through a series of adaptor proteins and kinases to activate NF-κB, a central transcriptional driver of innate immune responses [10]. A canonical activator of TLR4 is bacterial pathogens-derived lipopolysaccharide (LPS), and LPS stress has frequently been used as the prototypical model to study innate immune responses in CMs [11] and other cell types [12].

Recently, cross talks have been found between TLR signaling and the Hippo-YAP pathway [13], an integral pathway regulating organ growth [14]. The already defined communications between these pathways happen through post-translational mechanisms, with YAP being one of the crucial node molecules. As the terminal effector of the Hippo-YAP pathway, YAP interacts with several transcription factors, such as TEAD1, to potently stimulate proliferation and promote cell survival [15]. We have recently reported that cardiomyocyte-specific YAP depletion up-regulated *Tlr2* and *Tlr4* [16]; however, it is unknown whether YAP directly regulates these two TLR genes through TEAD1 and whether YAP/TEAD1 complex also regulates other TLR genes. Addressing these questions will add a new layer of regulation between Hippo-YAP and TLR pathways and provide novel insights towards understanding the pathogenesis of heart failure.

This study documented the expression patterns of nine murine TLR genes (*Tlr1–9*) in developing mouse hearts and disease-stressed adult mouse hearts, studied the relationship between YAP/TEAD1 complex and TLR genes, and investigated YAP's role in the regulation of cardiomyocyte innate immune signaling. Our results suggest that YAP/TEAD1 complex is a default repressor of cardiomyocyte TLR genes and indicate that YAP is required to restrain cardiomyocyte innate immune responses.

2. Results

2.1. The Expression of TLR Genes Increases with Age during Postnatal Heart Development

TLRs have been well-known as crucial players regulating cardiac inflammation [17], and recent data have also implicated TLR signaling in CM maturation [18]. Therefore, it is pivotal to delineate TLR gene expression patterns in the intact and disease-stressed heart. The TLRs have diverse cellular distributions and an array of ligands. TLR1, TLR2, TLR4, TLR5, and TLR6 are localized on the cell membrane and recognize bacterial and viral pathogen-associated molecular patterns in the extracellular matrix. TLR3, TLR7, TLR8, and TLR9 are intracellular TLRs associated with endosomes and recognize viral or bacterial genetic materials (Figure 1A) [19].

The molecular mechanisms regulating the expression of these TLR genes have not been sufficiently addressed. We analyzed whole-genome RNA sequencing data from fetal and adult murine hearts [20] for expression of the TLR genes (Figure 1B). The cardiac expression of *Tlr5* was higher in the fetal than in the adult heart, and the expression of *Tlr1*, *Tlr6*, *Tlr9*, *Tlr12* did not differ significantly between these two developmental stages. The remaining TLR genes all had lower expression levels in the fetal heart (Figure 1B). To further analyze the expression dynamics of these TLR genes during postnatal heart development, we selected four time points: embryonic day 18.5 (E18.5), postnatal day 5 (P5; Neonate), P14 (Juvenile), and P42 (Adult). Among the five genes encoding cell membrane TLRs, *Tlr2* and *Tlr4* increased with age, *Tlr5* decreased with age, and *Tlr1/Tlr6* fluctuated among the four developmental stages (Figure 1C). Different from the cell membrane TLRs, the expression of intracellular TLR genes, including *Tlr3*, *Tlr7*, *Tlr8*, and *Tlr9*, showed an increasing trend during heart development (Figure 1D). Here we summarized the expression level order of the four intracellular cardiac TLR genes: *Tlr3 > Tlr8, Tlr9 > Tlr7* (Figure 1D).

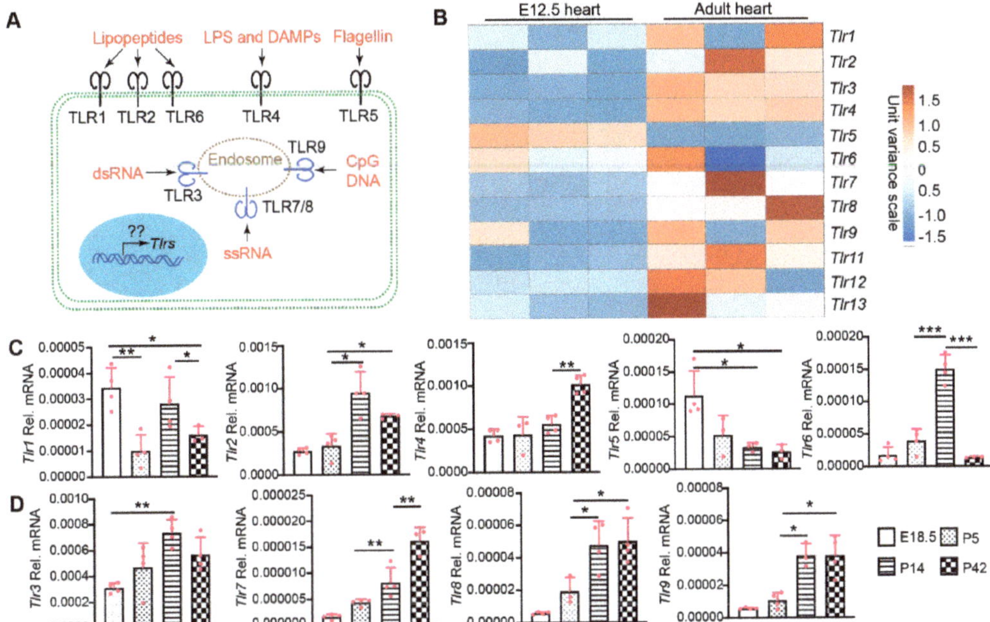

Figure 1. The expression of TLR genes in different age hearts. (**A**). Schematic view of TLR cellular localization and ligands. TLR1, 2, and 6 recognize bacteria- or mycoplasma-derived lipopeptides. TLR4 recognizes bacteria-derived lipopolysaccharide (LPS) and host-cell-derived death associated molecular patterns (DAMPs). Bacteria-derived Flagellin is the natural ligand of TLR5. Virus-derived double strand RNA (dsRNA) and single strand RNA (ssRNA) binds to and activates TLR3 and TLR7/8, respectively. TLR9 recognizes DNA containing unmethylated CpG motifs prevalent in microorganisms and mitochondrial DNA, but not in vertebrate genomic DNA. On the transcriptional level, mechanisms regulating the expression of these TLR genes are largely unknown. (**B**). Heat map of TLR gene expression levels in fetal and adult hearts. TLR gene expression values were retrieved from published RNA seq data (22). The heat map was generated with an online tool (ClustVis). (**C,D**). qRT-PCR measurement of TLR genes during heart development. Total RNA from hearts with the indicated ages was used for qRT-PCR. Gene expression level was normalized to *Gapdh*. E18.5, embryonic day 18.5; P5, P14, P42, postnatal days 5, 14, and 42. One-way ANOVA post hoc Tukey's multiple comparisons test, *, $p < 0.05$; **, $p < 0.01$, ***, $p < 0.001$. $n = 4$.

2.2. The Expression of TLR Genes in Disease-Stressed Hearts

To test whether non-pathogen stress affects TLR gene expression, we examined the mRNA levels of *Tlr1–9* in pressure overload (PO) and ischemia/reperfusion (IR) stressed mouse hearts. PO stress was induced by transverse aortic constriction (TAC) surgery, and hearts were collected ten days after TAC. IR stress was triggered by occluding the left anterior descending (LAD) coronary artery for 50 min, and hearts were collected two days after IR surgery. Compared with TLR genes in sham controls, *Tlr2, 4, 7, 8*, and *9* were significantly increased in the PO-stressed hearts (Figure 2A). In IR-stressed hearts, *Tlr2, 4, 5*, and *9* were significantly increased (Figure 2B). These data suggest that TLR genes are largely primed in disease-stressed hearts.

Because TLR4 is the best studied TLR and validated commercial TLR4 antibody is available, we corroborated our observation by examining the protein level of TLR4. In sham control myocardium, TLR4 protein was only detected in non-CMs (Figure 2C). However, in either IR- or PO-stressed myocardium, cardiomyocyte TLR4 was readily detectable

(Figure 2D). Taken together, these data suggest that TLR signaling pathways may be hyperactive in PO- or IR-stressed hearts.

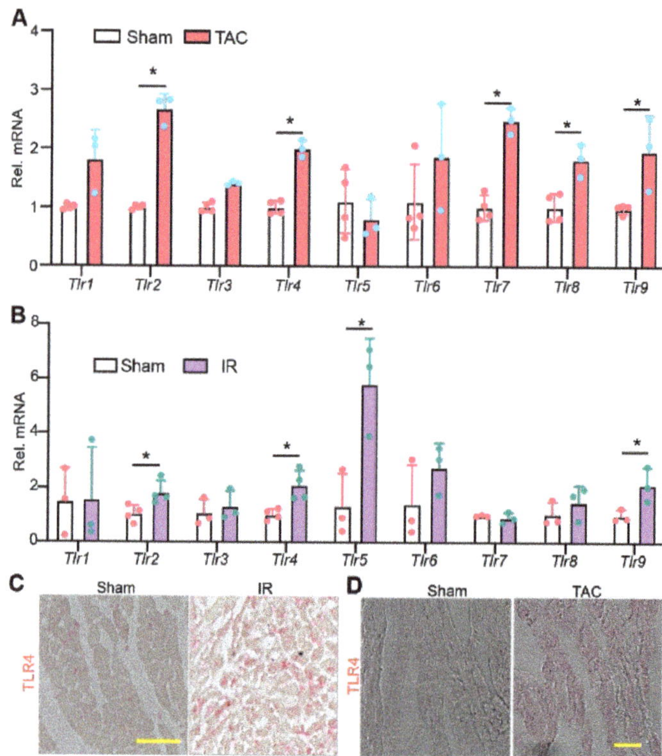

Figure 2. The expression pattern of TLR genes in stressed hearts. qRT-PCR measurement of TLR genes in pressure overload stressed hearts. 6–8 weeks old mice were stressed with TAC (transverse aortic constriction) surgery for ten days before being sacrificed for cardiac gene expression analysis. (**B**). qRT-PCR measurement of TLR genes in ischemia-reperfusion (IR) stressed hearts. Gene expression level was normalized to *Gapdh*. (**A,B**), student *t*-test, *, $p < 0.05$. $n = 3–4$. (**C,D**). Representative images of myocardium immunostained for TLR4. Scale bar = 50 μm. (**B,C**), hearts were collected for analysis two days after IR surgery.

2.3. Cardiomyocyte Specific YAP Depletion Does Not Induce Cardiac Hypertrophic Remodeling in the First 12 Days after Birth

Because TLRs play essential roles in regulating cardiac inflammation [8], it is pivotal to understand the molecular mechanisms that govern their expression. We previously reported that cardiomyocyte-specific YAP depletion increased the expression of *Tlr2* and *Tlr4* [16] and that cardiac YAP/TEAD1 complex decreased with age [21]. We then hypothesized that YAP/TEAD1 restrained the expression of cardiac TLR genes. To test this hypothesis, we generated cardiomyocyte-specific *Yap* knockout (Yap^{cKO}) mice by crossing *Myh6::Cre* transgenic mice [22] to *Yap* flox allele [23] (Figure 3A). Cardiomyocyte specific YAP depletion causes dilated cardiomyopathy in adult mice [24], and the associated hypertrophic remodeling process may confound TLR gene expression analysis. To avoid this potential culprit, we determined to find a developmental stage at which the hypertrophic remodeling process had not started in the Yap^{cKO} heart.

Compared to littermate control, Yap^{cKO} mice had a lower heart-to-body weight ratio trend at postnatal day 7 (P7), which reached significance at P12. Because YAP/TEAD complex is required for CM proliferation [25,26], and some CMs are still proliferative in the neonatal heart [27], it is possible that the lower heart-to-body weight ratio of P12 *Yap^cKO* mice is due to less CM proliferation. At P35, the heart-to-body weight ratio between control and *Yap^cKO* mice was not distinguishable (Figure 3B). These heart-to-body weight ratio data indicate that compensatory cardiac hypertrophy happens between P13 and P35. In support of this notion, no increase in cardiac fibrosis was observed in the *Yap^cKO* heart at P12 (Figure 3C). At this development stage, YAP knockdown was confirmed by western blot (Figure 3D), and the presence of residual YAP might be contributed by non-CMs [21] or due to the young age of *Yap^cKO* mice. We then checked YAP expression in 2-month-old hearts and found that YAP was efficiently knocked down at this age (Supplementary material Figure S1A).

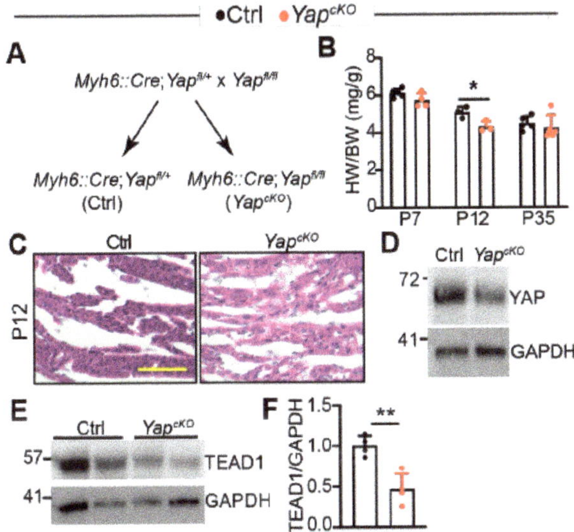

Figure 3. Cardiomyocyte-specific YAP depletion does not cause cardiac hypertrophy in mice less than 12 days of age. (**A**). Genetic strategy to knock out YAP in CMs. *Myh6::Cre; Yap^fl/+* mice were used as control (Ctrl) in the subsequent studies. (**B**). Heart-to-bodyweight ratio. Heart and body weight was measured at postnatal day 7, 12, and 35. *n* = 3–5 for each group. (**C**). H&E staining of P12 control and *Yap^cKO* hearts. Scale bar = 50 μm. (**D**). YAP immunoblot. (**E**). TEAD1 immunoblot. (**D,E**), total heart protein from P12 mice was used for immunoblot. **F**. Densitometry quantification of TEAD1. TEAD1 protein level was normalized to GAPDH. *n* = 3. (**B,F**), student's t test, *, *p* < 0.05; **, *p* < 0.01.

Interestingly, TEAD1 protein was decreased in *Yap^cKO* hearts (Figure 3E), suggesting that loss of YAP destabilizes TEAD1. Together, these data suggest that knocking out YAP does not induce cardiac hypertrophic remodeling in the first 12 days after birth. Therefore, we used P12 hearts for TLR gene expression analysis in the following studies.

2.4. YAP/TEAD1 Complex Regulates the Expression of TLR Genes

To determine whether YAP/TEAD1 complex directly regulates the expression of TLR genes, we analyzed recently reported fetal and adult cardiac TEAD1 chromatin immunoprecipitation and high throughput sequencing (ChIP-seq) data for TEAD1 binding to regions neighboring TLR genes [20]. In the fetal heart, TEAD1 directly bound to the promoters of *Tlr2, Tlr4, Tlr5, Tlr6, Tlr7*, and to regions neighboring *Tlr1, Tlr2, Tlr5, Tlr9* (Figure 4A). In the adult heart, TEAD1 bound to the promoters of *Tlr3, Tlr4, Tlr6*, and to regions neighboring *Tlr1, Tlr5*, and *Tlr9* (Figure 4A). For *Tlr2* and *Tlr7*, no TEAD1 ChIP seq peaks were detected in adult hearts. In addition, in both fetal and adult hearts, TEAD1 ChIP seq occupancy was not detected adjacent to *Tlr8* (Supplementary material Figure S1B).

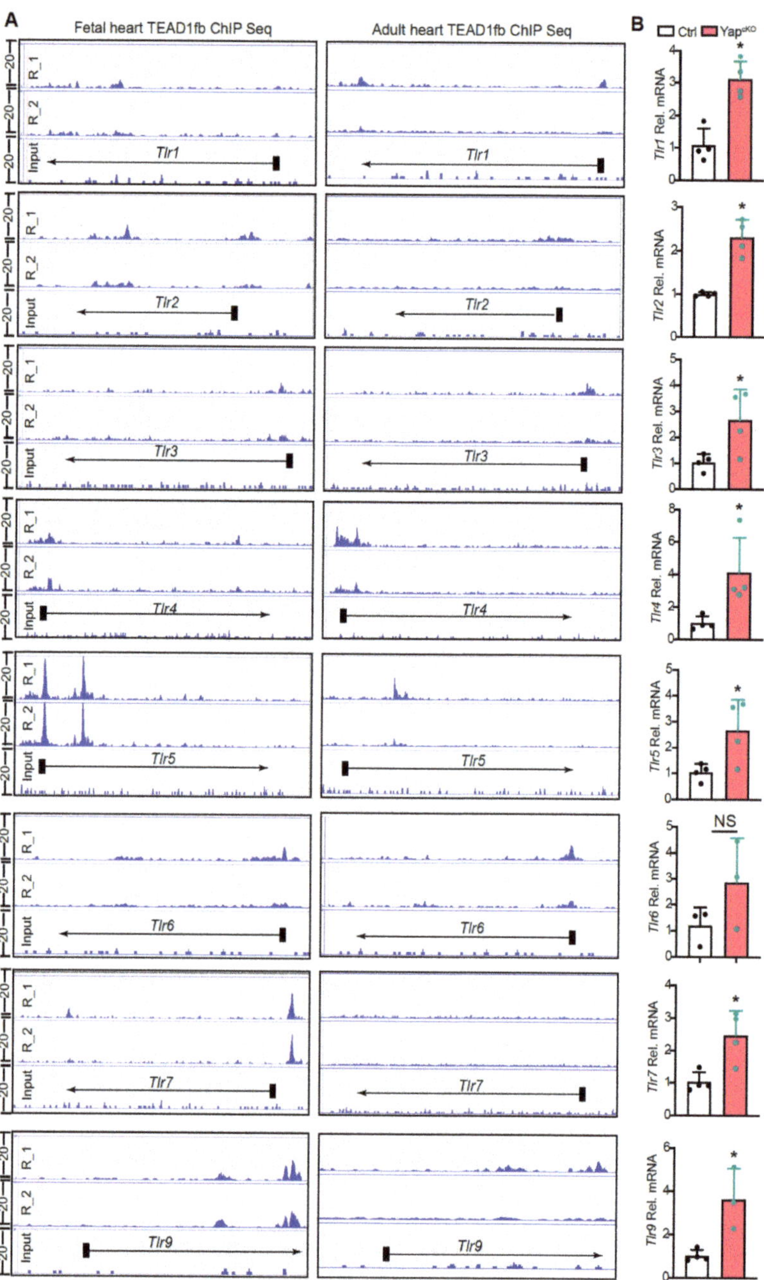

Figure 4. Cardiomyocyte-specific YAP depletion activates TLR genes expression. (**A**). Genome browser view showing chromatin immunoprecipitation and high throughput sequencing (ChIP-seq) of TEAD1 occupancy near the transcriptional start site (TSS) or gene body of the TLR genes. E12.5 fetal heart and adult hearts were used for TEAD1^{fb} ChIP-seq. ChIP-seq data from two biological replicates are sown. (**B**). qRT-PCR measurement of cardiac TLR genes. Gene expression level was normalized to *Gapdh*. $n = 4$. Student *t*-test, *, $p < 0.05$.

We then measured the expression of *Tlr1–9* with qRT-PCR. Our results showed that except for *Tlr6*, all the other eight TLR genes were significantly up-regulated in *Yap^cKO* hearts (Figure 4B, Supplementary material Figure S1C). Together, these data suggest that YAP/TEAD1 complex is a master suppressor of cardiac TLR genes.

2.5. YAP/TEAD1 Complex Directly Suppresses the Expression of Tlr4

The TEAD1 ChIP-seq and *Yap^cKO* gene expression data suggest that YAP/TEAD1 complex directly regulates the expression of TLR genes. To validate this hypothesis, we used *Tlr4* as a prototype gene for further analysis. The TEAD1 ChIP-seq peak of *Tlr4* spans its transcription start region (TSSR), which contains a conserved TEAD1 binding motif 104 bp downstream of the transcription start site (Figure 5A). We first did ChIP-qPCR to validate TEAD1 binding of this region. The TEAD1 ChIP products of adult hearts were amplified by two sets of primers: the first set of primers spanning the *Tlr4* TSSR and the second set of primers annealing to the region 2.5 kb upstream of *Tlr4* TSSR. qPCR amplicons from the first but not from the second primer set were enriched in the ChIP products (Figure 5B), confirming the direct binding of TEAD1 to *Tlr4* TSSR.

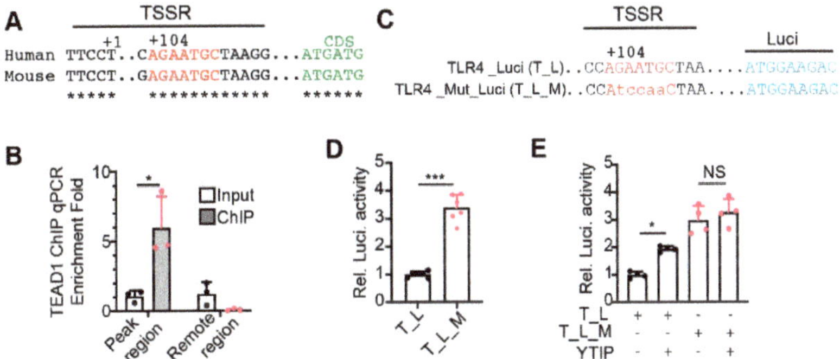

Figure 5. YAP/TEAD1 complex directly suppresses *Tlr4* expression. (**A**). Schematic view of the transcription start site (TSS, +1) regions of human and mouse TLR4 genes. The conserved TEAD1 binding motif was depicted with red letters, and the TLR4 coding sequences were indicated with green letters. (**B**). TEAD1 ChIP qPCR. A primer set spanning the mouse *Tlr4* TSS region was used to amplify ChIP products. Another primer set amplifying one fragment 2.5 kb away from the TSS was used as a negative control. Amplicon values from TEAD1 ChIP products were normalized with that of the input. *n* = 3. Student *t*-test, *, *p* < 0.05. (**C**). Depiction of the luciferase reporter constructs. TLR4_Luci (T_L) reporter contains the human TLR4 promoter driving luciferase. A conserved TEAD1 binding motif in the TLR4 promoter was mutated to generate TLR4_Mut_Luci (T_L_M). (**D,E**). Dual-luciferase reporter assay in HEK293T cells. Dual-luciferase assay was performed 24 h after transfection. YTIP: YAP and TEAD1 interference peptide. (**D**), student *t*-test, *, *p* < 0.05. (**E**), One-way ANOVA post hoc Tukey's multiple comparisons test, *, *p* < 0.05, ***, *p* < 0.001. (**D,E**), *n* = 4–6.

Next, we cloned the human TLR4 TSSR and its adjacent region into a luciferase reporter vector, TLR4_Luci. We created another reporter, TLR4_Mut_Luci, by mutating the TEAD1 binding motif sequence AGAATGC into AtccaaC (Figure 5C). If YAP/TEAD complex was required for suppressing TLR4 expression, disrupting YAP/TEAD interaction would activate TLR4_Luci but not TLR4_Mut_Luci reporter. In 293 T cells, TLR4_Mut_Luci reporter had significantly higher activity than TLR4_Luci (Figure 5D), indicating that the conserved TEAD1 binding motif is required for suppressing TLR4 expression. Consistently, disrupting YAP and TEAD interaction with a YAP-TEAD interfering peptide (YTIP) [26] significantly increased the activity of TLR4_Luci but had no effects on TLR4_Mut_Luci (Figure 5E), indicating that YAP/TEAD complex is required to repress TLR4 expression.

2.6. CM-Specific YAP Depletion Activates TLR4/NF-κB Signaling

In the heart, loss of either YAP or TEAD1 decreased CM survival and caused heart failure, probably due to the downregulation of anti-apoptosis genes and disruption of mitochondria structure [24,28]. Here, our data indicate another mechanism that the TLR genes are primed in the YAP or TEAD1 knockout hearts. This sensitizes the CMs to environmental stresses, such as extracellular damage-associated molecular patterns (DAMPs) released by damaged cells and intracellular mitochondrial DNA escaped from impaired mitochondria. Because TLR4 is one of the best-characterized DAMPs receptors, and TLR4/NF-κB signaling axis plays essential roles in the pathophysiology of heart failure [10], we tested our hypothesis by focusing on analyzing whether loss of YAP resulted in activation of TLR4/NF-κB pathway.

Consistent with our expectation, TLR4 and RelA protein levels were both increased in P12 *Yap^cKO* hearts (Figure 6A,B). We further measured the expression of four NF-κB target genes: *Ccl2* [29], *Il1b* [30], *Il12a* [31], and *Il12b* [32]. Except for *Ccl2*, the other three NF-κB target genes were significantly elevated in the *Yap^cKO* hearts (Figure 6C). In situ immunohistochemistry staining confirmed that the synthesis of both IL1β and IL12β was increased in the *Yap^cKO* myocardium (Figure 6D). Together, these data strongly suggest that cardiomyocyte-specific YAP depletion activates TLR4/NF-κB signaling pathway.

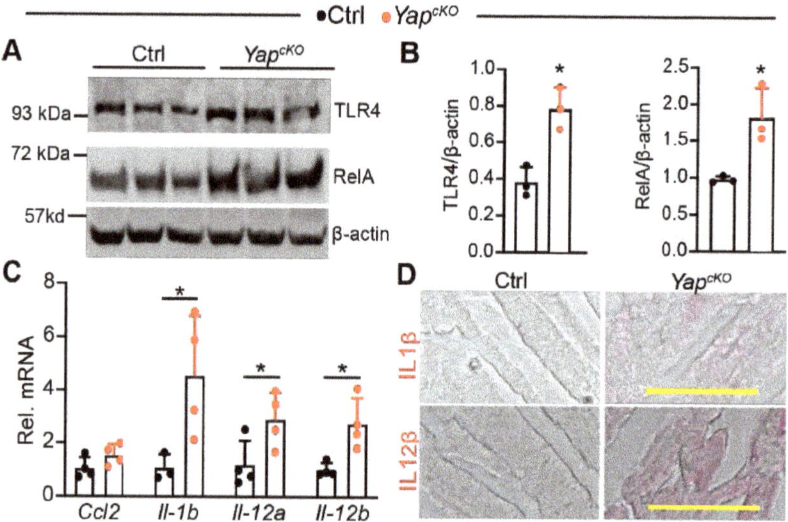

Figure 6. Cardiomyocyte-specific YAP depletion activates TLR4/NF-κB pathway. (**A**). Immunoblot of TLR4 and RelA. Total heart protein from postnatal day 12 (P12) mice was used for immunoblot. (**B**). Densitometry quantification of TLR4 and RelA. The protein levels of TLR4 and RelA were normalized to β-actin. $n = 3$ for each group. (**C**). qRT-PCR measurement of pro-inflammatory genes. Total RNA from P12 heart was used. Gene expression level was normalized to *Gapdh*. $n = 4$. (**B,C**), student *t*-test, *, $p < 0.05$. (**D**). Immunohistochemistry staining of P12 myocardium. Alkaline phosphatase-based detection system was used to visualize the expression of IL-1β and IL-12β. Scale bar = 50 μm.

2.7. Knocking Down YAP in the CMs Predisposes the Heart to Lipopolysaccharide (LPS) Stress

Our current data suggest that YAP is required to blunt CMs innate immune signaling. We tested whether knocking down YAP predisposed the heart to acute LPS stress to further validate this hypothesis. Because whole-heart YAP depletion caused heart failure [24], we used an Adeno-associated virus (AAV) system to knock out YAP in a portion of CMs. AAV9.cTnT.iCre (AAV.iCre) [33] was retro-orbitally delivered to three weeks old *Yap^fl/fl* mice at a dose of 5×10^9 virus genomes (vg)/gram (g) body weight, which is sufficient to transduce 50–60% of the CMs [34]. *Yap^{fl/fl}* mice receiving the same dose of AAV9.cTnT.GFP (AAV.GFP) were used as a control. At 21 days after AAV delivery, mice were treated with a sub-lethal dose of LPS (6 mg/kg) for 6 h [35]. Echocardiography measurements were performed before and after LPS treatment (Figure 7A).

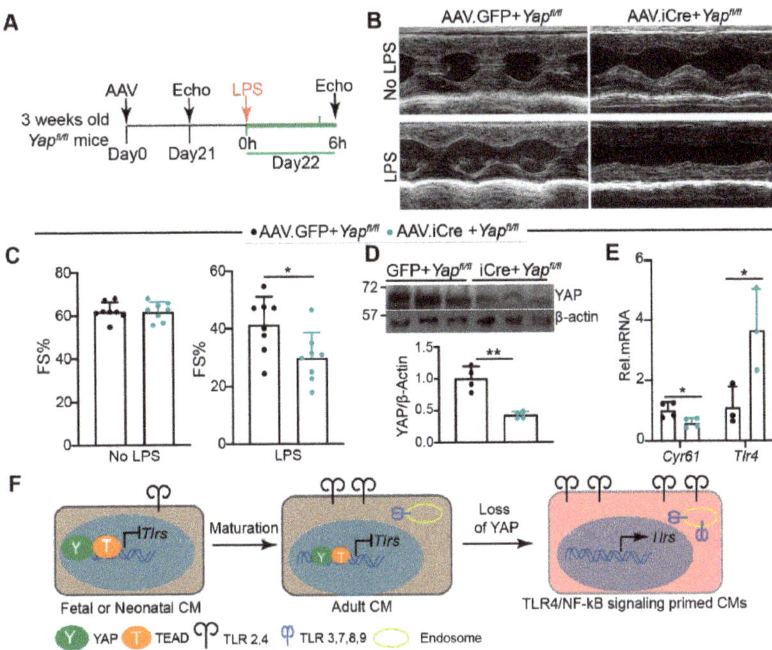

Figure 7. Knocking down YAP in CMs predisposes the heart to LPS stress. (**A**). Experimental design. Indicated AAVs were retro-orbitally delivered into 3 weeks old *Yap^{fl/fl}* mice. Twenty-one days later, echocardiography was performed to measure cardiac function. Twenty-two days after AAV delivery, 6 mg/kg LPS was intraperitoneally injected into AAV transduced mice. Echocardiography measurements were performed 6 h after LPS treatment. (**B**). Representative M mode cardiac echocardiograms before and after LPS treatment. (**C**). Fraction shortening before and after LPS treatment. AAV9.cTnT.GFP (AAV.GFP), *n* = 8; AAV9.cTnT.iCre (AAV.iCre), *n* = 8. (**D**). YAP Immunot Blot and densitometry quantification. YAP protein level was normalized to β-actin. *n* = 4 for each group. (**E**). qRT-PCR measurement. *N* = 3–4 for each group. Gene expression level was normalized to *Gapdh*. (**D,E**), 6 h after LPS treatment, hearts were collected for protein and RNA extraction. (**C–E**), student's t test, *. *p* < 0.05, **, *p* < 0.01. (**F**). Schematic summary of the current study. During heart development, the expression of *Tlr2* and *Tlr4* and intracellular TLR genes increases with age. YAP/TEAD1 complex is a default repressor of TLR genes in intact hearts. Cardiomyocyte-specific YAP depletion unleashes the expression of TLR genes and primes the CMs for generating inflammatory cytokine peptides. Brown color background in fetal/neonatal and adult CMs indicates a healthy status; pink color background in YAP depleted CMs indicates an innate immune signaling active status.

Three weeks after AAV delivery, AAV.iCre + *Yap^{fl/fl}* mice had similar fraction shortening with AAV.GFP + *Yap^{fl/fl}* mice (Figure 7B,C). After LPS treatment, both AAV.iCre + *Yap^{fl/fl}* and AAV.GFP + *Yap^{fl/fl}* mice had reduced systolic heart function (Figure 7B,C). Compared to AAV.GFP + *Yap^{fl/fl}* mice, AAV.iCre + *Yap^{fl/fl}* mice were more vulnerable to LPS stress, as evidenced by significantly lower fraction shortening (Figure 7B,C). Nevertheless, the heart rate was not distinguishable between these two groups of mice before and after LPS treatment (Supplementary material Figure S2A). At the molecular level, we confirmed that YAP was knocked down, and TLR4 and RelA were up-regulated in the AAV.iCre + *Yap^{fl/fl}* hearts (Figure 7D, Supplementary material Figure S2B). *Cyr61* is a well-defined YAP target [36]. Consistent with YAP knockdown, *Cyr61* was significantly reduced in AAV.iCre + *Yap^{fl/fl}* hearts (Figure 7E). Additionally, *Tlr4* was significantly up-regulated in AAV.iCre + *Yap^{fl/fl}* hearts (Figure 7E).

Activation of YAP has been reported to reduce LPS-induced CM apoptosis [37]. We then examined whether knocking down YAP increased LPS-induced CM apoptosis. Unlike the published study [37], we treated mice with LPS for 6 h instead of two days. In this condition, we did not detect CM apoptosis in both AAV.GFP + *Yap^{fl/fl}* and AAV.iCre + *Yap^{fl/fl}* hearts (supplementary material Figure S2C). Additionally, 6 h LPS treatment did not result in macrophages and neutrophils infiltration in either AAV.GFP + *Yap^{fl/fl}* or AAV.iCre + *Yap^{fl/fl}* hearts (Supplementary material Figure S2D). These observations align with previous reports that acute LPS-TLR4 signaling decreases heart function by impairing CM contractility but not by inducing either CM loss [38,39] or leukocytes infiltration [9]. Additionally, together with Figure 6, our data suggest that knocking down YAP in CMs increases TLR4 expression, which predisposes the heart to LPS stress by sensitizing CMs to LPS-TLR4 signaling.

3. Discussion

This study investigated the expression patterns of nine cardiac TLR genes and studied the transcriptional relationship between YAP/TEAD1 complex and these genes. We found that the expression levels of most cardiac TLR genes were associated with age and activated by pathological stress. Furthermore, our data suggest that YAP/TEAD1 complex directly suppresses the expression of most cardiac TLR genes, and that loss of YAP primes CM innate immune gene expression programs by activating TLR4/NF-κB signaling (Figure 7F).

3.1. Expression of Cardiac TLR Genes Is Associated with Age and Activated by Pathological Stress

During postnatal heart development, CMs experience a maturation process [40], and non-CMs gradually expand their population [41]. Here, we found that the expression levels of most cardiac TLR genes were low in neonatal mice but significantly higher in adults. Three possible mechanisms underlie these observations: first, the expression of cardiac TLR genes increases due to CM maturation; second, the increase of cardiac TLR genes is due to the expansion of non-CMs; third, both CM maturation and non-CM expansion contribute to the upregulation of cardiac TLR genes. Because the percentage of non-CMs does not increase between E18.5 and postnatal day 14 [41], the upregulation of cardiac TLR genes should be mainly due to CM maturation during this growth period. Interestingly, the expression of all the intracellular TLR genes increases during this development phase, suggesting that upregulation of intracellular TLR genes is positively correlated with CM maturation.

In the adult heart, we examined the expression of TLR genes in two different heart disease models: PO-induced chronic cardiac hypertrophy and IR-triggered acute cardiac injury. Our data demonstrate that the expression of many cardiac TLR genes was activated by pathological stress. In this study, we used RNA from the whole heart to do gene expression analysis. In a more detailed study, after myocardial infarction (MI), cardiac tissues from the infarct zone and the remote zone were separated for TLR gene expression analysis. The results showed that TLR genes were increased in the infarct zone but kept unchanged in the remote zone [6], suggesting that upregulation of cardiac TLR genes may be mainly due to the expansion of non-CMs in disease-stressed adult hearts.

3.2. YAP/TEAD1 Complex Is a Default Repressor of Cardiomyocyte TLR Genes

Recently, we and others have shown that activating YAP in CMs suppressed the expression of inflammatory genes [16,42]. This study extended our understanding of YAP's role in cardiac homeostasis by systemically interrogating the relationship between YAP and cardiac TLR genes. Our data demonstrated that TEAD1 is directly bound to the regions neighboring *TLR genes*, and CM-specific YAP depletion increased TLR gene expression. It is therefore conceivable that YAP/TEAD1 complex serves as a default suppressor of cardiomyocyte TLR genes. Of note, YAP might be required but not sufficient for suppressing TLR genes. For instance, YAP activation did not suppress *Tlr4* expression in either LPS-treated CMs or IR-stressed hearts [16]. Furthermore, some TLR genes were increased in IR-stressed hearts (Figure 2B), whereas nuclear YAP was enriched in the CMs of diseased hearts [24,43]. Therefore, YAP activation is likely insufficient to suppress TLR genes, and pathological stress may activate additional molecular mechanisms that overcome YAP/TEAD-mediated default repression of TLR genes.

Despite that activating YAP represses a subset of target genes by recruiting nucleosome remodeling and histone deacetylase (NuRD) complex [44] in MCF10A mammary epithelial cells, our current data suggest that this is unlikely the case for YAP/TEAD1 regulation of *Tlr4*. Interestingly, TEAD1 protein was decreased in *Yap^{cKO}* hearts (Figure 3), suggesting that loss of YAP increases *Tlr4* expression through its effect on TEAD1 expression. Future studies are secured to elucidate the mechanism of how YAP/TEAD1 complex regulates TLR genes.

3.3. YAP Is Required for Blunting CM Innate Immune Signaling

Ischemic or non-ischemic pathological stress activates CM innate immune signaling pathways [45], which stimulate pro-inflammatory cytokine release and reactive oxygen species (ROS) production [7,46]. These innate immune responses are beneficial for defending CMs against pathogen invasion and tissue repair, but also cause cardiac inflammation and myocardial damage, including CM dysfunction or death. Our published [16] and current data have identified YAP as a crucial suppressor of TLR4-mediated innate immune responses in CMs. Consistent with published data [9], we showed that acute LPS stress reduced heart function before the onset of cardiac inflammation. We further showed that knocking down YAP in the heart activated TLR4/NF-κB signaling (Figure 6) and exacerbated LPS-induced cardiac shock (Figure 7). These data suggest that YAP in the CMs cell-autonomously protects the heart against LPS stress by blunting TLR4/NF-κB pathway.

Although we found that loss of YAP increased TLR4 and RelA expression, we did not reveal the underlying mechanism of how YAP suppresses TLR4/NF-κB signaling. The TLR4/NF-κB signaling axis comprises multiple adaptor proteins and kinases, such as MyD88, TRAF6, and TAK1 [10]. Recently, in a non-cardiac context, YAP has been shown to inhibit intracellular innate immune signaling independent of its transcriptional activity. Instead, YAP directly impedes the activity of crucial innate immune signaling components, such as IRF3 [47], TAK1 [48], TBK1 [49], and TRAF6 [50]. More work needs to be done to dissect YAP's role in CM innate immune signaling, including a determination of whether YAP regulates CM innate immune responses through its transcriptional or non-transcriptional activity, or both.

4. Material and Methods

Supplemental information provides expanded material and experimental procedures.

4.1. Experimental Animals

All animal procedures were approved by the Institute Animal Care and Use Committees of Masonic Medical Research Institute and Boston Children's Hospital. All experiments were performed in accordance with NIH guidelines and regulations. C57BL/6J mice aged 6–8 weeks were obtained from Jackson Labs. Swiss Webster (CFW) mice aged 6–8 weeks were obtained from Charles River Laboratories. *Myh6::Cre* [22] and *Yap* flox alleles [23] were reported previously.

4.2. LPS Treatment

E. coli O55:B5 LPS (Sigma, St. Louis, MO, USA, #L2880) was dissolved in saline and sterile-filtered. LPS was intraperitoneally delivered at a dose of 6 mg/kg body weight.

Then, 6 h after LPS delivery, mice were tested for cardiac function. The mice were sacrificed for tissue harvesting after echocardiography measurements.

4.3. Gene Expression

Total RNA was isolated using Trizol. For qRT-PCR, RNA was reverse transcribed (Applied Biological Materials Inc., Richmond, BC, Canada, G454), and specific transcripts were measured using Sybr Green chemistry (Lifesct., Rockville, MD, USA, LS01131905Y) and normalized to *Gapdh*. Primer sequences were provided in Table S1. Primary antibodies used for immunoblot and immunohistochemistry staining were listed in Table S2.

4.4. Statistics

Values were expressed as mean $\pm$ SD. In addition, student's *t*-test or ANOVA with Tukey's honestly significant difference post hoc test was used to test for statistical significance involving two or more than two groups, respectively.

Supplementary Materials: The following are available online at https://www.mdpi.com/article/10.3390/ijms22136649/s1, Figure S1: YAP/TEAD1 complex regulation of cardiac TLR genes; Figure S2: Short period of LPS treatment does not induce CM apoptosis. Table S1: qRT-PCR primers; Table S2: Primary antibody list.

Author Contributions: Y.G., Y.S., A.G.E.-S. and J.S.K. carried out the experiments. Q.M. performed IR and TAC surgery. B.N.A. and W.T.P. provided the Heart RNA-Seq data and TEAD1 ChIP-seq data. A.G.E.-S., W.T.P. and M.I.K. reviewed and revised the manuscript. Z.L. designed and performed part of the experiments, interpreted the data and wrote the manuscript. All authors have read and agreed to the published version of the manuscript.

Funding: Z.L. was supported by the Masonic Medical Research Institute. A. Ercan-Sencicek was supported by the Grand Lodge of Free and Accepted Masons of the State of New York-Autism.

Institutional Review Board Statement: The study was conducted according to the guidelines of the Declaration of Helsinki, and approved by the Institutional Review Board (or Ethics Committee) Masonic Medical Research Institute (protocol code 2018-8-30-ZL01, approved in September 2020).

Informed Consent Statement: Not applicable.

Data Availability Statement: The RNA-seq and ChIP-seq data presented in this study are openly available at NCBI GEO database (GSE124008), Reference doi: 10.1038/s41467-019-12812-3.

Acknowledgments: We thank Ye Sun from Boston Children's Hospital for critically reviewing the manuscript.

Conflicts of Interest: The authors declare no conflict of interest.

References

1. Virani, S.S.; Alonso, A.; Aparicio, H.J.; Benjamin, E.J.; Bittencourt, M.S.; Callaway, C.W.; Carson, A.P.; Chamberlain, A.M.; Cheng, S.; Delling, F.N.; et al. Heart Disease and Stroke Statistics—2021 Update: A Report from the American Heart Association. *Circulation* **2021**, *143*, e254–e743. [CrossRef]
2. Yu, L.; Feng, Z. The Role of Toll-Like Receptor Signaling in the Progression of Heart Failure. *Mediat. Inflamm.* **2018**, *2018*, 1–11. [CrossRef] [PubMed]
3. Moresco, E.M.Y.; LaVine, D.; Beutler, B. Toll-like receptors. *Curr. Biol.* **2011**, *21*, R488–R493. [CrossRef] [PubMed]
4. Boyd, J.H.; Mathur, S.; Wang, Y.; Bateman, R.M.; Walley, K.R. Toll-like receptor stimulation in cardiomyoctes decreases contractility and initiates an NF-kappaB dependent inflammatory response. *Cardiovasc. Res.* **2006**, *72*, 384–393. [CrossRef]
5. Frantz, S.; Kobzik, L.; Kim, Y.-D.; Fukazawa, R.; Medzhitov, R.; Lee, R.T.; Kelly, R.A. Toll4 (TLR4) expression in cardiac myocytes in normal and failing myocardium. *J. Clin. Investig.* **1999**, *104*, 271–280. [CrossRef] [PubMed]
6. Becher, P.M.; Hinrichs, S.; Fluschnik, N.; Hennigs, J.K.; Klingel, K.; Blankenberg, S.; Westermann, D.; Lindner, D. Role of Toll-like receptors and interferon regulatory factors in different experimental heart failure models of diverse etiology: IRF7 as novel cardiovascular stress-inducible factor. *PLoS ONE* **2018**, *13*, e0193844. [CrossRef] [PubMed]
7. Ninh, V.K.; Brown, J.H. The contribution of the cardiomyocyte to tissue inflammation in cardiomyopathies. *Curr. Opin. Physiol.* **2021**, *19*, 129–134. [CrossRef]
8. Wagner, K.B.; Felix, S.B.; Riad, A. Innate immune receptors in heart failure: Side effect or potential therapeutic target? *World J. Cardiol.* **2014**, *6*, 791–801. [CrossRef] [PubMed]

9. Fallach, R.; Shainberg, A.; Avlas, O.; Fainblut, M.; Chepurko, Y.; Porat, E.; Hochhauser, E. Cardiomyocyte Toll-like receptor 4 is involved in heart dysfunction following septic shock or myocardial ischemia. *J. Mol. Cell. Cardiol.* **2010**, *48*, 1236–1244. [CrossRef] [PubMed]

10. Yang, Y.; Lv, J.; Jiang, S.; Ma, Z.; Wang, D.; Hu, W.; Deng, C.; Fan, C.; Di, S.; Sun, Y.; et al. The emerging role of Toll-like receptor 4 in myocardial inflammation. *Cell Death Dis.* **2016**, *7*, e2234. [CrossRef]

11. Lin, L.; Knowlton, A.A. Innate immunity and cardiomyocytes in ischemic heart disease. *Life Sci.* **2014**, *100*, 1–8. [CrossRef]

12. Lu, Y.-C.; Yeh, W.-C.; Ohashi, P.S. LPS/TLR4 signal transduction pathway. *Cytokine* **2008**, *42*, 145–151. [CrossRef]

13. Wang, S.; Zhou, L.; Ling, L.; Meng, X.; Chu, F.; Zhang, S.; Zhou, F. The Crosstalk Between Hippo-YAP Pathway and Innate Immunity. *Front. Immunol.* **2020**, *11*, 323. [CrossRef]

14. Pan, D. The Hippo Signaling Pathway in Development and Cancer. *Dev. Cell* **2010**, *19*, 491–505. [CrossRef] [PubMed]

15. Mauviel, A.; Nalletstaub, F.; Varelas, X. Integrating developmental signals: A Hippo in the (path)way. *Oncogene* **2011**, *31*, 1743–1756. [CrossRef] [PubMed]

16. Chen, J.; Ma, Q.; King, J.S.; Sun, Y.; Xu, B.; Zhang, X.; Zohrabian, S.; Guo, H.; Cai, W.; Li, G.; et al. aYAP modRNA reduces cardiac inflammation and hypertrophy in a murine ischemia-reperfusion model. *Life Sci. Alliance* **2020**, *3*, e201900424. [CrossRef]

17. Feng, Y.; Chao, W. Toll-Like Receptors and Myocardial Inflammation. *Int. J. Inflamm.* **2011**, *2011*, 1–21. [CrossRef] [PubMed]

18. Hodgkinson, C.P.; Pratt, R.E.; Kirste, I.; Dal-Pra, S.; Cooke, J.P.; Dzau, V.J. Cardiomyocyte Maturation Requires TLR3 Activated Nuclear Factor Kappa B. *Stem Cells* **2018**, *36*, 1198–1209. [CrossRef]

19. Newton, K.; Dixit, V.M. Signaling in Innate Immunity and Inflammation. *Cold Spring Harb. Perspect. Biol.* **2012**, *4*, a006049. [CrossRef]

20. Akerberg, B.N.; Gu, F.; VanDusen, N.J.; Zhang, X.; Dong, R.; Li, K.; Zhang, B.; Zhou, B.; Sethi, I.; Ma, Q.; et al. A reference map of murine cardiac transcription factor chromatin occupancy identifies dynamic and conserved enhancers. *Nat. Commun.* **2019**, *10*, 1–16. [CrossRef]

21. Lin, Z.; Guo, H.; Cao, Y.; Zohrabian, S.; Zhou, P.; Ma, Q.; VanDusen, N.; Guo, Y.; Zhang, J.; Stevens, S.M.; et al. Acetylation of VGLL4 Regulates Hippo-YAP Signaling and Postnatal Cardiac Growth. *Dev. Cell* **2016**, *39*, 466–479. [CrossRef]

22. Agah, R.; Frenkel, P.A.; French, B.A.; Michael, L.H.; Overbeek, P.; Schneider, M.D. Gene recombination in postmitotic cells. Targeted expression of Cre recombinase provokes cardiac-restricted, site-specific rearrangement in adult ventricular muscle in vivo. *J. Clin. Investig.* **1997**, *100*, 169–179. [CrossRef]

23. Zhang, N.; Bai, H.; David, K.K.; Dong, J.; Zheng, Y.; Cai, J.; Giovannini, M.; Liu, P.; Anders, R.A.; Pan, D. The Merlin/NF2 Tumor Suppressor Functions through the YAP Oncoprotein to Regulate Tissue Homeostasis in Mammals. *Dev. Cell* **2010**, *19*, 27–38. [CrossRef]

24. Del Re, D.P.; Yang, Y.; Nakano, N.; Cho, J.; Zhai, P.; Yamamoto, T.; Zhang, N.; Yabuta, N.; Nojima, H.; Pan, D.; et al. Yes-associated Protein Isoform 1 (Yap1) Promotes Cardiomyocyte Survival and Growth to Protect against Myocardial Ischemic Injury. *J. Biol. Chem.* **2013**, *288*, 3977–3988. [CrossRef] [PubMed]

25. Liu, R.; Jagannathan, R.; Li, F.; Lee, J.; Balasubramanyam, N.; Kim, B.S.; Yang, P.; Yechoor, V.K.; Moulik, M. Tead1 is required for perinatal cardiomyocyte proliferation. *PLoS ONE* **2019**, *14*, e0212017. [CrossRef]

26. von Gise, A.; Lin, Z.; Schlegelmilch, K.; Honor, L.B.; Pan, G.M.; Buck, J.N.; Ma, Q.; Ishiwata, T.; Zhou, B.; Camargo, F.D.; et al. YAP1, the nuclear target of Hippo signaling, stimulates heart growth through cardiomyocyte proliferation but not hypertrophy. *Proc. Natl. Acad. Sci. USA* **2012**, *109*, 2394–2399. [CrossRef] [PubMed]

27. Porrello, E.R.; Mahmoud, A.I.; Simpson, E.; Hill, J.A.; Richardson, J.A.; Olson, E.N.; Sadek, H. Transient Regenerative Potential of the Neonatal Mouse Heart. *Science* **2011**, *331*, 1078–1080. [CrossRef]

28. Liu, J.; Wen, T.; Dong, K.; He, X.; Zhou, H.; Shen, J.; Fu, Z.; Hu, G.; Ma, W.; Li, J.; et al. TEAD1 protects against necroptosis in postmitotic cardiomyocytes through regulation of nuclear DNA-encoded mitochondrial genes. *Cell Death Differ.* **2021**, 1–15. [CrossRef]

29. Goebeler, M.; Gillitzer, R.; Kilian, K.; Utzel, K.; Bröcker, E.B.; Rapp, U.R.; Ludwig, S. Multiple signaling pathways regulate NF-kappaB-dependent transcription of the monocyte chemoattractant protein-1 gene in primary endothelial cells. *Blood* **2001**, *97*, 46–55. [CrossRef] [PubMed]

30. Hiscott, J.; Marois, J.; Garoufalis, J.; D'Addario, M.; Roulston, A.; Kwan, I.; Pepin, N.; Lacoste, J.; Nguyen, H.; Bensi, G. Characterization of a functional NF-kappa B site in the human interleukin 1 beta promoter: Evidence for a positive autoregulatory loop. Mol. *Cell Biol.* **1993**, *13*, 6231–6240. [CrossRef]

31. Homma, Y.; Cao, S.; Shi, X.; Ma, X. The Th2 transcription factor c-Maf inhibits IL-12p35 gene expression in activated macrophages by targeting NF-kappaB nuclear translocation. *J. Interferon Cytokine Res.* **2007**, *27*, 799–808.

32. Murphy, T.L.; Cleveland, M.G.; Kulesza, P.; Magram, J.; Murphy, K.M. Regulation of interleukin 12 p40 expression through an NF-kappa B half-site. *Mol. Cell. Biol.* **1995**, *15*, 5258–5267. [CrossRef] [PubMed]

33. Lin, Z.; Von Gise, A.; Zhou, P.; Gu, F.; Ma, Q.; Jiang, J.; Yau, A.L.; Buck, J.N.; Gouin, K.A.; Van Gorp, P.R.R.; et al. Cardiac-Specific YAP Activation Improves Cardiac Function and Survival in an Experimental Murine MI Model. *Circ. Res.* **2014**, *115*, 354–363. [CrossRef]

34. Guo, Y.; VanDusen, N.J.; Zhang, L.; Gu, W.; Sethi, I.; Guatimosim, S.; Ma, Q.; Jardin, B.D.; Ai, Y.; Zhang, D.; et al. Analysis of Cardiac Myocyte Maturation Using CASAAV, a Platform for Rapid Dissection of Cardiac Myocyte Gene Function In Vivo. *Circ. Res.* **2017**, *120*, 1874–1888. [CrossRef] [PubMed]

35. Li, N.; Zhou, H.; Wu, H.; Wu, Q.; Duan, M.; Deng, W.; Tang, Q. STING-IRF3 contributes to lipopolysaccharide-induced cardiac dysfunction, inflammation, apoptosis and pyroptosis by activating NLRP3. *Redox Biol.* **2019**, *24*, 101215. [CrossRef] [PubMed]

36. Zhang, H.; Pasolli, H.A.; Fuchs, E. Yes-associated protein (YAP) transcriptional coactivator functions in balancing growth and differentiation in skin. *Proc. Natl. Acad. Sci. USA* **2011**, *108*, 2270–2275. [CrossRef]

37. Yu, W.; Mei, X.; Zhang, Q.; Zhang, H.; Zhang, T.; Zou, C. Yap overexpression attenuates septic cardiomyopathy by inhibiting DRP1-related mitochondrial fission and activating the ERK signaling pathway. *J. Recept. Signal Transduct.* **2019**, *39*, 175–186. [CrossRef]

38. Yang, J.; Zhang, R.; Jiang, X.; Lv, J.; Li, Y.; Ye, H.; Liu, W.; Wang, G.; Zhang, C.; Zheng, N.; et al. Toll-like receptor 4–induced ryanodine receptor 2 oxidation and sarcoplasmic reticulum Ca2+ leakage promote cardiac contractile dysfunction in sepsis. *J. Biol. Chem.* **2018**, *293*, 794–807. [CrossRef]

39. Bai, T.; Hu, X.; Zheng, Y.; Wang, S.; Kong, J.; Cai, L. Resveratrol protects against lipopolysaccharide-induced cardiac dysfunction by enhancing SERCA2a activity through promoting the phospholamban oligomerization. *Am. J. Physiol. Circ. Physiol.* **2016**, *311*, H1051–H1062. [CrossRef] [PubMed]

40. Guo, Y.; Pu, W.T. Cardiomyocyte Maturation: New Phase in Development. *Circ. Res.* **2020**, *126*, 1086–1106. [CrossRef]

41. Banerjee, I.; Fuseler, J.W.; Price, R.L.; Borg, T.K.; Baudino, T.A. Determination of cell types and numbers during cardiac development in the neonatal and adult rat and mouse. *Am. J. Physiol. Circ. Physiol.* **2007**, *293*, H1883–H1891. [CrossRef]

42. Guo, H.; Lu, Y.W.; Lin, Z.; Huang, Z.-P.; Liu, J.; Wang, Y.; Seok, H.Y.; Hu, X.; Ma, Q.; Li, K.; et al. Intercalated disc protein Xinβ is required for Hippo-YAP signaling in the heart. *Nat. Commun.* **2020**, *11*, 1–11. [CrossRef]

43. Hou, N.; Wen, Y.; Yuan, X.; Xu, H.; Wang, X.; Li, F.; Ye, B. Activation of Yap1/Taz signaling in ischemic heart disease and dilated cardiomyopathy. *Exp. Mol. Pathol.* **2017**, *103*, 267–275. [CrossRef]

44. Kim, M.; Kim, T.; Johnson, R.L.; Lim, D.-S. Transcriptional Co-repressor Function of the Hippo Pathway Transducers YAP and TAZ. *Cell Rep.* **2015**, *11*, 270–282. [CrossRef]

45. Adamo, L.; Rocha-Resende, C.; Prabhu, S.D.; Mann, D.L. Reappraising the role of inflammation in heart failure. *Nat. Rev. Cardiol.* **2020**, *17*, 269–285. [CrossRef] [PubMed]

46. Hobai, I.A.; Morse, J.C.; Siwik, D.A.; Colucci, W.S. Lipopolysaccharide and cytokines inhibit rat cardiomyocyte contractility in vitro. *J. Surg. Res.* **2015**, *193*, 888–901. [CrossRef] [PubMed]

47. Wang, S.; Xie, F.; Chu, F.; Zhang, Z.; Yang, B.; Dai, T.; Gao, L.; Wang, L.; Ling, L.; Jia, J.; et al. YAP antagonizes innate antiviral immunity and is targeted for lysosomal degradation through IKKε-mediated phosphorylation. *Nat. Immunol.* **2017**, *18*, 733–743. [CrossRef]

48. Deng, Y.; Lu, J.; Li, W.; Wu, A.; Zhang, X.; Tong, W.; Ho, K.K.; Qin, L.; Song, H.; Mak, K.K. Reciprocal inhibition of YAP/TAZ and NF-κB regulates osteoarthritic cartilage degradation. *Nat. Commun.* **2018**, *9*, 4564. [CrossRef]

49. Zhang, Q.; Meng, F.; Chen, S.; Plouffe, S.W.; Wu, S.; Liu, S.; Li, X.; Zhou, R.; Wang, J.; Zhao, B.; et al. Hippo signalling governs cytosolic nucleic acid sensing through YAP/TAZ-mediated TBK1 blockade. *Nat. Cell Biol.* **2017**, *19*, 362–374. [CrossRef] [PubMed]

50. Lv, Y.; Kim, K.; Sheng, Y.; Cho, J.; Qian, Z.; Zhao, Y.-Y.; Hu, G.; Pan, D.; Malik, A.B.; Hu, G. YAP Controls Endothelial Activation and Vascular Inflammation Through TRAF6. *Circ. Res.* **2018**, *123*, 43–56. [CrossRef]

International Journal of
Molecular Sciences

Article

Implications of the Wilms' Tumor Suppressor Wt1 in Cardiomyocyte Differentiation

Nicole Wagner [1,*,†], Marina Ninkov [1,†], Ana Vukolic [1,2], Günseli Cubukcuoglu Deniz [3], Minoo Rassoulzadegan [1], Jean-François Michiels [4] and Kay-Dietrich Wagner [1,*]

1 CNRS, INSERM, iBV, Université Côte d'Azur, 06107 Nice, France; Marina.Ninkov@univ-cotedazur.fr (M.N.); ana.vukolic@roche.ch (A.V.); minoo.rassoulzadegan@unice.fr (M.R.)
2 Roche Glycart AG, 8952 Schlieren, Switzerland
3 Stem Cell Institute, Ankara University, 06520 Ankara, Turkey; gunselicubukcu@gmail.com
4 Department of Pathology, CHU Nice, 06107 Nice, France; michiels.jf@chu-nice.fr
* Correspondence: nwagner@unice.fr (N.W.); kwagner@unice.fr (K.-D.W.); Tel.: +33-493-377665 (N.W. & K.-D.W.)
† These authors contributed equally to this work.

Abstract: The Wilms' tumor suppressor Wt1 is involved in multiple developmental processes and adult tissue homeostasis. The first phenotypes recognized in Wt1 knockout mice were developmental cardiac and kidney defects. Wt1 expression in the heart has been described in epicardial, endothelial, smooth muscle cells, and fibroblasts. Expression of Wt1 in cardiomyocytes has been suggested but remained a controversial issue, as well as the role of Wt1 in cardiomyocyte development and regeneration after injury. We determined cardiac Wt1 expression during embryonic development, in the adult, and after cardiac injury by quantitative RT-PCR and immunohistochemistry. As in vitro model, phenotypic cardiomyocyte differentiation, i.e., the appearance of rhythmically beating clones from mouse embryonic stem cells (mESCs) and associated changes in gene expression were analyzed. We detected Wt1 in cardiomyocytes from embryonic day (E10.5), the first time point investigated, until adult age. Cardiac Wt1 mRNA levels decreased during embryonic development. In the adult, Wt1 was reactivated in cardiomyocytes 48 h and 3 weeks following myocardial infarction. Wt1 mRNA levels were increased in differentiating mESCs. Overexpression of Wt1(-KTS) and Wt1(+KTS) isoforms in ES cells reduced the fraction of phenotypically cardiomyocyte differentiated clones, which was preceded by a temporary increase in c-kit expression in Wt1(-KTS) transfected ES cell clones and induction of some cardiomyocyte markers. Taken together, Wt1 shows a dynamic expression pattern during cardiomyocyte differentiation and overexpression in ES cells reduces their phenotypical cardiomyocyte differentiation.

Keywords: Wilms' tumor suppressor 1; cardiomyocyte differentiation; mouse embryonic stem cells; myocardial infarction

Citation: Wagner, N.; Ninkov, M.; Vukolic, A.; Cubukcuoglu Deniz, G.; Rassoulzadegan, M.; Michiels, J.-F.; Wagner, K.-D. Implications of the Wilms' Tumor Suppressor Wt1 in Cardiomyocyte Differentiation. *Int. J. Mol. Sci.* **2021**, *22*, 4346. https://doi.org/10.3390/ijms22094346

Academic Editor: Lih Kuo

Received: 2 April 2021
Accepted: 20 April 2021
Published: 21 April 2021

Publisher's Note: MDPI stays neutral with regard to jurisdictional claims in published maps and institutional affiliations.

1. Introduction

The Wilms' tumor 1 (Wt1) gene encodes a zinc finger protein that has multiple roles in embryonic development, adult health, and disease. Wt1 is an important regulator during embryogenesis [1–4] but is also involved in pathological processes, such as carcinogenesis. Originally proposed as a tumor suppressor, Wt1 is nowadays considered as an oncogene [5–12]. Wt1 is an evolutionary conserved transcription factor [13] with a high specificity for GC-rich regions [14]. Alternative RNA splicing results in formation of numerous protein isoforms, which can be classified into two major groups: Wt1(+KTS) and Wt1(-KTS), depending on the presence (+) or the absence (−) of three amino acids (lysine, threonine, and serine/KTS) in the linker sequence between zinc fingers 3 and 4 in exon 9 [15]. In general, it is thought that Wt1(-KTS) isoforms bind DNA with high affinity and regulate gene transcription [16], while Wt1(+KTS) isoforms have a higher affinity for RNA and might play a role in mRNA processing [17,18].

It is known that Wt1 regulates development and maintenance of various tissues of the cardiovascular, urogenital, nervous, hematopoietic, and immune system [19–29] through regulation of genes involved in proliferation, differentiation, and apoptosis—the essential processes for establishing early cellular fates within the embryo [23,24,30–34]. The crucial role of Wt1 in heart formation became clear when it was shown that homozygous deletion of Wt1 in mouse embryos was lethal, due to disturbed cardiac development [23,33]. Moreover, Wt1 cardiac conditional knockout mice died between E16.5 and E18.5 [35], when the heart should have achieved its definitive prenatal configuration [21,33,36]. Additionally, Wt1-deficient embryoid bodies failed to differentiate towards cardiac progenitor cells in vitro [35]. Wt1 is not only relevant for embryonic heart development but might also be involved in adult heart regeneration. Wt1 re-expression was noted in adult hearts following myocardial injury [37,38]. The role of Wt1 in adult cardiomyocytes is still controversial.

In development, the earliest recognizable structure in the growing heart is the primitive heart tube, which is formed at embryonic day 8.5 (E8.5), in the mouse [39]. Wt1 expression was first observed in a transitory cluster of cells—the proepicardium and the coelomic epithelium at E9.5. Wt1-positive proepicardial cells migrate across the pericardial cavity, proliferate, and spread over the surface of the myocardium to form the epicardial layer by E12.5 [40–42]. The highest proliferation levels and migratory capacity of epicardial cells correlate with elevated Wt1 expression during epicardial development [43]. Between E11.5 and E12.5, Wt1- expressing cells begin to migrate from the epicardium into the subepicardial zone to form a layer of subepicardial mesenchymal cells (SEMCs) [21]. Around E13.5, a subset of epicardial cells undergoes epithelial-to-mesenchymal transition (EMT), which induces the formation of epicardial-derived cells (EPDCs), a population of multipotent mesenchymal cardiac progenitor cells, which differentiate into the major cardiovascular cell types—cardiomyocytes, fibroblasts, smooth muscle, and endothelial cells [42,44]. The expression of Wt1 is essential for EMT and resulting differentiation of EPDCs and their derivates, through repression of the epithelial phenotype in epicardial cells [35,45]. However, additional consequences of Wt1 expression in cardiovascular progenitor cells are largely unknown. Generally, cardiovascular progenitors are defined by distinct combinations of cardiac-specific and stem cell associated genetic markers (Isl1, Nkx 2–5, c-kit, Oct3/4, Nanog). They maintain proliferative potential and are the main source of cardiomyocytes during development [46]. However, the impact of Wt1 expression on cardiomyocyte terminal differentiation was not studied in detail. Therefore, the purpose of this study was to examine how Wt1 affects the course of cardiomyocyte differentiation from progenitor cells during embryogenesis and adult life.

In the present study, we demonstrate Wt1 expression in cardiomyocytes during embryonic development, in the adult, and in response to injury *in vivo*. We show that transient Wt1 overexpression reduces phenotypic cardiomyocyte differentiation of ES cell clones *in vitro*, which is associated with modified expression levels of stem cell and cardiomyocyte marker genes.

2. Results

2.1. Wt1 Expression in Developing and Adult Hearts

To analyze systematically cardiac Wt1 expression during embryonic development and after birth, we measured its expression in heart tissue at different time points by quantitative RT-PCR (Figure 1). For this purpose, RNA was isolated from hearts including epicardium, myocardium, and endocardium, but excluding atria and outflow tract. In the developing heart of mouse embryos, we detected the highest Wt1 expression levels between E10.5 and E12.5, when the covering of myocardium with Wt1 positive progenitors should be accomplished [41]. In our study, Wt1 mRNA levels gradually decreased from E14.5, followed by a sharp drop after birth. Nevertheless, we were able to detect cardiac Wt1 mRNA expression until the end of observation period, at postnatal day (P) 21.

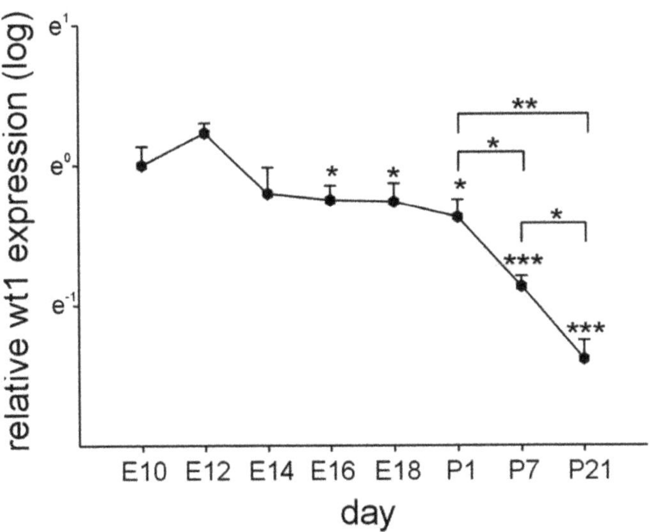

Figure 1. Cardiac Wt1 expression during embryonic development and after birth. Quantitative RT-PCRs for Wt1 in mouse hearts at different time points of embryonic development and after birth (n = 4 each, the four samples for E10 were each pooled from seven different organs, at E12 and E14 the four samples were pooled from four organs each). Expression of Wt1 was normalized to the mean of the respective Gapdh, actin, and Rplp0 expression. Next, the average of all samples at E10 was calculated. Individual samples were then normalized against this average value. Significance was tested between E10 and P21. Data are represented as means ± SEM. * $p < 0.05$, ** $p < 0.01$, *** $p < 0.001$.

Next, we performed immunohistochemical analyses of embryonic tissues and postnatal mouse hearts. Interestingly, we identified Wt1 positive cardiomyocytes in the heart from E10.5, the first time point analyzed, until adulthood (Figure 2). For Wt1 immunostaining of embryonic stages, we used paraffin sections of our mouse embryo collection. Rectangles in the scanned slides indicate the position of the higher magnifications in Figure 2. Only Wt1-positive cardiomyocytes are indicated by arrows as epicardial, endothelial, smooth muscle, and fibroblast expression of Wt1 had been reported already extensively [9,11,12,23,24,37,41,47–56]. At postnatal age, cardiac Wt1 expression diminished compared to the embryonic stages (Figures 1 and 2), which corresponds to data reported in the literature [38]. However, in contrast to this report we show that Wt1 is not restricted to epicardium and endothelial cells, but it is still expressed in some cardiomyocytes after birth and in the adult (Figure 2). Interestingly, Wt1 expression in cardiomyocytes presents in a speckled manner, eventually suggesting a role of Wt1(+KTS). To confirm the histomorphologically observed expression of Wt1 in cardiomyocytes on a molecular level, we performed immunofluorescence double-labelling of Wt1 and cardiac troponin T, followed by confocal imaging for the different developmental stages (Figure 2c).

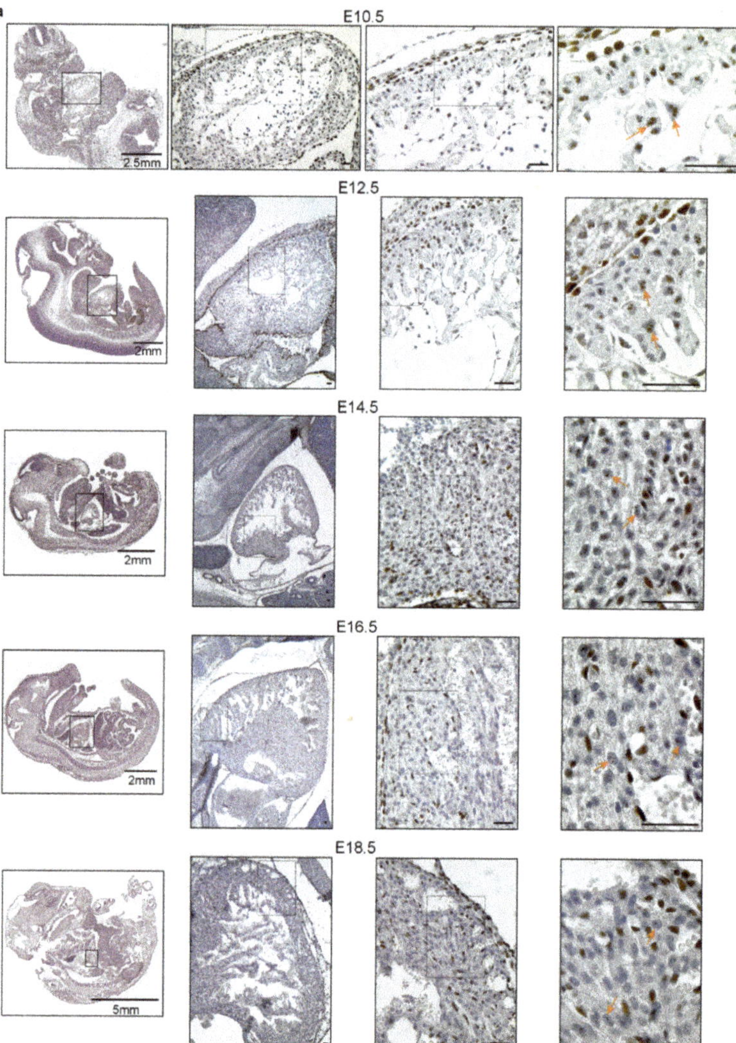

Figure 2. *Cont.*

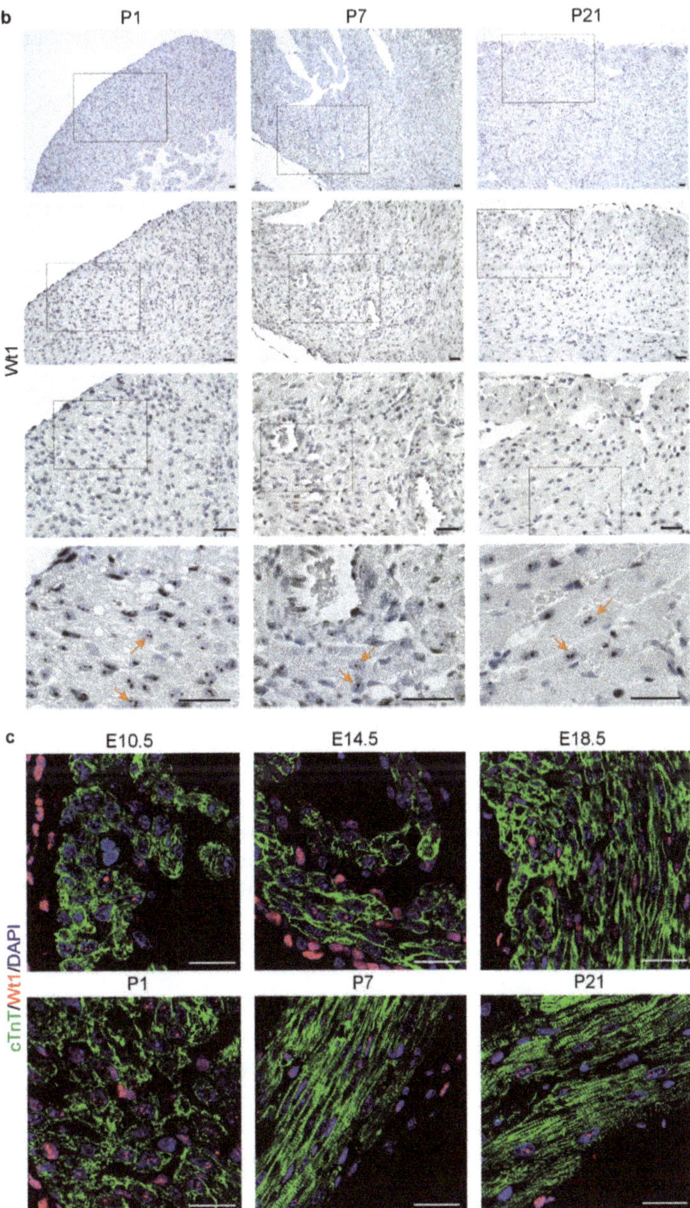

Figure 2. Wt1 is highly expressed in different cell types, including cardiomyocytes, during development of the mouse embryonic heart and persists in some cardiomyocytes after birth. Representative photomicrographs of Wt1 immunostaining on sections of mouse embryos (3,3′ diaminobenzidine (DAB) substrate, brown, hematoxylin counterstain) at different stages before birth (**a**) and of heart sections (**b**) after birth. Rectangles indicate the position of the magnification. Arrows mark Wt1 positive cardiomyocytes. (**c**) Confocal images of Wt1 (red)/cardiac troponin T (green) double-labeling on cardiac tissues at different time points of embryonic development and after birth. Nuclei were counterstained with DAPI (blue). Unless otherwise indicated, scale bars represent 50 μm.

2.2. Wt1 Expression in Infarcted Hearts

Next, we aimed at identifying the relevance of Wt1 expression under pathophysiological conditions in adult hearts. As similar processes are employed during organ development and regeneration, we hypothesized that Wt1 might contribute directly to cardiomyocyte cellular and functional differentiation in adult hearts following myocardial infarction (MI). Wt1 mRNA was determined by qRT-PCRs in hearts 48 h (acute phase) or 3 weeks (chronic phase) following left anterior descending coronary artery (LAD) ligation and in sham-operated controls without LAD ligation. Compared to control mice, a tenfold increase in Wt1 expression was measured 48 h following MI and a fivefold increase 3 weeks after ligation of the LAD (Figure 3a). Next, we employed immunohistochemistry to localize Wt1 expression in control and infarcted hearts of mice. In control hearts, only a few Wt1 positive cardiomyocytes were detected as described above, while the frequency was notably increased in acute MI samples especially in the border zone of the myocardial infarction, and more Wt1-positive cardiomyocytes, compared to controls, were still detected 3 weeks after MI (Figure 3b). The identity of a subset of Wt1 positive cells as cardiomyocytes was confirmed by colocalization of Wt1 (red) and cardiac troponin T (green) as cardiomyocytes markers within infarcted mouse hearts 48 h or 3 weeks after LAD ligation (Figure 3c).

2.3. Cardiomyocyte Differentiation In Vitro

To get additional insights into the process of cardiomyocyte development, we used mouse embryonic stem cell (mESC) differentiation *in vitro*. mESCs have the potential to differentiate spontaneously into cardiomyocytes and represent a validated model for cardiac developmental investigations [57] as in vivo and in vitro cardiac cell differentiation employs the same signaling pathways [58]. Therefore, mESCs were differentiated as embryoid bodies (EBs) using the hanging drop method and their phenotype and temporal gene expression profiles were investigated. The first evidence of cardiomyocyte differentiation was the emergence of spontaneously beating clones from day 2 of EBs culture until day 21, when approximately 90% of clones exhibited rhythmic beating (Supplementary Figure S1). In line with our in vivo data, qRT-PCR analyses showed elevated Wt1 mRNA levels in ESC clones during cardiac differentiation. Wt1 mRNA levels increased until day 6 of culture. Afterwards average values decreased but remained above the basal levels measured on day 0 (Figure 4a). Immunocytochemistry for Wt1 in differentiating mESCs followed by confocal imaging confirmed, quantitatively, the results of the qRT-PCR analyses (Figure 4b).

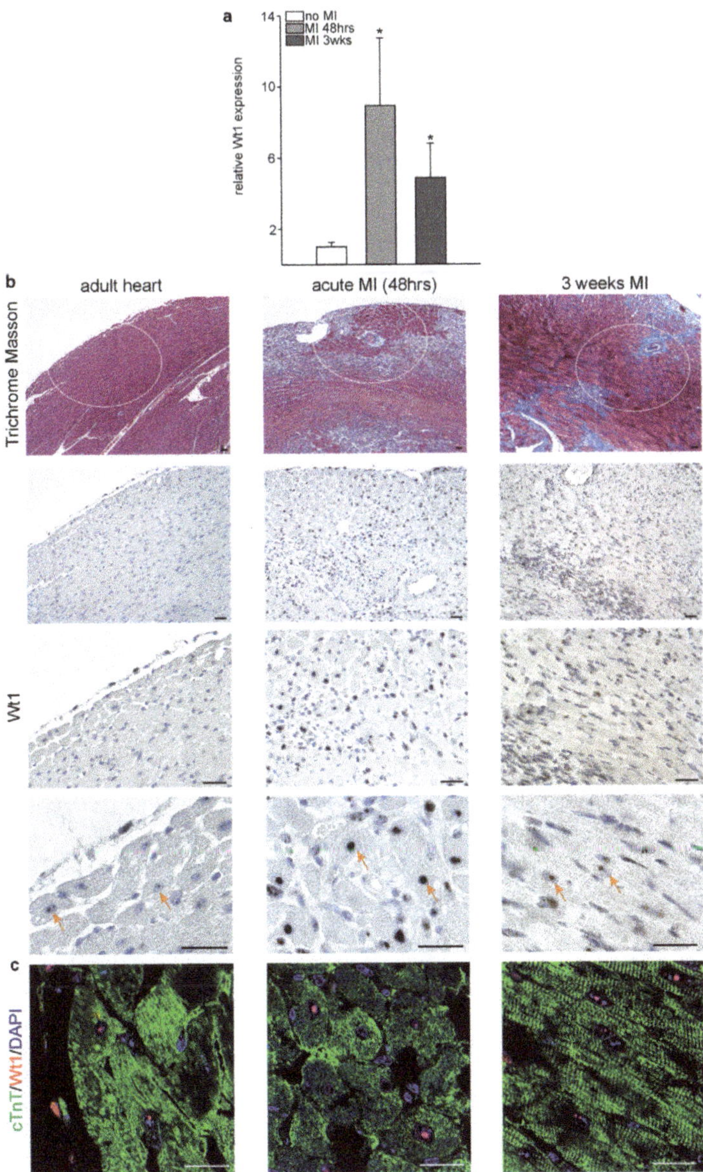

Figure 3. Wt1 is upregulated in cardiomyocytes after myocardial infarction. (**a**) Quantitative RT-PCRs for Wt1 in normal mouse hearts and hearts after 48 h or 3 weeks following LAD ligation (*n* = 4 for each group). (**b**) Upper panel: Trichrome Masson stainings for adult mouse heart and hearts 48 h or 3 weeks after myocardial infarction. Ellipses indicate the regions where subsequent photomicrographs of Wt1 immunostaining (panels below) for the adult mouse heart (3,3′ diaminobenzidine (DAB) substrate, brown, hematoxylin counterstain) were taken. Arrows mark Wt1 positive cardiomyocytes. (**c**) Confocal images of Wt1 (red)/cardiac troponin T (green) double-labelling of normal and infarcted mouse hearts after 48 h or 3 weeks after LAD ligation. Nuclei were counterstained with DAPI (blue). Data are presented as means ± SEM. * *p* < 0.05. Scale bars represent 50 μm.

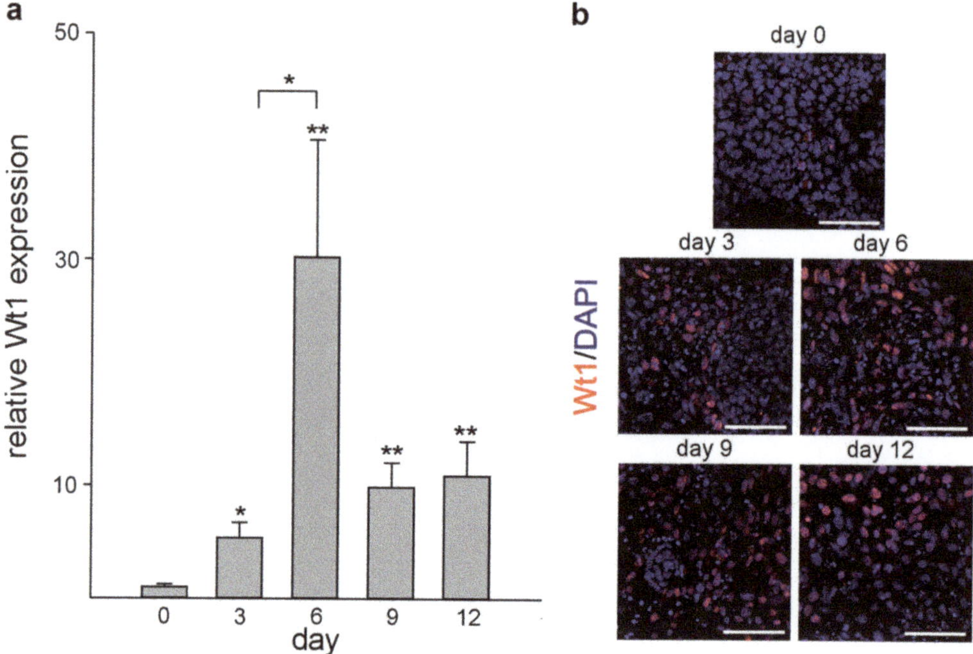

Figure 4. Wt1 is expressed in cardiomyocyte precursors and differentiated clones. (**a**) Quantitative RT-PCRs for Wt1 Randomly selected EBs were harvested on days: 0, 3, 6, 9, and 12 of culture. Expression of Wt1 was normalized to the mean of the respective Gapdh, actin, and Rplp0 expression. The Wt1 values are relative to Wt1 mRNA from day 0. The data from three independent experiments are represented as means $\pm$ S.E.M. * $p < 0.05$, ** $p < 0.01$ vs. day 0. (**b**) Confocal images of Wt1 -labelling (red) of clones at the indicated time points. Cells were counterstained with DAPI (blue). Scale bars indicate 50 µm.

As developmental changes are based on downregulation of embryonic genes and up-regulation of those required for the adult differentiated phenotype, we analyzed temporal expression of stem cell (cKit, Sox2, Oct4, Nanog, Myc) (Supplementary. Figure S2) and cardiomyocyte markers (Nkx 2–5, Myh6, Myh7, Kdr, Pdgfra) (Supplementary. Figure S3) in randomly selected EBs on days: 0, 3, 6, 9, and 12 by qRT-PCR. As expected, the mRNA levels of the majority of stem cell markers started to decline from the earliest time point analyzed in differentiating mESC clones (Supplementary Figure S2). Expression of c-kit, however, did not diminish and followed the temporary expression pattern of Wt1 (Figure 4) without reaching statistical significance. Although the Myc gene has been described to be crucial for maintenance of pluripotency and self-renewal of mESC [59], expression of this stem cell factor was significantly downregulated only on day 9 of EBs culture (Supplementary Figure S2). Concerning cardiomyocyte markers, only Myh6 was significantly upregulated after 12 days of culture in randomly selected clones in this set of experiments (Supplementary Figure S3).

2.4. Wt1 Overexpression Affects the Course of Cardiomyocyte Differentiation

As Wt1 was highly expressed in embryonic hearts and differentiating EBs, we finally intended to define its impact on cardiomyocyte differentiation, on a cellular and molecular level. For this purpose, we transitionally transfected mESC with plasmids containing Wt1(-KTS), Wt1(+KTS), or empty vectors as control. The transfection efficacy was validated 1 day after electroporation. Wt1 expression was moderately increased in samples electroporated with Wt1(-KTS) and Wt1(+KTS) containing plasmids compared to controls at this time point (Figure 5). Regarding stem cell markers, Sox2 expression was higher in Wt1(-KTS) and Wt1(+KTS) transfected cells at this time point, while Nanog expression was significantly increased only in Wt1(-KTS) expressing mESCs. No significant differences in cardiomyocyte marker expression between Wt1 overexpressing cells and the respective controls could be detected at this early time point (Figure 6).

Next, we analyzed the time course of Wt1 expression in empty vector control, Wt1(-KTS), and Wt1(+KTS) expressing clones (Figure 5). Wt1 expression increased on days 6, 8, and 10 of differentiation in the empty vector control group compared to the 1-day post-transfection time point. Significantly higher Wt1 mRNA values were observed on day 5, 6, 8, 9, and 10 compared to 24 h in the Wt1(-KTS) group, while significantly higher Wt1 mRNA levels were detected only on days 6, 8, and 10 in the Wt1(+KTS) group. The upregulation of Wt1 expression during cardiomyocyte differentiation in all groups resembles the situation in non-transfected cells, mentioned above, without an additional significant effect of the initial transient overexpression of the Wt1(-KTS) and Wt1(+KTS) constructs.

The time course of stem cell marker expression during cardiac differentiation of the transfected mESCs also resembled the expression pattern of non-transfected cells mentioned above. Beside the increase in Sox2 expression in Wt1(-KTS) and Wt1(+KTS) overexpressing cells and of Nanog in Wt1(-KTS) cells after 1 day, Oct4, Nanog, and Sox2 expression levels dropped rapidly in clones at day 3 of differentiation in all groups compared to day 1 controls. Surprisingly, Sox2, Oct4, and Nanog expression was higher in Wt1(+KTS) clones compared to empty vector control clones at the same time point. The decrease in Myc expression was less pronounced and reached statistical significance only at days 8 and 9 of differentiation in all transfected groups. Interestingly, c-kit expression showed a more dynamic pattern with significant increases on days 5 and 6 in Wt1(-KTS) and Wt1(+KTS) overexpressing clones and empty vector controls, when compared to the initial control values. A significant increase in c-kit in WT1(-KTS) clones on day 5 compared to empty vector transfected cells on the same time point is in agreement with our recent description of c-kit as a direct transcriptional target of Wt1 [9]. Afterwards, c-kit RNA levels returned to control values in all groups (Figure 5).

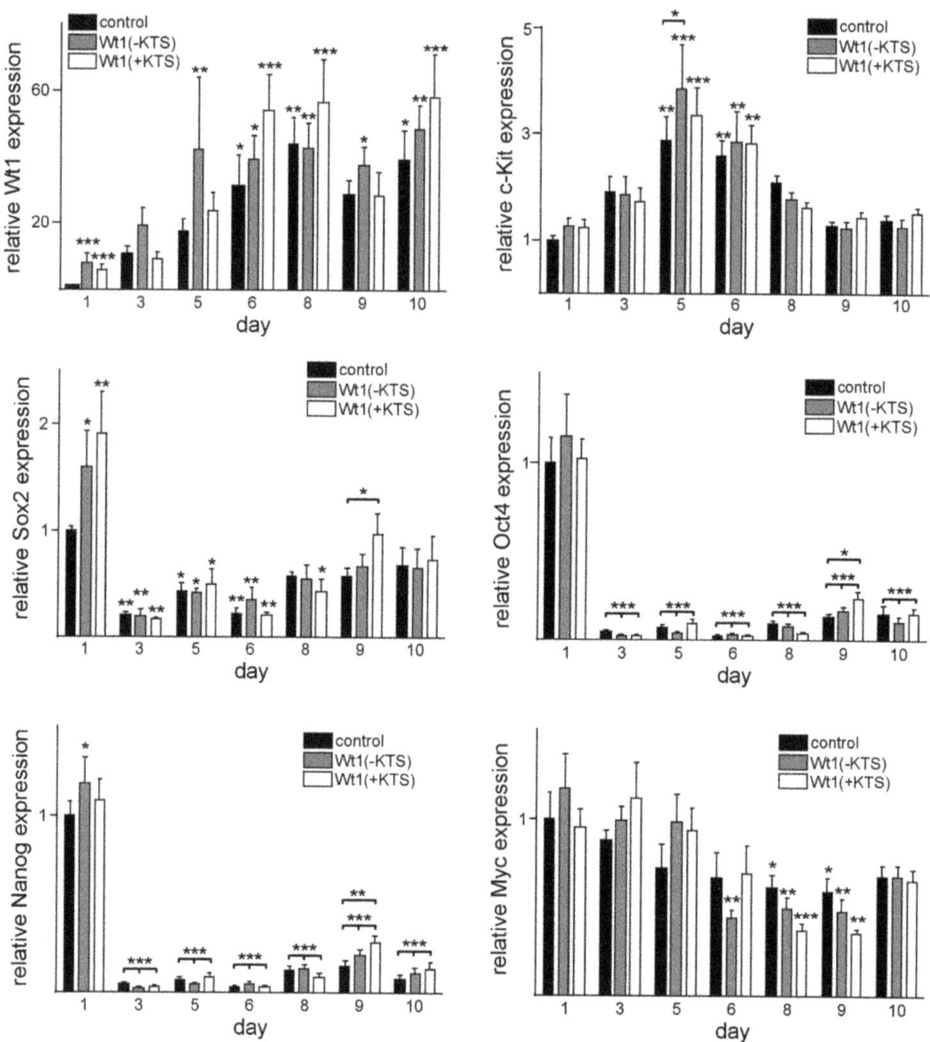

Figure 5. Time course of Wt1 and stem cell marker expression after electroporation of Wt1(-KTS), Wt1(+KTS), or empty vector control. Gene expression was quantified by qRT-PCR. Expression of each gene was normalized to the mean of the respective Gapdh, actin, and Rplp0 expression. The gene expression values are relative to respective control mRNA values (empty vector expression) at day 1. Data from four independent experiments including 12 independent randomly selected clones for each time point are represented as means ± S.E.M. and analyzed by Two-way ANOVA (Fisher's LSD test). * $p < 0.05$, ** $p < 0.01$, *** $p < 0.001$.

Next, we analyzed the effects of Wt1 overexpression on cardiomyocyte markers including Myh6, Myh7, Nkx2–5, Kdr, and Pdgfra. Kdr and Pdgfra represent factors, which are involved in cardiomyocyte differentiation, but not limited to cardiomyocyte progenitors [60,61]. Regarding the gene expression time course, qRT-PCR analyses showed the highest mRNA levels of all genes investigated on day 3 of culture, compared to 24 h for all three groups of clones. In addition, there were statistically significant increases in Nkx 2–5, Myh6, Myh7, and Pdgfra mRNAs in Wt1(-KTS) clones on day 3, compared to control clones at the same day, while in Wt1(+KTS) clones, only Kdr expression was significantly higher on day 3, compared to controls (Figure 6).

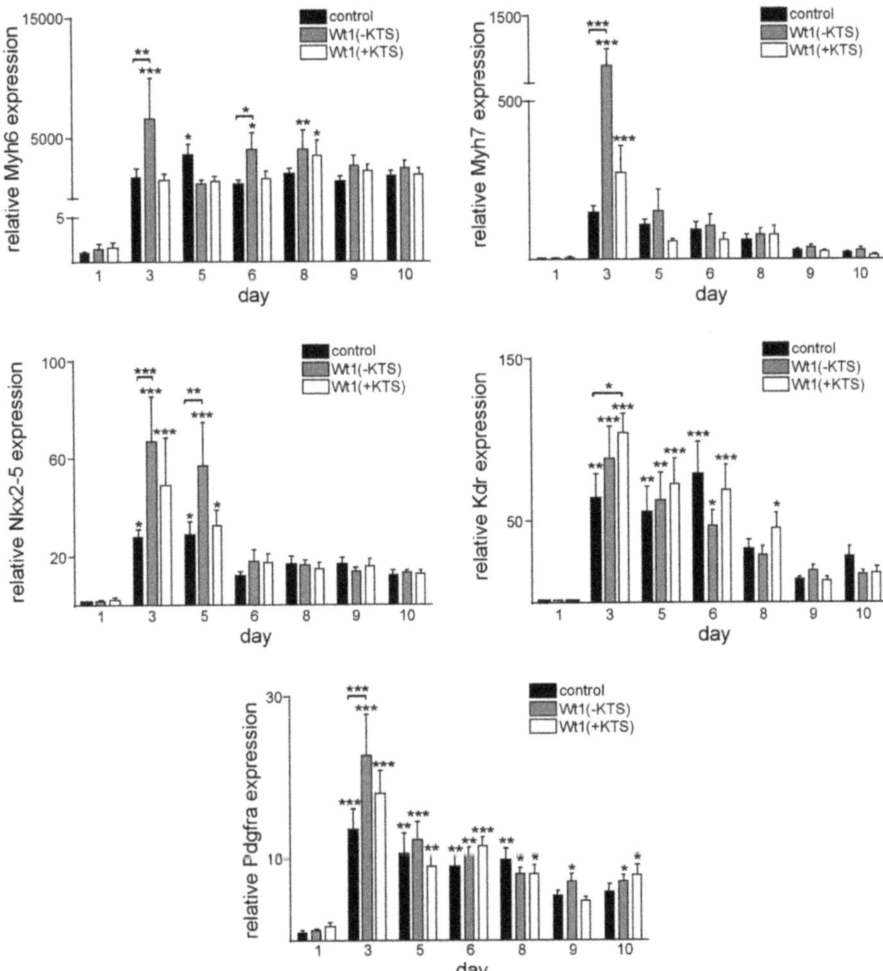

Figure 6. Time course of cardiomyocyte marker mRNA expression. Quantitative RT-PCRs for cardiomyocyte marker mRNA in control, Wt1(-KTS) and Wt1(+KTS) groups. Gene expression was normalized to the mean of the respective Gapdh, actin, and Rplp0 expression. Individual gene expression values are relative to 1 day post-transfection time point control mRNA values. The data from four independent experiments including 12 independent randomly selected clones for each time point are represented as means ± S.E.M. and analyzed by Two-way ANOVA (Fisher's LSD test). * $p < 0.5$, ** $p < 0.01$, *** $p < 0.001$.

Phenotype analysis of the embryoid bodies revealed that Wt1 overexpression (both Wt1(+KTS) and Wt1(-KTS) forms) resulted in lower rate of cardiomyocyte functional differentiation, compared to the controls (Figure 7, Supplementary Video S1). Beating clones were evident at day 2 in the control groups and at day 3 in Wt1(+KTS) and Wt1(-KTS) groups. Moreover, in the control group, the percentage of differentiated clones reached a plateau after day 10 and more than 90 % of EBs contained beating clones by the end of the observation period. At the same time, differentiation rate in Wt1(+KTS) and Wt1(-KTS) groups followed a slower dynamic with around 80 % of contractile clones by the end point of the experiments. The most pronounced drop in cardiomyocyte functional differentiation dynamics was found in the Wt1(+KTS) group, compared to controls. The percentage of phenotypically differentiated beating clones was significantly lower from day 6 until day 10 of the observation period in Wt1(-KTS) and Wt1(+KTS) clones (Figure 7).

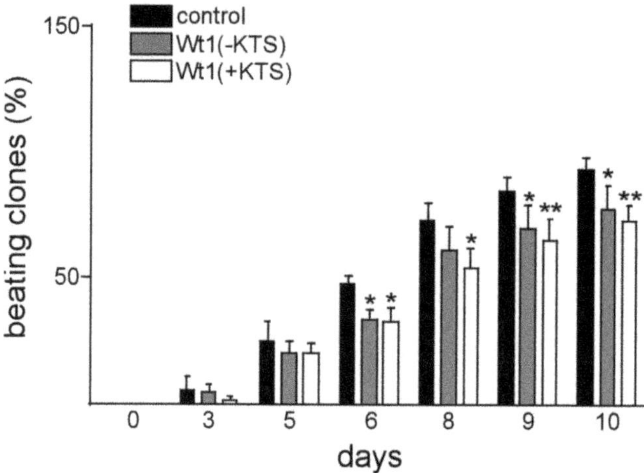

Figure 7. Wt1 overexpression delays functional cardiomyocyte differentiation. Wt1(-KTS), Wt1(+KTS), or empty vector controls were electroporated in mESCs. Hundred embryonic bodies were established per experiment and the number of beating clones counted at the indicated time points. Data from four independent experiments are represented as means $\pm$ S.E.M. * $p < 0.05$, ** $p < 0.01$.

3. Discussion

The expansion of different cardiac cell types in a timely and spatiotemporal pattern is required for normal heart development (for review see [62]). It has been noted earlier that the Wilms' tumor suppressor, Wt1, is required for murine heart development as Wt1 knockout mice have severely hypoplastic hearts and die during mid-gestation, most likely due to heart failure [21,33]. There is a general consensus that Wt1 expressing cells contribute to the development of endothelium, smooth muscle cells, and fibroblasts in the heart [12,23,24,35,37,38,47,48,62,63], but the contribution of Wt1-positive cells to cardiomyocytes during development and in cardiac repair still remains controversial. Earlier, a significant contribution of epicardial-derived cells to the cardiomyocyte lineage in the developing heart has been described based on lineage tracing experiments [42,44,56], which was questioned later [55,64,65]. Regarding a possible role of epicardial-derived Wt1 expressing progenitor cells for cardiac repair, the situation is comparable with some studies postulating an important role after myocardial infarction [52,66–69], while others did not confirm these results [70]. These contradictory results could be explained by different experimental approaches, staining procedures, limitations of the Wt1-Cre mouse models used [65], and by the fact that the re-activated epicardium is heterogenous and different from developmental epicardial cells [71] and only a fraction of cells in adult epicardium expresses Wt1 and is reliable targeted by the Wt1Cre lines [72]. For these reasons, we measured endogenous Wt1 mRNA levels and used a sensitive immunohistochemistry approach to characterize Wt1 expressing cells during cardiac development, in the adult, and during repair after myocardial infarction. The highest Wt1 mRNA expression was observed at E12.5. During this time window, Wt1 plays an important role in epithelial-mesenchymal transition (EMT) and mesenchymal epicardial-derived cell (EPDC) development through downregulation of E-cadherin, upregulation of Snail, and regulation of the Wnt/β-catenin signaling [35,53]. During later stages of embryonic and postnatal development, we observed a decrease in Wt1 mRNA expression, which is in agreement with previous results using a reporter system [21]. Nevertheless, some cardiomyocytes remained Wt1-positive even in the hearts of adult mice. It is conceivable that Wt1 regulates some of the cardiac progenitors by preventing terminal cardiomyocyte differentiation [73,74]. These progenitors could contribute to the cardiomyocyte lineage during development and give rise to the sparse de novo cardiomyocytes formation in adulthood [52].

Wt1 upregulation after myocardial infarction has already been reported several years ago [37]. However, earlier we focused mainly on the angiogenic response after myocardial infarction. It has been shown that Wt1 upregulation in adult heart vasculature was the response to local ischemia and hypoxia in rodents. It is thought that endothelial Wt1 expression is associated with neovascularization and recovery following MI [37,38]. The hypoxic conditions during development and after MI induced vascular formation via hypoxia-inducible factors, which directly upregulated Wt1 [49,75]. Wt1, in turn, regulates the expression of several angiogenic factors and receptors, positively [9–12,23,24,48,76–78]. De novo cardiomyocytes were reported to develop in adjacent areas of a myocardial infarction [79,80] and hypoxic regulation of Wt1 re-expression might also function to promote the tissue regeneration by cardiomyocytes differentiation from an activated progenitor pool [49,56,81]. Additionally, proinflammatory cytokines, such as TNF-α, IL-1β, and IL-6, could favor Wt1 activation after MI through activation of NF-κB. Furthermore, WT1 expression after injury might also be induced by soluble factors secreted by the myocardium [82]. This might induce progenitor cell proliferation and cell survival [83,84]. Given the distance to the epicardium, where we detected Wt1-positive cardiomyocytes already 48 h after myocardial infarction, it is unlikely that these cells are directly epicardium-derived. As the epicardium containing Wt1-positive cells promotes immune cell recruitment, neovascularization, and re-entry of cardiomyocytes into the cell cycle via mitogen secretion in response to injury (for review see [65,85]), both Wt1 expressing cell types might interact in repair. Interestingly, Tyser et al. [86] recently identified a common progenitor pool of the epicardium and myocardium by single cell transcriptomic analyses, most of the clusters expressing Wt1, which could explain expression in some cardiomyocytes and epicardium later in life. Developmental cardiac Wt1 expression diminished at the termination of heart formation, but we assume that low levels of Wt1 expression are sufficient to maintain a cardiac progenitor subset from terminal differentiation. This would support cardiac tissue regeneration by Wt1 reactivation when stimuli, such as hypoxia/inflammation, occur. However, more detailed examination of this pre-cardiomyocyte subset is necessary, as they could be employed as a valuable therapeutic tool for repair following myocardial infarction.

To obtain additional insights into the role of Wt1 in cardiomyocyte differentiation, we used cardiac differentiation of mESCs, a well-established model [57,58,87,88]. Wt1 was detectable in some undifferentiated mESCs and its expression levels increased along with cardiac differentiation until day 6. Wt1 expression in mESCs seems to be necessary for cardiomyocyte differentiation as *Wt1* null mESCs cells failed to differentiate towards the cardiomyocyte lineage [35]. To further characterize the role of Wt1 for cardiomyocyte differentiation of mESCs, we transiently overexpressed Wt1(-KTS) or Wt1(+KTS) constructs. Interestingly, this transient and moderate Wt1 overexpression reduced phenotypical cardiomyocyte differentiation, i.e., the percentage of beating clones throughout the observation period. At the onset of differentiation, signaling pathways regulating pluripotency of mESCs are inhibited through downregulation of stem cell genes, such as Sox2, Oct4, and Nanog [89]. This corresponds to our findings with significant downregulation of these genes at day 3 of differentiation when the first clones started beating. Whether the increase in Sox2 mRNA one day after Wt1(-KTS) and Wt1(+KTS) overexpression is directly related to activation by Wt1 remains to be clarified. Nevertheless, it might contribute to reduced phenotypic differentiation. The higher expression of Sox2, Oct4, and Nanog in Wt1(+KTS) transfected clones compared to control at day 9 reflects, most likely, an indirect effect of reduced differentiation. In contrast to these pluripotency factors, c-kit showed an increasing expression with a peak at day 5 of differentiation with a significantly higher value in Wt1(-KTS) transfected cells, compared to controls. This might be in agreement with c-kit representing a direct transcriptional target of Wt1 [9]. Increasing c-kit expression during mESCs differentiation is compatible with the induction of a cardiovascular progenitor phenotype [90]. Also increased expression of cardiac (Myh6, Myh7, Nkx2–5) and cardiovascular progenitor markers (Kdr, Pdgfra) [91] coincided with the onset of phenotypic differentiation of mESCs. Surprisingly, Myh6, Myh7, Nkx2–5, and Pdgfra were all expressed significantly higher in Wt1(-KTS) clones on day 3, while KDR mRNA was increased in Wt1(+KTS) overexpressing clones. Direct activation of Kdr by Wt1 has been documented already in murine developing gonads although this was mainly attributed to the Wt1(-KTS) isoform [76]. Whether the other differentially expressed genes identified here represent bona fide Wt1 target genes remains subject of further study. The temporary increases in Myh6, Myh7, Nkx2–5, and Pdgfra seem to not be sufficient to support long-term phenotypic cardiomyocyte differentiation, as clones with transient Wt1 overexpression showed less contractility. Whether this is related to short term increases in Sox2 and c-kit or other factors not identified in the present study remains an open question.

Although the mESC cardiac differentiation model is well established, the heterogeneity of clones, and cells within a clone, limit the use for further molecular and transcriptomic studies. The effects of transient Wt1 isoform overexpression and spontaneous increase in Wt1 expression upon cardiac differentiation of mESCs might, in addition, result in mixed outcomes. Regarding a potential role of Wt1 expressing progenitors for cardiac repair in vivo, a major limitation is currently the lack of techniques to isolate and expand these cells.

4. Materials and Methods

4.1. Mice and Tissue Preparation

All animal work was conducted according to National and international guidelines and was approved by the local ethics committee (Nice, France, 09.01.2013) (PEA-NCE/2013/106).

Timed pregnant mice (NMRI) were purchased from Janvier Labs (Le Genest-Saint-Isle, France). Pregnant mice were sacrificed by cervical dislocation. Embryonic hearts were dissected, and tissues were used to prepare RNA. For immunohistochemistry, collections of paraffin-embedded whole embryos were used up to E18.5; for later stages, hearts were dissected.

Myocardial infarctions were induced by ligation of the left coronary artery (LAD), as described elsewhere [92]. In brief, anaesthetized mice were endotracheally intubated, a thoracotomy between the third and fourth rib was performed, and the LAD was closed permanently with a 7–0 suture 2 mm distal to the left auricle. The thoracotomy and the skin wound were closed with 4–0 sutures and the mice remained intubated until spontaneous respiration was re-established. Animals were sacrificed at the indicated time points after infarction, the apex dissected for RNA preparation, and the remaining heart tissues used for paraffin-embedding followed by histological and immunohistological analyses.

4.2. Cell Culture

4.2.1. Mouse Embryonic Stem Cell Culture

The mouse embryonic stem cell line, R1, was used. In order to stimulate proliferation and prevent differentiation, mESCs were cultured on a layer of mitotically inactivated primary mouse embryonic fibroblasts (MEFs)—feeder cells. The MEFs were prepared following the well-established protocol available online (http://www.ispybio.com/search/protocols/MEF_protocol.pdf, accessed on 4 January 2020). In short, a pregnant female mice was sacrificed by cervical dislocation around day 13.5 post coitum. Embryos were separated from the placenta. The head and dark red organs were removed, and the remaining tissue was minced with razor blades and then trypsinised until a single-cell suspension was obtained. Isolated MEFs were expanded in MEF medium (DMEM Glutamax/Gibco 61965-026 supplemented with 10% FBS/Gibco 10270-106; 1/100 L-glutamine/200 mM: Gibco 25030-024; 1/100 penicillin/streptomycin/10,000 U/mL, Gibco 15140-122, ThermoScientific, Cergy Pontoise, France) in 10 cm tissue culture dishes (Corning, NY, USA), until 90–100% confluence was reached. Then, cells were split at 1:4 ratio and MEFs from passages 3–5 at 80% of confluence were inactivated with mitomycin C (10 µg/mL, BML-GR311–0010, Enzo Life Sciences, Farmingdale, NY, USA) 3 h at 37 °C to generate feeder layers. Inactivated MEFs (iMEFs) were ready to use 24 h following mitomycin C treatment.

Mouse ESCs were grown on a feeder layer in an ESC medium composed of DMEM "Knock-out" (10829018) medium supplemented with 15% ES cell grade FBS (16141–002), 1/100 MEM non-essential amino acids (11140–035), 1/100 L-glutamine, 1/1000 2 mercapthoethanol (31350–010), 1/100 sodium pyruvate (11360–039), 1/100 penicillin/streptom ycin—all from Thermo Scientific-Gibco. Prior to ESCs seeding, 10 μg/mL of mLIF was added to the medium (ESGRO Leukemia Inhibitory Factor supplement for mouse cell culture 10^7 U/mL. Hemicon International, Inc., Temecula, CA, USA ESG 1107). After 48 h of culture, when mESCs formed multiple large colonies (100 % confluence), they were trypsinised and used further for differentiation or/and electroporation.

4.2.2. mESC Differentiation by the Hanging Drop Method

The differentiation of mESCs was carried out, according to the protocol described by Wang and Yang [93]. Briefly, mESCs were resuspended in differentiation medium (ESC medium without mLIF supplementation). A tissue culture dish (10 cm) was filled with 10 mL of sterile PBS. Then, 50 drops of differentiation medium containing 500 cells/drop were placed to the lid of the dish and cultured for 72 h in order to form embryoid bodies (EBs). The drops with embryoid bodies (EBs) were then transferred to 96 well plates and incubated for the next 72 h. Following this incubation, EBs were transferred from 96 to 24 well plates coated with 0.2 % gelatin to enhance EBs attachment. The first day of EBs culture in 24 well plates was defined as day 0. The differentiation medium was changed daily and EBs were monitored for morphological and functional changes (contractility). At indicated time points, random samples were harvested for quantitative RT-PCR analysis.

4.3. Electroporation

For transient transfection of confluent undifferentiated mESCs cultured as mentioned in Section 4.2.1, plasmids containing either Wt1(−KTS) or Wt1(+KTS) expression vectors (Wt1 cDNA in pCB6+ plasmid), or empty vector as a control were used. For each group, 1 μg of plasmid was incubated with 3×10^6 mESCs in 0.8 mL PBS 30 min on ice. Following the incubation, the electroporation was performed (400 V, 250 μF) using the Bio-Rad Gene pulser (Bio-Rad, Richmond, CA, USA). Once the electroporation was conducted, cells were quickly resuspended in ESC medium, plated on fresh feeder layers and incubated for 24 h to enable plasmid baseline expression prior to hanging drop culture, described in Section 4.2.2.

4.4. Quantitative RT-PCR

Total RNA was extracted from mESCs, EBs, and organs using the Trizol reagent (Invitrogen, Carlsbad, CA, USA). The RNA pellet was dissolved in diethyl pyrocarbonate-treated H_2O and RNA concentration was assessed spectrophotometrically. For reverse transcription, 0.5 μg of total RNA from mESCs and EBs was transcribed to cDNA using Maxima First Strand cDNA Synthesis kit (Thermo Scientific). The reverse transcription products were diluted $10\times$ and 1 μl of diluted cDNA was used for quantitative PCRs. Detection of PCR products in real time was performed on the LightCycler Instrument (Roche Diagnostics, Mannheim, Germany) using the PowerUp SYBR Green Master Mix kit (Thermo Fisher Scientific, Waltham, MA, USA). For organs, first-strand cDNA synthesis was performed with 0.5 μg of total RNA using oligo(dT) primers and Superscript III reverse transcriptase (Invitrogen) (Table 1). One μl of the reaction product was taken for real time RT-PCR amplification (ABI Prism 7000, Applied Biosystems, Foster City, CA, USA) using a commercial SYBR® Green kit (Eurogentec, Angers, France). Expression of each gene was normalized to the respective Gapdh, actin, and Rplp0 expression. For the in vivo part of the study, the average mRNA values of all samples at E10.5 were calculated; individual samples were then normalized against this average value. Data for mRNA from non-transfected cells were expressed as relative value to mRNA from day 0 (value normalized to 1); mRNA data for transfected cells were expressed as relative value to mRNA empty vector control 24 h post-transfection (value normalized to 1).

Table 1. Primer Sequences.

Name	Sequence
Wt1 forward	CCA GCT CAG TGA AAT GGA CA [11]
Wt1 reverse	CTG TAC TGG GCA CCA CAG AG [11]
Kit forward	GCC TGA CGT GCA TTG ATC C [94]
Kit reverse	AGT GGC CTC GGC TTT TTC C [94]
Sox2 forward	CGC CCA GTA GAC TGC ACA
Sox2 reverse	CCC TCA CAT GTG CGA CAG
Oct4 forward	TGG GCG TTC TCT TTG GAA
Oct4 reverse	GTT GTC GGC TTC CTC CAC
Nanog forward	CAG GTT TCA GAA GCA GAA GTA CC
Nanog reverse	GGT TTT GAA ACC AGG TCT TAA CC
Myh6 forward	CCA AGA CTG TCC GGA ATG A
Myh6 reverse	TCC AAA GTG GAT CCT GAT GA
Myh7 forward	GCC TCC ATT GAT GAC TCT G
Myh7 reverse	CGC CTG TCA GCT TGT AAA TG
Nkx2–5 forward	ATT TTA CCC GGG AGC CTA CG
Nkx2–5 reverse	CAG CGC GCA CAG CTC TTT T
Kdr forward	AGT GGT ACT GGC AGC TAG AAG [94]
Kdr reverse	ACA AGC ATA CGG GCT TGT TT [94]
Pdgfra forward	ATG AGA GTG AGA TCG AAG GCA [94]
Pdgfra reverse	CGG CAA GGT ATG ATG GCA GAG [94]
Rplp0 forward	CAC TGG TCT AGG ACC CGA GAA G [95]
Rplp0 reverse	GGT GCC TCT GGA GAT TTT CG [95]
Gapdh forward	CCA ATG TGT CCG TCG TGG ATC T [48,95]
Gapdh reverse	GTT GAA GTC GCA GGA GAC AAC C [48,95]
Actb forward	CTT CCT CCC TGG AGA AGA GC [48,95]
Actb reverse	ATG CCA CAG GAT TCC ATA CC [48,95]

4.5. Mouse Tissue Samples, Histology and Immunohistology

Samples from at least three different animals per time point were analyzed. Three μm paraffin sections were used for histological and immunohistological procedures. Haematoxylin-Eosin staining was routinely performed on all tissue samples. For Wt1 immunohistology, after heat-mediated antigen retrieval and quenching of endogenous peroxidase activity, the antigen was detected after antibody application (Wt1 rabbit monoclonal antibody, clone CAN-R9(IHC)-56-2, Abcam, Cambridge, UK,) using the EnVisionTM Peroxidase/DAB Detection System from Dako (Trappes, France). Sections were counterstained with Hematoxylin (Sigma, St. Louis, MO, USA). Omission of the first antibody served as a negative control. Additionally, some slides were incubated with an IgG Isotype Control (1:100, rabbit monoclonal, clone SP137, Abcam) as a negative control. Slides were photographed using a slide scanner (Leica Microsystems, Nanterre, France) or an epifluorescence microscope (DMLB, Leica, Germany) connected to a digital camera (Spot RT Slider, Diagnostic Instruments, Scotland). For immunofluorescence double-labelling of mouse hearts, anti-Wt1 rabbit monoclonal antibody from Abcam was combined with a mouse monoclonal anti-cardiac troponin antibody (clone 4C2, Abcam) using Dylight 594 donkey anti rabbit and Dylight 488 donkey anti mouse secondary antibodies (Jackson ImmunoResearch, Newmarket, Suffolk, UK). Cells were stained using anti-Wt1 rabbit monoclonal antibody from Abcam and Dylight 594 donkey anti rabbit secondary antibody. Negative controls were obtained by omission of first antibodies. Images were taken using a confocal ZEISS LSM Exciter microscope (Zeiss, Jena, Germany).

4.6. Statistical Analysis

Statistical analyses were performed using the GraphPad Prism software (version 6.2; GraphPad Software Inc, San Diego, CA, USA). Data are expressed as means ± standard error of the mean (S.E.M.). Statistical differences between mean values were assessed by analysis of variance (one-way or two-way ANOVA) followed by the Bonferroni post-hoc,

Mann–Whitney, or Fisher's test as indicated. A p value < 0.05 was considered to reflect statistical significance.

5. Conclusions

Here we show that Wt1 is expressed in cardiomyocytes during heart development and in adult life. Wt1 expressing cardiomyocytes might represent a subset of pre-differentiated cells that are able to contribute to regeneration of damaged heart tissue. On a cellular level, Wt1 kept a population of ESC derived cardiomyocytes from their final mature differentiated state through modulation of stem cell markers' expression. A detailed analysis of the molecular mechanisms by which Wt1 regulates cardiomyocyte maturation will be subject of further studies.

Supplementary Materials: The following are available online at https://www.mdpi.com/article/10.3390/ijms22094346/s1.

Author Contributions: Conceptualization, K.-D.W. and N.W., M.R.; methodology, K.-D.W., N.W., M.R. and J.-F.M.; software, K.-D.W., N.W. and M.N.; validation, K.-D.W., N.W. and M.N.; formal analysis, K.-D.W., N.W., G.C.D., M.R., J.-F.M., A.V. and M.N.; investigation, K.-D.W., N.W., A.V. and M.N.; resources, K.-D.W. and M.R.; data curation, K.-D.W., N.W. and M.N.; writing—original draft preparation, K.-D.W., N.W. and M.N.; writing—review and editing, K.-D.W., N.W., A.V., G.C.D., M.R., J.-F.M. and M.N.; visualization, K.-D.W., N.W. and M.N.; supervision, K.-D.W.; project administration, K.-D.W., N.W. and M.N.; funding acquisition, K.-D.W. and N.W. All authors have read and agreed to the published version of the manuscript.

Funding: This research was funded by Fondation pour la Recherche Medicale, grant number FRM DPC20170139474 (Kay-Dietrich Wagner), Fondation ARC pour la recherche sur le cancer", grant number n°PJA 20161204650 (N.W.), Gemluc (Nicole Wagner), and Plan Cancer INSERM (Kay-Dietrich Wagner).

Institutional Review Board Statement: Not applicable.

Informed Consent Statement: Not applicable.

Data Availability Statement: Data is contained within the article or Supplementary Materials.

Acknowledgments: The authors thank J. Jiang, M. Audrey, A. Borderie, S. Destree, A. Biancardini, A. Martres A. Loubat, and M. Cutajar-Bossert for technical assistance.

Conflicts of Interest: The authors declare no conflict of interest.

Abbreviations

DMEM	Dulbecco's Modified Eagle Medium
E	Embryonic day
EBs	Embryoid bodies
EPDC	Epicardial-derived cells
FBS	Fetal bovine serum
LAD	Left anterior descending coronary artery
LIF	Leukemia inhibitory factor
MEF	Mouse embryonic fibroblasts
mESCs	Mouse embryonic stem cells
MI	Myocardial infarction
P	Postnatal day
pt	Post transfection
qRT-PCR	Quantitative reverse-transcription polymerase chain reaction
S.E.M.	Standard error of the mean
Wt1	Wilms' tumor Suppressor 1

References

1. Rackley, R.R.; Flenniken, A.M.; Kuriyan, N.P.; Kessler, P.M.; Stoler, M.H.; Williams, B.R. Expression of the Wilms' tumor suppressor gene WT1 during mouse embryogenesis. *Cell Growth Differ.* **1993**, *4*, 1023–1031.
2. Hastie, N.D. Wilms' tumour 1 (WT1) in development, homeostasis and disease. *Development* **2017**, *144*, 2862–2872. [CrossRef] [PubMed]
3. Wagner, K.D.; Wagner, N.; Schedl, A. The complex life of WT1. *J. Cell Sci.* **2003**, *116*, 1653–1658. [CrossRef]
4. Hohenstein, P.; Hastie, N.D. The many facets of the Wilms' tumour gene, WT1. *Hum. Mol. Genet.* **2006**, *15*, R196–R201. [CrossRef]
5. Oji, Y.; Miyoshi, S.; Maeda, H.; Hayashi, S.; Tamaki, H.; Nakatsuka, S.; Yao, M.; Takahashi, E.; Nakano, Y.; Hirabayashi, H.; et al. Overexpression of the Wilms' tumor gene WT1 in de novo lung cancers. *Int. J. Cancer* **2002**, *100*, 297–303. [CrossRef] [PubMed]
6. Rampal, R.; Figueroa, M.E. Wilms tumor 1 mutations in the pathogenesis of acute myeloid leukemia. *Haematologica* **2016**, *101*, 672–679. [CrossRef] [PubMed]
7. Huff, V.; Miwa, H.; Haber, D.A.; Call, K.M.; Housman, D.; Strong, L.C.; Saunders, G.F. Evidence for WT1 as a Wilms tumor (WT) gene: Intragenic germinal deletion in bilateral WT. *Am. J. Hum. Genet.* **1991**, *48*, 997–1003.
8. Haber, D.A.; Buckler, A.J.; Glaser, T.; Call, K.M.; Pelletier, J.; Sohn, R.L.; Douglass, E.C.; Housman, D.E. An internal deletion within an 11p13 zinc finger gene contributes to the development of Wilms' tumor. *Cell* **1990**, *61*, 1257–1269. [CrossRef]
9. Wagner, K.D.; Cherfils-Vicini, J.; Hosen, N.; Hohenstein, P.; Gilson, E.; Hastie, N.D.; Michiels, J.F.; Wagner, N. The Wilms' tumour suppressor Wt1 is a major regulator of tumour angiogenesis and progression. *Nat. Commun.* **2014**, *5*, 5852. [CrossRef]
10. Belali, T.; Wodi, C.; Clark, B.; Cheung, M.K.; Craig, T.J.; Wheway, G.; Wagner, N.; Wagner, K.D.; Roberts, S.; Porazinski, S.; et al. WT1 activates transcription of the splice factor kinase SRPK1 gene in PC3 and K562 cancer cells in the absence of corepressor BASP1. *Biochim. Biophys. Acta Gene Regul. Mech.* **2020**, *1863*, 194642. [CrossRef]
11. Wagner, K.D.; El Maï, M.; Ladomery, M.; Belali, T.; Leccia, N.; Michiels, J.F.; Wagner, N. Altered VEGF Splicing Isoform Balance in Tumor Endothelium Involves Activation of Splicing Factors Srpk1 and Srsf1 by the Wilms' Tumor Suppressor Wt1. *Cells* **2019**, *8*, 41. [CrossRef]
12. Wagner, N.; Michiels, J.F.; Schedl, A.; Wagner, K.D. The Wilms' tumour suppressor WT1 is involved in endothelial cell proliferation and migration: Expression in tumour vessels in vivo. *Oncogene* **2008**, *27*, 3662–3672. [CrossRef]
13. Eisermann, K.; Tandon, S.; Bazarov, A.; Brett, A.; Fraizer, G.; Piontkivska, H. Evolutionary conservation of zinc finger transcription factor binding sites in promoters of genes co-expressed with WT1 in prostate cancer. *BMC Genom.* **2008**, *9*, 337. [CrossRef]
14. Rauscher, F.J.; Morris, J.F.; Tournay, O.E.; Cook, D.M.; Curran, T. Binding of the Wilms' tumor locus zinc finger protein to the EGR-1 consensus sequence. *Science* **1990**, *250*, 1259–1262. [CrossRef]
15. Haber, D.A.; Sohn, R.L.; Buckler, A.J.; Pelletier, J.; Call, K.M.; Housman, D.E. Alternative splicing and genomic structure of the Wilms tumor gene WT1. *Proc. Natl. Acad. Sci. USA* **1991**, *88*, 9618–9622. [CrossRef]
16. Toska, E.; Roberts, S.G. Mechanisms of transcriptional regulation by WT1 (Wilms' tumour 1). *Biochem. J.* **2014**, *461*, 15–32. [CrossRef]
17. Laity, J.H.; Dyson, H.J.; Wright, P.E. Molecular basis for modulation of biological function by alternate splicing of the Wilms' tumor suppressor protein. *Proc. Natl. Acad. Sci. USA* **2000**, *97*, 11932–11935. [CrossRef] [PubMed]
18. Larsson, S.H.; Charlieu, J.P.; Miyagawa, K.; Engelkamp, D.; Rassoulzadegan, M.; Ross, A.; Cuzin, F.; van Heyningen, V.; Hastie, N.D. Subnuclear localization of WT1 in splicing or transcription factor domains is regulated by alternative splicing. *Cell* **1995**, *81*, 391–401. [CrossRef]
19. Alberta, J.A.; Springett, G.M.; Rayburn, H.; Natoli, T.A.; Loring, J.; Kreidberg, J.A.; Housman, D. Role of the WT1 tumor suppressor in murine hematopoiesis. *Blood* **2003**, *101*, 2570–2574. [CrossRef]
20. Herzer, U.; Crocoll, A.; Barton, D.; Howells, N.; Englert, C. The Wilms tumor suppressor gene wt1 is required for development of the spleen. *Curr. Biol.* **1999**, *9*, 837–840. [CrossRef]
21. Moore, A.W.; McInnes, L.; Kreidberg, J.; Hastie, N.D.; Schedl, A. YAC complementation shows a requirement for Wt1 in the development of epicardium, adrenal gland and throughout nephrogenesis. *Development* **1999**, *126*, 1845–1857.
22. Wagner, N.; Wagner, K.D.; Hammes, A.; Kirschner, K.M.; Vidal, V.P.; Schedl, A.; Scholz, H. A splice variant of the Wilms' tumour suppressor Wt1 is required for normal development of the olfactory system. *Development* **2005**, *132*, 1327–1336. [CrossRef] [PubMed]
23. Wagner, N.; Wagner, K.D.; Theres, H.; Englert, C.; Schedl, A.; Scholz, H. Coronary vessel development requires activation of the TrkB neurotrophin receptor by the Wilms' tumor transcription factor Wt1. *Genes Dev.* **2005**, *19*, 2631–2642. [CrossRef] [PubMed]
24. Wagner, N.; Wagner, K.D.; Scholz, H.; Kirschner, K.M.; Schedl, A. Intermediate filament protein nestin is expressed in developing kidney and heart and might be regulated by the Wilms' tumor suppressor Wt1. *Am. J. Physiol. Regul. Integr. Comp. Physiol.* **2006**, *291*, R779–R787. [CrossRef] [PubMed]
25. Wagner, K.D.; Wagner, N.; Vidal, V.P.; Schley, G.; Wilhelm, D.; Schedl, A.; Englert, C.; Scholz, H. The Wilms' tumor gene Wt1 is required for normal development of the retina. *EMBO J.* **2002**, *21*, 1398–1405. [CrossRef] [PubMed]
26. Schnerwitzki, D.; Hayn, C.; Perner, B.; Englert, C. Wt1 Positive dB4 Neurons in the Hindbrain Are Crucial for Respiration. *Front. Neurosci.* **2020**, *14*, 529487. [CrossRef]
27. Weiss, A.C.; Rivera-Reyes, R.; Englert, C.; Kispert, A. Expansion of the renal capsular stroma, ureteric bud branching defects and cryptorchidism in mice with Wilms tumor 1 gene deletion in the stromal compartment of the developing kidney. *J. Pathol.* **2020**, *252*, 290–303. [CrossRef] [PubMed]

28. Schnerwitzki, D.; Perry, S.; Ivanova, A.; Caixeta, F.V.; Cramer, P.; Günther, S.; Weber, K.; Tafreshiha, A.; Becker, L.; Vargas Panesso, I.L.; et al. Neuron-specific inactivation of. *Life Sci. Alliance* **2018**, *1*, e201800106. [CrossRef]
29. Nathan, A.; Reinhardt, P.; Kruspe, D.; Jörß, T.; Groth, M.; Nolte, H.; Habenicht, A.; Herrmann, J.; Holschbach, V.; Toth, B.; et al. The Wilms tumor protein Wt1 contributes to female fertility by regulating oviductal proteostasis. *Hum. Mol. Genet.* **2017**, *26*, 1694–1705. [CrossRef]
30. Bharathavikru, R.; Dudnakova, T.; Aitken, S.; Slight, J.; Artibani, M.; Hohenstein, P.; Tollervey, D.; Hastie, N. Transcription factor Wilms' tumor 1 regulates developmental RNAs through 3′ UTR interaction. *Genes Dev.* **2017**, *31*, 347–352. [CrossRef]
31. Hartwig, S.; Ho, J.; Pandey, P.; Macisaac, K.; Taglienti, M.; Xiang, M.; Alterovitz, G.; Ramoni, M.; Fraenkel, E.; Kreidberg, J.A. Genomic characterization of Wilms' tumor suppressor 1 targets in nephron progenitor cells during kidney development. *Development* **2010**, *137*, 1189–1203. [CrossRef] [PubMed]
32. Kirschner, K.M.; Wagner, N.; Wagner, K.D.; Wellmann, S.; Scholz, H. The Wilms tumor suppressor Wt1 promotes cell adhesion through transcriptional activation of the alpha4integrin gene. *J. Biol. Chem.* **2006**, *281*, 31930–31939. [CrossRef] [PubMed]
33. Kreidberg, J.A.; Sariola, H.; Loring, J.M.; Maeda, M.; Pelletier, J.; Housman, D.; Jaenisch, R. WT-1 is required for early kidney development. *Cell* **1993**, *74*, 679–691. [CrossRef]
34. Lee, S.B.; Huang, K.; Palmer, R.; Truong, V.B.; Herzlinger, D.; Kolquist, K.A.; Wong, J.; Paulding, C.; Yoon, S.K.; Gerald, W.; et al. The Wilms tumor suppressor WT1 encodes a transcriptional activator of amphiregulin. *Cell* **1999**, *98*, 663–673. [CrossRef]
35. Martínez-Estrada, O.M.; Lettice, L.A.; Essafi, A.; Guadix, J.A.; Slight, J.; Velecela, V.; Hall, E.; Reichmann, J.; Devenney, P.S.; Hohenstein, P.; et al. Wt1 is required for cardiovascular progenitor cell formation through transcriptional control of Snail and E-cadherin. *Nat. Genet.* **2010**, *42*, 89–93. [CrossRef]
36. Miquerol, L.; Kelly, R.G. Organogenesis of the vertebrate heart. *Wiley Interdiscip. Rev. Dev. Biol.* **2013**, *2*, 17–29. [CrossRef] [PubMed]
37. Wagner, K.D.; Wagner, N.; Bondke, A.; Nafz, B.; Flemming, B.; Theres, H.; Scholz, H. The Wilms' tumor suppressor Wt1 is expressed in the coronary vasculature after myocardial infarction. *FASEB J.* **2002**, *16*, 1117–1119. [CrossRef] [PubMed]
38. Duim, S.N.; Kurakula, K.; Goumans, M.J.; Kruithof, B.P. Cardiac endothelial cells express Wilms' tumor-1: Wt1 expression in the developing, adult and infarcted heart. *J. Mol. Cell. Cardiol.* **2015**, *81*, 127–135. [CrossRef]
39. Männer, J.; Wessel, A.; Yelbuz, T.M. How does the tubular embryonic heart work? Looking for the physical mechanism generating unidirectional blood flow in the valveless embryonic heart tube. *Dev. Dyn.* **2010**, *239*, 1035–1046. [CrossRef]
40. Armstrong, J.F.; Pritchard-Jones, K.; Bickmore, W.A.; Hastie, N.D.; Bard, J.B. The expression of the Wilms' tumour gene, WT1, in the developing mammalian embryo. *Mech. Dev.* **1993**, *40*, 85–97. [CrossRef]
41. Vicente-Steijn, R.; Scherptong, R.W.; Kruithof, B.P.; Duim, S.N.; Goumans, M.J.; Wisse, L.J.; Zhou, B.; Pu, W.T.; Poelmann, R.E.; Schalij, M.J.; et al. Regional differences in WT-1 and Tcf21 expression during ventricular development: Implications for myocardial compaction. *PLoS ONE* **2015**, *10*, e0136025. [CrossRef] [PubMed]
42. Zhou, B.; Ma, Q.; Rajagopal, S.; Wu, S.M.; Domian, I.; Rivera-Feliciano, J.; Jiang, D.; von Gise, A.; Ikeda, S.; Chien, K.R.; et al. Epicardial progenitors contribute to the cardiomyocyte lineage in the developing heart. *Nature* **2008**, *454*, 109–113. [CrossRef] [PubMed]
43. Wu, M.; Smith, C.L.; Hall, J.A.; Lee, I.; Luby-Phelps, K.; Tallquist, M.D. Epicardial spindle orientation controls cell entry into the myocardium. *Dev. Cell* **2010**, *19*, 114–125. [CrossRef] [PubMed]
44. Cai, C.L.; Martin, J.C.; Sun, Y.; Cui, L.; Wang, L.; Ouyang, K.; Yang, L.; Bu, L.; Liang, X.; Zhang, X.; et al. A myocardial lineage derives from Tbx18 epicardial cells. *Nature* **2008**, *454*, 104–108. [CrossRef] [PubMed]
45. Velecela, V.; Torres-Cano, A.; García-Melero, A.; Ramiro-Pareta, M.; Müller-Sánchez, C.; Segarra-Mondejar, M.; Chau, Y.Y.; Campos-Bonilla, B.; Reina, M.; Soriano, F.X.; et al. Epicardial cell shape and maturation are regulated by Wt1 via transcriptional control of. *Development* **2019**, *146*. [CrossRef]
46. Sereti, K.I.; Nguyen, N.B.; Kamran, P.; Zhao, P.; Ranjbarvaziri, S.; Park, S.; Sabri, S.; Engel, J.L.; Sung, K.; Kulkarni, R.P.; et al. Analysis of cardiomyocyte clonal expansion during mouse heart development and injury. *Nat. Commun.* **2018**, *9*, 754. [CrossRef] [PubMed]
47. Cano, E.; Carmona, R.; Ruiz-Villalba, A.; Rojas, A.; Chau, Y.Y.; Wagner, K.D.; Wagner, N.; Hastie, N.D.; Muñoz-Chápuli, R.; Pérez-Pomares, J.M. Extracardiac septum transversum/proepicardial endothelial cells pattern embryonic coronary arterio-venous connections. *Proc. Natl. Acad. Sci. USA* **2016**, *113*, 656–661. [CrossRef]
48. Wagner, N.; Morrison, H.; Pagnotta, S.; Michiels, J.F.; Schwab, Y.; Tryggvason, K.; Schedl, A.; Wagner, K.D. The podocyte protein nephrin is required for cardiac vessel formation. *Hum. Mol. Genet.* **2011**, *20*, 2182–2194. [CrossRef]
49. Wagner, K.D.; Wagner, N.; Wellmann, S.; Schley, G.; Bondke, A.; Theres, H.; Scholz, H. Oxygen-regulated expression of the Wilms' tumor suppressor Wt1 involves hypoxia-inducible factor-1 (HIF-1). *FASEB J.* **2003**, *17*, 1364–1366. [CrossRef]
50. Carmona, R.; Barrena, S.; López Gambero, A.J.; Rojas, A.; Muñoz-Chápuli, R. Epicardial cell lineages and the origin of the coronary endothelium. *FASEB J.* **2020**, *34*, 5223–5239. [CrossRef]
51. Ikeda, N.; Nakazawa, N.; Kurata, Y.; Yaura, H.; Taufiq, F.; Minato, H.; Yoshida, A.; Ninomiya, H.; Nakayama, Y.; Kuwabara, M.; et al. Tbx18-positive cells differentiated from murine ES cells serve as proepicardial progenitors to give rise to vascular smooth muscle cells and fibroblasts. *Biomed. Res.* **2017**, *38*, 229–238. [CrossRef]
52. Smart, N.; Bollini, S.; Dubé, K.N.; Vieira, J.M.; Zhou, B.; Davidson, S.; Yellon, D.; Riegler, J.; Price, A.N.; Lythgoe, M.F.; et al. De novo cardiomyocytes from within the activated adult heart after injury. *Nature* **2011**, *474*, 640–644. [CrossRef]

53. Von Gise, A.; Zhou, B.; Honor, L.B.; Ma, Q.; Petryk, A.; Pu, W.T. WT1 regulates epicardial epithelial to mesenchymal transition through β-catenin and retinoic acid signaling pathways. *Dev. Biol.* **2011**, *356*, 421–431. [CrossRef]
54. Wilm, B.; Ipenberg, A.; Hastie, N.D.; Burch, J.B.; Bader, D.M. The serosal mesothelium is a major source of smooth muscle cells of the gut vasculature. *Development* **2005**, *132*, 5317–5328. [CrossRef] [PubMed]
55. Rudat, C.; Kispert, A. Wt1 and epicardial fate mapping. *Circ. Res.* **2012**, *111*, 165–169. [CrossRef] [PubMed]
56. Christoffels, V.M.; Grieskamp, T.; Norden, J.; Mommersteeg, M.T.; Rudat, C.; Kispert, A. Tbx18 and the fate of epicardial progenitors. *Nature* **2009**, *458*, E8–E9. [CrossRef] [PubMed]
57. Maltsev, V.A.; Rohwedel, J.; Hescheler, J.; Wobus, A.M. Embryonic stem cells differentiate in vitro into cardiomyocytes representing sinusnodal, atrial and ventricular cell types. *Mech. Dev.* **1993**, *44*, 41–50. [CrossRef]
58. Boheler, K.R.; Czyz, J.; Tweedie, D.; Yang, H.T.; Anisimov, S.V.; Wobus, A.M. Differentiation of pluripotent embryonic stem cells into cardiomyocytes. *Circ. Res.* **2002**, *91*, 189–201. [CrossRef]
59. Varlakhanova, N.V.; Cotterman, R.F.; de Vries, W.N.; Morgan, J.; Donahue, L.R.; Murray, S.; Knowles, B.B.; Knoepfler, P.S. myc maintains embryonic stem cell pluripotency and self-renewal. *Differentiation* **2010**, *80*, 9–19. [CrossRef]
60. Yang, L.; Soonpaa, M.H.; Adler, E.D.; Roepke, T.K.; Kattman, S.J.; Kennedy, M.; Henckaerts, E.; Bonham, K.; Abbott, G.W.; Linden, R.M.; et al. Human cardiovascular progenitor cells develop from a KDR+ embryonic-stem-cell-derived population. *Nature* **2008**, *453*, 524–528. [CrossRef]
61. Kim, B.J.; Kim, Y.H.; Lee, Y.A.; Jung, S.E.; Hong, Y.H.; Lee, E.J.; Kim, B.G.; Hwang, S.; Do, J.T.; Pang, M.G.; et al. Platelet-derived growth factor receptor-alpha positive cardiac progenitor cells derived from multipotent germline stem cells are capable of cardiomyogenesis in vitro and in vivo. *Oncotarget* **2017**, *8*, 29643–29656. [CrossRef] [PubMed]
62. Meilhac, S.M.; Buckingham, M.E. The deployment of cell lineages that form the mammalian heart. *Nat. Rev. Cardiol.* **2018**, *15*, 705–724. [CrossRef] [PubMed]
63. Wessels, A.; van den Hoff, M.J.; Adamo, R.F.; Phelps, A.L.; Lockhart, M.M.; Sauls, K.; Briggs, L.E.; Norris, R.A.; van Wijk, B.; Perez-Pomares, J.M.; et al. Epicardially derived fibroblasts preferentially contribute to the parietal leaflets of the atrioventricular valves in the murine heart. *Dev. Biol.* **2012**, *366*, 111–124. [CrossRef]
64. Zeng, B.; Ren, X.F.; Cao, F.; Zhou, X.Y.; Zhang, J. Developmental patterns and characteristics of epicardial cell markers Tbx18 and Wt1 in murine embryonic heart. *J. Biomed. Sci.* **2011**, *18*, 67. [CrossRef] [PubMed]
65. Redpath, A.N.; Smart, N. Recapturing embryonic potential in the adult epicardium: Prospects for cardiac repair. *Stem Cells Transl. Med.* **2021**, *10*, 511–521. [CrossRef] [PubMed]
66. Marín-Juez, R.; El-Sammak, H.; Helker, C.S.M.; Kamezaki, A.; Mullapuli, S.T.; Bibli, S.I.; Foglia, M.J.; Fleming, I.; Poss, K.D.; Stainier, D.Y.R. Coronary Revascularization During Heart Regeneration Is Regulated by Epicardial and Endocardial Cues and Forms a Scaffold for Cardiomyocyte Repopulation. *Dev. Cell* **2019**, *51*, 503–515. [CrossRef]
67. Liu, Y.H.; Lai, L.P.; Huang, S.Y.; Lin, Y.S.; Wu, S.C.; Chou, C.J.; Lin, J.L. Developmental origin of postnatal cardiomyogenic progenitor cells. *Future Sci. OA* **2016**, *2*, FSO120. [CrossRef]
68. Huang, G.N.; Thatcher, J.E.; McAnally, J.; Kong, Y.; Qi, X.; Tan, W.; DiMaio, J.M.; Amatruda, J.F.; Gerard, R.D.; Hill, J.A.; et al. C/EBP transcription factors mediate epicardial activation during heart development and injury. *Science* **2012**, *338*, 1599–1603. [CrossRef]
69. Van Wijk, B.; Gunst, Q.D.; Moorman, A.F.; van den Hoff, M.J. Cardiac regeneration from activated epicardium. *PLoS ONE* **2012**, *7*, e44692. [CrossRef]
70. Zhou, B.; Honor, L.B.; Ma, Q.; Oh, J.H.; Lin, R.Z.; Melero-Martin, J.M.; von Gise, A.; Zhou, P.; Hu, T.; He, L.; et al. Thymosin beta 4 treatment after myocardial infarction does not reprogram epicardial cells into cardiomyocytes. *J. Mol. Cell. Cardiol.* **2012**, *52*, 43–47. [CrossRef]
71. Bollini, S.; Vieira, J.M.; Howard, S.; Dubè, K.N.; Balmer, G.M.; Smart, N.; Riley, P.R. Re-activated adult epicardial progenitor cells are a heterogeneous population molecularly distinct from their embryonic counterparts. *Stem Cells Dev.* **2014**, *23*, 1719–1730. [CrossRef]
72. Zhou, B.; Honor, L.B.; He, H.; Ma, Q.; Oh, J.H.; Butterfield, C.; Lin, R.Z.; Melero-Martin, J.M.; Dolmatova, E.; Duffy, H.S.; et al. Adult mouse epicardium modulates myocardial injury by secreting paracrine factors. *J. Clin. Investig.* **2011**, *121*, 1894–1904. [CrossRef] [PubMed]
73. Urbanek, K.; Torella, D.; Sheikh, F.; De Angelis, A.; Nurzynska, D.; Silvestri, F.; Beltrami, C.A.; Bussani, R.; Beltrami, A.P.; Quaini, F.; et al. Myocardial regeneration by activation of multipotent cardiac stem cells in ischemic heart failure. *Proc. Natl. Acad. Sci. USA* **2005**, *102*, 8692–8697. [CrossRef] [PubMed]
74. Beltrami, A.P.; Barlucchi, L.; Torella, D.; Baker, M.; Limana, F.; Chimenti, S.; Kasahara, H.; Rota, M.; Musso, E.; Urbanek, K.; et al. Adult cardiac stem cells are multipotent and support myocardial regeneration. *Cell* **2003**, *114*, 763–776. [CrossRef]
75. Krueger, K.; Catanese, L.; Sciesielski, L.K.; Kirschner, K.M.; Scholz, H. Deletion of an intronic HIF-2α binding site suppresses hypoxia-induced WT1 expression. *Biochim. Biophys. Acta Gene Regul. Mech.* **2019**, *1862*, 71–83. [CrossRef]
76. Kirschner, K.M.; Sciesielski, L.K.; Krueger, K.; Scholz, H. Wilms tumor protein-dependent transcription of VEGF receptor 2 and hypoxia regulate expression of the testis-promoting gene. *J. Biol. Chem.* **2017**, *292*, 20281–20291. [CrossRef] [PubMed]
77. El Maï, M.; Wagner, K.D.; Michiels, J.F.; Ambrosetti, D.; Borderie, A.; Destree, S.; Renault, V.; Djerbi, N.; Giraud-Panis, M.J.; Gilson, E.; et al. The Telomeric Protein TRF2 Regulates Angiogenesis by Binding and Activating the PDGFRβ Promoter. *Cell Rep.* **2014**, *9*, 1047–1060. [CrossRef]

78. McCarty, G.; Awad, O.; Loeb, D.M. WT1 protein directly regulates expression of vascular endothelial growth factor and is a mediator of tumor response to hypoxia. *J. Biol. Chem.* **2011**, *286*, 43634–43643. [CrossRef]
79. Senyo, S.E.; Steinhauser, M.L.; Pizzimenti, C.L.; Yang, V.K.; Cai, L.; Wang, M.; Wu, T.D.; Guerquin-Kern, J.L.; Lechene, C.P.; Lee, R.T. Mammalian heart renewal by pre-existing cardiomyocytes. *Nature* **2013**, *493*, 433–436. [CrossRef]
80. Hsieh, P.C.; Segers, V.F.; Davis, M.E.; MacGillivray, C.; Gannon, J.; Molkentin, J.D.; Robbins, J.; Lee, R.T. Evidence from a genetic fate-mapping study that stem cells refresh adult mammalian cardiomyocytes after injury. *Nat. Med.* **2007**, *13*, 970–974. [CrossRef]
81. Scholz, H.; Kirschner, K.M. Oxygen-Dependent Gene Expression in Development and Cancer: Lessons Learned from the Wilms' Tumor Gene, WT1. *Front. Mol. Neurosci.* **2011**, *4*, 4. [CrossRef] [PubMed]
82. Sanz, R.L.; Mazzei, L.; Manucha, W. Implications of the transcription factor WT1 linked to the pathologic cardiac remodeling post-myocardial infarction. *Clin. Investig. Arterioscler.* **2019**, *31*, 121–127. [CrossRef] [PubMed]
83. Balbi, C.; Lodder, K.; Costa, A.; Moimas, S.; Moccia, F.; van Herwaarden, T.; Rosti, V.; Campagnoli, F.; Palmeri, A.; De Biasio, P.; et al. Reactivating endogenous mechanisms of cardiac regeneration via paracrine boosting using the human amniotic fluid stem cell secretome. *Int. J. Cardiol.* **2019**, *287*, 87–95. [CrossRef] [PubMed]
84. Zhang, Y.; Sivakumaran, P.; Newcomb, A.E.; Hernandez, D.; Harris, N.; Khanabdali, R.; Liu, G.S.; Kelly, D.J.; Pébay, A.; Hewitt, A.W.; et al. Cardiac Repair With a Novel Population of Mesenchymal Stem Cells Resident in the Human Heart. *Stem Cells* **2015**, *33*, 3100–3113. [CrossRef]
85. Quijada, P.; Trembley, M.A.; Small, E.M. The Role of the Epicardium During Heart Development and Repair. *Circ. Res.* **2020**, *126*, 377–394. [CrossRef]
86. Tyser, R.C.V.; Ibarra-Soria, X.; McDole, K.; Arcot Jayaram, S.; Godwin, J.; van den Brand, T.A.H.; Miranda, A.M.A.; Scialdone, A.; Keller, P.J.; Marioni, J.C.; et al. Characterization of a common progenitor pool of the epicardium and myocardium. *Science* **2021**, *371*. [CrossRef] [PubMed]
87. Hescheler, J.; Fleischmann, B.K.; Lentini, S.; Maltsev, V.A.; Rohwedel, J.; Wobus, A.M.; Addicks, K. Embryonic stem cells: A model to study structural and functional properties in cardiomyogenesis. *Cardiovasc. Res.* **1997**, *36*, 149–162. [CrossRef]
88. Maltsev, V.A.; Wobus, A.M.; Rohwedel, J.; Bader, M.; Hescheler, J. Cardiomyocytes differentiated in vitro from embryonic stem cells developmentally express cardiac-specific genes and ionic currents. *Circ. Res.* **1994**, *75*, 233–244. [CrossRef]
89. Whitmill, A.; Liu, Y.; Timani, K.A.; Niu, Y.; He, J.J. Tip110 Deletion Impaired Embryonic and Stem Cell Development Involving Downregulation of Stem Cell Factors Nanog, Oct4, and Sox2. *Stem Cells* **2017**, *35*, 1674–1686. [CrossRef]
90. Tallini, Y.N.; Greene, K.S.; Craven, M.; Spealman, A.; Breitbach, M.; Smith, J.; Fisher, P.J.; Steffey, M.; Hesse, M.; Doran, R.M.; et al. c-kit expression identifies cardiovascular precursors in the neonatal heart. *Proc. Natl. Acad. Sci. USA* **2009**, *106*, 1808–1813. [CrossRef]
91. Noseda, M.; Harada, M.; McSweeney, S.; Leja, T.; Belian, E.; Stuckey, D.J.; Abreu Paiva, M.S.; Habib, J.; Macaulay, I.; de Smith, A.J.; et al. PDGFRα demarcates the cardiogenic clonogenic Sca1+ stem/progenitor cell in adult murine myocardium. *Nat. Commun.* **2015**, *6*, 6930. [CrossRef] [PubMed]
92. Van Laake, L.W.; Passier, R.; Monshouwer-Kloots, J.; Nederhoff, M.G.; Ward-van Oostwaard, D.; Field, L.J.; van Echteld, C.J.; Doevendans, P.A.; Mummery, C.L. Monitoring of cell therapy and assessment of cardiac function using magnetic resonance imaging in a mouse model of myocardial infarction. *Nat. Protoc.* **2007**, *2*, 2551–2567. [CrossRef] [PubMed]
93. Wang, X.; Yang, P. In vitro differentiation of mouse embryonic stem (mES) cells using the hanging drop method. *J. Vis. Exp.* **2008**. [CrossRef]
94. Wagner, K.D.; Du, S.; Martin, L.; Leccia, N.; Michiels, J.F.; Wagner, N. Vascular PPARβ/δ Promotes Tumor Angiogenesis and Progression. *Cells* **2019**, *8*, 1623. [CrossRef] [PubMed]
95. Keber, R.; Motaln, H.; Wagner, K.D.; Debeljak, N.; Rassoulzadegan, M.; Ačimovič, J.; Rozman, D.; Horvat, S. Mouse knockout of the cholesterogenic cytochrome P450 lanosterol 14alpha-demethylase (Cyp51) resembles Antley-Bixler syndrome. *J. Biol. Chem.* **2011**, *286*, 29086–29097. [CrossRef] [PubMed]

International Journal of
Molecular Sciences

Review

Recent Advances in Gene Therapy for Cardiac Tissue Regeneration

Yevgeniy Kim, Zharylkasyn Zharkinbekov , Madina Sarsenova, Gaziza Yeltay and Arman Saparov *

Department of Medicine, School of Medicine, Nazarbayev University, Nur-Sultan 010000, Kazakhstan; yevgeniy.kim@nu.edu.kz (Y.K.); zharylkasyn.zharkinbekov@nu.edu.kz (Z.Z.); madina.sarsenova@nu.edu.kz (M.S.); gaziza.yeltay@nu.edu.kz (G.Y.)
* Correspondence: asaparov@nu.edu.kz; Tel.: +7-717-270-6140

Abstract: Cardiovascular diseases (CVDs) are responsible for enormous socio-economic impact and the highest mortality globally. The standard of care for CVDs, which includes medications and surgical interventions, in most cases, can delay but not prevent the progression of disease. Gene therapy has been considered as a potential therapy to improve the outcomes of CVDs as it targets the molecular mechanisms implicated in heart failure. Cardiac reprogramming, therapeutic angiogenesis using growth factors, antioxidant, and anti-apoptotic therapies are the modalities of cardiac gene therapy that have led to promising results in preclinical studies. Despite the benefits observed in animal studies, the attempts to translate them to humans have been inconsistent so far. Low concentration of the gene product at the target site, incomplete understanding of the molecular pathways of the disease, selected gene delivery method, difference between animal models and humans among others are probable causes of the inconsistent results in clinics. In this review, we discuss the most recent applications of the aforementioned gene therapy strategies to improve cardiac tissue regeneration in preclinical and clinical studies as well as the challenges associated with them. In addition, we consider ongoing gene therapy clinical trials focused on cardiac regeneration in CVDs.

Keywords: gene therapy; cardiovascular diseases; cardiac regeneration; cardiac reprogramming; therapeutic angiogenesis; growth factors; reactive oxygen species; apoptosis

Citation: Kim, Y.; Zharkinbekov, Z.; Sarsenova, M.; Yeltay, G.; Saparov, A. Recent Advances in Gene Therapy for Cardiac Tissue Regeneration. *Int. J. Mol. Sci.* **2021**, 22, 9206. https://doi.org/10.3390/ijms22179206

Academic Editors: Nicole Wagner and Kay-Dietrich Wagner

Received: 31 July 2021
Accepted: 16 August 2021
Published: 26 August 2021

Publisher's Note: MDPI stays neutral with regard to jurisdictional claims in published maps and institutional affiliations.

1. Introduction

Cardiovascular diseases (CVDs) remain the principal cause of mortality and morbidity worldwide. In 2019 alone, it was estimated that there were approximately 18.6 million deaths due to CVDs, which accounted for almost one third of all global mortalities [1]. At the same time, the prevalence of CVDs exceeded half a billion cases in 2019 [1]. A more concerning fact about morbidity and mortality related to CVDs is that they are continuing to grow. According to the Global Burden of Disease 2019 Study, the number of deaths caused by CVDs increased by approximately 6.5 million for the period between 1990–2019, while the amount of prevalent cases of CVDs almost doubled for the same period [1]. In addition to the huge impact on people's health and quality of life, CVDs are blameworthy for huge economic losses. In the countries of the European Union in 2017, total spending associated with CVDs were estimated to be €210 billion euro [2]. In the USA in 2017, CVD-related healthcare and productivity loss costs exceeded $320 billion in 2017 [3]. The estimated healthcare costs are predicted to increase more than 2.5-fold, reaching $818 billion, by 2030 [3].

The standard treatments for CVDs include pharmacologic agents, therapeutic devices and surgical interventions. Although these therapies improve survival, reduce symptoms, and improve quality of life, they do not reverse the pathologic processes associated with CVDs and because of this, the health condition of most patients deteriorate with time [4].

Advances in cellular and molecular biology shed light on the pathogenetic and pathophysiologic mechanisms of CVDs and give promising opportunities for the use of novel therapies to combat heart and vascular diseases. Since these therapies are targeted towards the molecular cause of a specific CVD, they could hold a promising approach to cure the disease. Such treatments include cell-based therapies, therapies with growth factors and other bioactive molecules, biomaterials, and others [5–8]. Another therapeutic strategy that could likely improve the outcome of CVDs is gene therapy.

Gene therapy has been investigated for the treatment of a multitude of CVDs and CVD-related conditions such as atherosclerosis, coronary artery disease (CAD), myocardial infarction (MI), peripheral arterial disease (PAD), hypertension, various types of arrhythmias, heart failure (HF), restenosis of coronary stents, failure of vein grafts, and others [9–13]. Preclinical trials involving gene therapy for the aforementioned conditions were largely successful, however, there are a number of challenges associated with the translation of this therapeutic modality to clinical use [9]. Although there were several clinical trials for gene therapy, which showed positive results in terms of safety and certain therapeutic efficiency (specifically, trials for CAD, PAD, and HF), many other randomized-controlled studies failed to show the benefits of gene therapy over standard treatments [9]. This lack of consistent results in clinical trials can be attributed to multiple reasons including low concentration of the transferred gene or its product at the target site, incomplete knowledge of the disease mechanism and hence wrong gene therapy strategy, significant differences between animal models and human subjects, and others [4,9].

In this review, we discuss the latest preclinical findings and clinical trials related to the application of gene therapy for the treatment of CVDs. Specifically, we consider how cardiac gene therapy can be used to enhance angiogenesis, remodel scar tissue, alleviate production of reactive oxygen species, and prevent apoptosis.

2. Cardiac Tissue Regeneration in Adult Heart

2.1. Mechanisms of Cardiac Tissue Regeneration

The myocardium of adult mammals including humans has an extremely low regenerative capacity due to low proliferation rates of cardiomyocytes and scarcity of cardiac stem cells. In fact, for a long time, it was believed that cardiomyocytes in the adult heart are completely unable to proliferate. However, several studies using carbon-14 dating have detected a slow but persistent turnover of cardiomyocytes throughout life [14–16]. According to the American Heart Association's 2017 Consensus Statement, the annual turnover rate was reported to be between 0.5% and 2.0% [17]. Moreover, after injury to the heart muscle, the cardiomyocyte proliferation rate increases several times [18]. In contrast to adults, the neonatal mammalian heart possesses an outstanding capacity for regeneration. In the last decade, there have been multiple studies that demonstrated a robust regenerative response following apical resection and ischemic injury in mice shortly after birth [19–23]. The ability of neonatal myocardium to completely regenerate was also established in porcine and rat models of MI [24–26]. Importantly, neonates exhibit this remarkable regenerative capacity only for a brief period of time after birth, i.e., it is rapidly lost two to seven days postnatally [19,25,27].

Cardiac tissue regeneration is a very complex process that involves multiple genes and signaling pathways. Many reports showed that myocardial recovery is mainly mediated by the proliferation of existing cardiomyocytes with little contribution from cardiac progenitor cells [19]. Cardiomyocyte proliferation in turn is regulated by multiple cells and signaling interactions that is demonstrated by recent gene expression and chromatin modification profiling studies [28,29]. Using single-cell RNA sequencing, Wang and colleagues have identified 22 non-cardiomyocyte cell types involved in cardiac regeneration, including endothelial cells, fibroblasts, epicardial cells, pericytes, smooth muscle cells, and a number of immune cells—namely, macrophages, monocytes, dendritic cell-like cells, granulocytes, T cells, B cells, and glial cells [29]. In another recent transcriptome profiling study, two novel regulators of neonatal heart regeneration were found, specifically, C-C motif chemokine ligand 24 (CCL24) and insulin-like growth factor 2 messenger RNA-binding protein 3 (Igf2bp3) [28].

The main players associated with cardiac tissue regeneration are growth factors, cell cycle regulators, and miRNAs. In addition, several intrinsic signaling pathways were found to be essential in cardiomyocyte proliferation and heart regeneration. Neuregulin 1 (NRG1) is probably the most studied growth factor involved in heart development and regeneration [30]. It signals via Erb-B2 receptor tyrosine kinases 2-4 (ERBB2-4) and mediates the formation of a normal heart during embryonic development as well as promotes proliferation of cardiomyocytes after cardiac injuries [31]. Other growth factors reported to have a role in cardiac regeneration are follistatin-like 1 and insulin-like growth factor 2, which can mediate cell cycle re-entry of cardiomyocytes [30]. Another group of molecules that participates in cardiac regeneration are cell cycle regulators, i.e., cyclins and cyclin-dependent kinases (CDKs). Earlier studies found that cyclins A2, D1, and D2 can promote cardiomyocyte proliferation in adult hearts [30]. A recent study by Mohamed and colleagues discovered that in addition to the cell cycle regulators mentioned above, cyclins B1 and D1 as well as CDK 1 and CDK 4 are capable of inducing cell division in adult cardiac tissue [32]. This effect was evidenced by the overexpression studies as well as inhibition of Wee1 and TGFβ, which directly or indirectly suppress the activity of the aforementioned cell cycle regulators. Yet another class of molecules involved in heart regeneration are miRNAs. Recently, a large-scale screen of miRNAs revealed that there are 96 miRNAs that can mediate cardiomyocyte proliferation [33]. The same study discovered that the majority of these miRNAs acted via Hippo pathway and depended on the pathway's effector YAP. Interestingly, individual silencing of the miRNA did not suppress cell division indicating that none of these miRNAs on its own is essential for cardiomyocyte proliferation. The Hippo pathway is likely the most investigated intrinsic pathway that regulates cardiac tissue regeneration and cardiomyocyte proliferation. This evolutionary conserved pathway controls heart size by inhibiting cardiomyocyte division [30]. In addition, the Hippo pathway was shown to suppress cardiac tissue regeneration since the deletion of some of its components enhanced heart repair. Hippo pathway's downstream effector YAP, by contrast, was found to have pro-proliferative effects on cardiomyocytes. Thus, Monroe and colleagues recently demonstrated that YAP5SA causes expression of fetal genes and promotes proliferation in mouse cardiomyocytes [34].

The loss of cardiac regenerative capacity shortly after birth is caused by an exit of cardiomyocytes from the cell cycle and an entering into quiescent state. It is suggested that a change in extracellular matrix (ECM) and cytoskeleton architecture is a main factor that is culpable for the transition of cardiomyocytes to a senescent state. In particular, as cardiac cytoskeleton and ECM become stiffer and more organized and stable, it becomes more difficult for cell division to occur [35]. This was confirmed in a recent study by Notari and colleagues [27], in which the transcriptome of mice on the first and second postnatal days were compared. The study revealed that P2 murine neonates could not regenerate myocardium after resection of about 15% of the apex and responded with fibrosis. Importantly, cardiomyocytes of P2 mice did not lose their proliferative capacity, which indicates that the loss of regenerative ability was not associated with quiescence of the cells. On the other hand, a transcriptome analysis found a significant difference in the expression of genes related to ECM and cytoskeleton between P1 and P2 neonates. Specifically, mice on the second postnatal day overexpressed the aforementioned genes [27]. These findings support the theory that ECM and cytoskeleton growth and development are responsible for the loss of cardiac regenerative capacity shortly after birth.

2.2. Overview of Strategies to Enhance Cardiac Tissue Regeneration

Given the fact that poor cardiac regeneration lies in the root of the pathogenesis and pathophysiology of many CVDs, multiple therapeutic strategies have been suggested in order to enhance the regenerative capacity of the adult heart. A multitude of approaches that were shown to improve cardiac repair and regeneration can be roughly divided into three main categories, namely, bioactive molecules and secretory factors, cell-based strategies, and biomaterials. Paracrine effects of growth factors and cytokines were found to improve the regeneration of cardiac tissue. For instance, vascular endothelial factor A (VEGF-A) and fibroblast growth factor 2 (FGF2) enhanced cardiac repair by reducing scar size and mediating angiogenesis in animal models of MI [36]. In addition to growth factors and cytokines, microRNAs (miRNAs) have also been considered for regenerative therapy. MiRNAs are highly conserved, short, single-stranded non-coding RNA molecules, which control gene expression at the post-transcriptional level [18]. Since miRNAs are important regulators of cardiomyocyte proliferation, it was proposed that their application could promote cardiac regeneration in the adult heart. Specifically, this could be done by the activation of miRNAs that upregulate cardiomyocyte proliferation such as miR-199a and miR-590, or by suppression of miRNAs that inhibit cardiomyocyte division, namely, miR-195, miR-15a, miR-15b, miR-16, and miR-497 [18,36,37]. However, the wide application of growth factors and other secreted molecules for cardiac regeneration is limited by their short half-life.

Early studies involving cell-based approaches for cardiac regeneration utilized skeletal myocytes, resident cardiac-derived cells, bone marrow-derived cells, and mesenchymal stem cells [36,38,39]. Despite certain success achieved by some research groups, the results of preclinical studies with these cells were either inconsistent or failed to further translate to clinical trials [36]. The next stage in the development of cellular therapy for cardiac repair was the application of human embryonic stem cells (hESCs) and human induced pluripotent stem cells (hiPSCs) [40]. In both cases, multiple animal studies reported positive results, i.e., hESCs and hiPSCs differentiated into cardiomyocytes in vitro, survived after transplantation to the injured heart and significantly enhanced cardiac function [38]. In addition to direct benefits of cardiomyocyte replacement, hESCs and hiPSCs contributed to the regeneration of cardiac tissue via paracrine mechanisms secreting factors that promoted cardiomyocyte proliferation and angiogenesis, and inhibited apoptosis and fibrosis [8]. Despite the promising results of cell-based therapy observed in animals, there are a number of challenges associated with its translation to human studies. Firstly, cell transplantation could induce tumorigenesis and life-threatening arrhythmias [36,41]. In addition, it was found that the efficiency of a cell-based approach is significantly limited because of low survival and poor engraftment of the transplanted cells [38,41]. Rapid cell death following transplantation was related to preparation techniques, harsh environment of the injured heart and immune response [36,38]. Current studies involving cell-based strategies are, therefore, focused on approaches that could address the hurdles mentioned above.

The major approaches utilized to optimize stem cell therapy for cardiac regeneration are preconditioning, genetic manipulations, and use of biomaterials. Stem cell preconditioning using hypoxia, chemical and pharmacologic agents, and bioactive molecules was shown to significantly improve survival of the cells after transplantation as well as enhance their beneficial effects on the heart [8,42,43]. The positive effects of preconditioning on stem cells are related to induction of pro-survival and inhibition of apoptotic genes [42]. For instance, hypoxia activates hypoxia-inducible factor-1α (HIF-1α), which in turn promotes the expression of VEGF and angiotensin, two molecules that stimulate angiogenesis [44,45]. At the same time, it suppresses the expression of apoptosis-related proteins Bcl-2 and Bcl-xL [8]. Biomaterials such as nanoparticles, hydrogels, cryogels, coacervate, and scaffolds were also utilized to deliver various cytokines and growth factors for improving tissue regeneration [46–49] and to enhance the efficiency of cell therapy for cardiac diseases [50]. In multiple preclinical studies on MI, the use of biomaterials combined with stem cells was associated with enhanced survival and engraftment of the cell [7]. Moreover, stem cells combined with biomaterials stimulated cardiomyogenesis and angiogenesis as well as the release of factors that are important for cardiomyocyte survival and proliferation [7,51]. Another strategy to improve cellular therapy for cardiac regeneration is genetic manipulation, which will be further discussed in the upcoming sections of the paper.

Overall, adult cardiac tissue in humans has an extremely low ability to regenerate and because of this, injury to myocardium results in the formation of non-contractile fibrotic tissue eventually leading to higher morbidity and mortality. Cardiac regeneration strategies involving bioactive molecules, cells, and biomaterials attempt to improve the regenerative capacity of adult myocardium.

3. Gene Therapy for Cardiac Tissue Regeneration

3.1. Gene Therapy with Growth Factors for Improving Angiogenesis

Therapeutic angiogenesis can be a potential treatment option for improving cardiac tissue functioning by stimulating blood vessel growth, increasing tissue perfusion and recovery [52]. In this regard, gene-based therapy with a qualitative gene delivery system (a plasmid or viral origin) in combination with adequate pro-angiogenic genes can serve as a promising tool for successful cardiac tissue regeneration (Figure 1). A well-developed gene-based system must include the following requirements: sustained long-term therapeutic effect, ability to target specific cell types and a decreased risk of systemic side-effects as compared to regular pharmacotherapy. To date, gene transfer systems include two approaches such as direct tissue injection and intravascular infusion with or without surgical or catheter-mediated interventions [53]. In addition, for gene therapy in heart diseases, the goal in most cases is to deliver genetic material directly to the myocardium. Expression of the targeted gene needs to be sustained over an extended period of time unless therapy is meant to repair a specific structural defect. Most importantly, the gene must encode a molecule that plays a critical role in the disease pathogenesis such that by altering expression of that gene alone, cardiac function will be improved sufficiently or will favorably alter the disease flow [54]. Thus, gene therapy using appropriate delivery systems, which meet the aforementioned requirements, and pro-angiogenic factors that act in processes such as blood flow, metabolic activity, and cardiac functioning are being applied for cardiac angiogenesis [55]. Among these factors are VEGF, FGF, and hepatocyte growth factor (HGF). In this section, we will focus on the insights from recent growth factor gene-based therapy approaches.

Among the pro-angiogenic factors, VEGF has been the most extensively studied. It is a 45-kD homodimeric glycoprotein that has four main isoforms: VEGF-A (possesses the ability to bind heparin), -B, -C, and -D. There are additional isoforms in VEGF-A: VEGF121, VEGF165, which is the most biologically active, VEGF189 and VEGF206 [56]. Many other known angiogenic factors act at least partly via VEGF-A. VEGF-B has some exceptional features compared to other VEGFs since it induces myocardium-specific angiogenesis without the risk of hyperpermeability and edema. Interestingly, it also seems to alter myocardial energy metabolism and fatty acid uptake and promote cell survival and compensatory hypertrophy [57]. The receptors for VEGF are FLT-1 and FLK-1, which activate intracellular tyrosine kinase. Neuropilin 1 (NP-1) is another receptor for VEGF that binds VEGF165. NP-1 and FLK-1 are key mediators of the phosphoinositide-3-kinase, and Akt (PI3K/Akt) and mitogen-activated protein kinase (MAPK) kinase pathways [58].

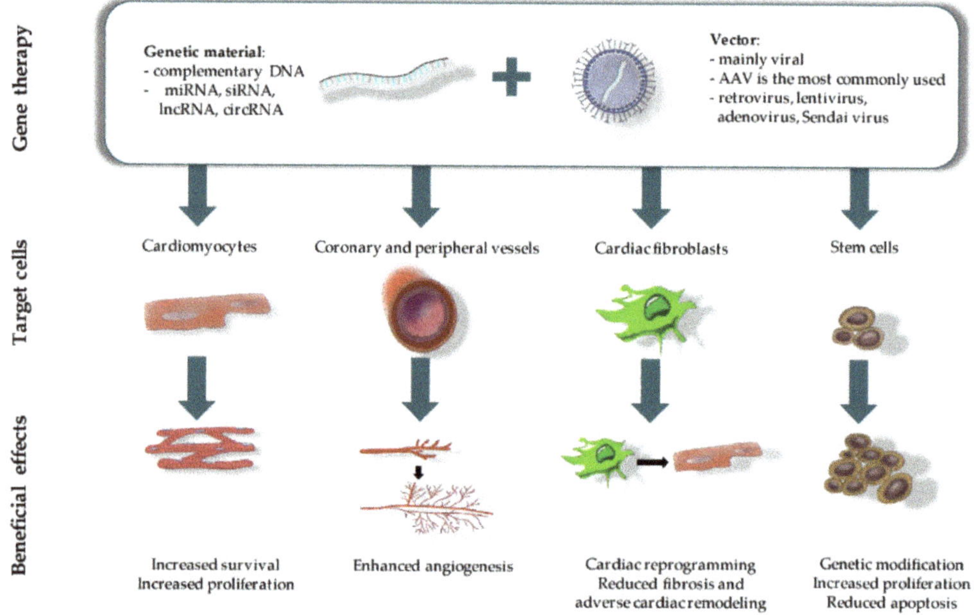

Figure 1. Gene therapy and therapeutic outcomes in CVDs treatment.

VEGF-A isoform has been well studied in a number of preclinical studies and several clinical trials. However, VEGF-B and VEGF-D isoforms have attracted much attention from scientists in the past few years [59]. In this regard, Nurro and colleagues demonstrated an angiogenic capacity of new members of the VEGF family, VEGF-B186 and VEGF-DΔNΔC, in porcine myocardium [60]. Both adenovirus-mediated VEGF-B186 (AdVEGF-B186) and AdVEGF-DΔNΔC gene transfers induced efficient angiogenesis in the myocardium, possibly due to their high solubility in tissues since they do not bind effectively to heparan sulfates, suggesting that they could be suitable candidates for the induction of therapeutic angiogenesis for the treatment of refractory angina. In a study by Huusko and colleagues, it was demonstrated that VEGF-C and VEGF-D are associated with the compensatory ventricular hypertrophy and that adeno-associated virus subtype 9 (AAV9)-VEGF-B186 gene transfer can rescue the function of the failing heart and postpone the transition towards HF in mice [61]. Another strategy of gene therapy is based on the concept of a combined approach using a simultaneous delivery of VEGF165 and HGF genes to alleviate the symptoms of MI in rats [62]. Combined application of these genes resulted in an increase of the number of cardiac stem cells in the peri-infarct area and sporadic proliferation of mature cardiomyocytes. These effects may be explained by the activation of VEGFRs and c-Met (HGF receptor) that initiate mitogenic signaling cascades. Furthermore, it is reported that VEGF-D stimulates both angiogenesis and lymphangiogenesis, which has been also tested in animal models and confirmed by a phase I/IIa clinical study with one-year follow-up [63]. Another clinical study investigated a novel VEGF-D, a new member of the VEGF family called VEGF-DΔNΔC, in 30 patients with refractory angina. This phase I/IIa randomized clinical trial used increasing doses of endocardial adenoviral (ad)-injections with electroanatomical targeting of injections using a NOGA catheter system. AdVEGF-DΔNΔC gene therapy is considered to be safe and well tolerated, and confirmed by a consistent report that lipoprotein (a) level in 50% of patients with refractory angina was significantly decreased [64].

FGF is another angiogenic factor that has been studied in CVDs. FGF is related to a family of 22 identified members of pleiotropic proteins in human and mouse and regulates crucial functions in the heart, ranging from development to homeostasis and disease, and is considered a cardiomyokine [65,66]. The mechanism of action of FGF proteins is based on FGF/FGFR signaling, which plays an important role in angiogenesis and lymphangiogenesis. FGF signaling can influence the whole process of angiogenesis and cardiomyocyte mitosis. Activation of FGFR1 or FGFR2 has been demonstrated to have a positive effect on vascular endothelial proliferation. The first step of angiogenesis is ECM degradation. For instance, FGF1, FGF2, and FGF4 promote the expression of matrix metallo-proteinases (MMPs) in endothelium cells. The second step in angiogenesis is endothelium migration. FGF1, FGF2, FGF8, and FGF10 were demonstrated to stimulate endothelium chemotaxis [67]. In clinical studies, FGF4 showed positive results in patients with stable angina following a series of AGENT-trials representing a single intracoronary infusion of adenovirus type 5 vector (Ad5FGF-4) [68]. In an ASPIRE trial, a single intracoronary infusion of Ad5FGF-4 using catheter-based administration was compared to standard care without a placebo group. It was planned to recruit 100 participants; however, the study was terminated apparently due to a low number of patients [69].

Another promising pro-angiogenic factor is HGF. It participates in mediation of angiogenesis by inducing endothelial cell proliferation, migration and survival. The HGF receptor c-Met is reported to be expressed on vascular endothelial cells. HGF/c-Met signaling induces endothelial cell proliferation through the MAPK/ERK and STAT3 pathways. In addition to the direct effect on stimulating endothelial cells to form tubular structures, HGF can recruit bone marrow endothelial progenitor cells to participate in angiogenesis in the ischemic area [70]. For cardiac tissue regeneration, HGF either in plasmid or adenoviral constructs has been tested but only in a small number of studies. Preclinical studies demonstrated the therapeutic effects of adenovirus carrying the HGF gene (Ad-HGF) in a minipig model of chronic myocardium ischemia in which an Ameroid constrictor was placed around the left circumflex branch of the coronary artery. The data showed that Ad-HGF significantly improves the heart function and blood supply in chronic myocardium ischemia models [71]. Furthermore, a catheter-based intramyocardial delivery (NavX) of Ad-HGF was proved to be safe and feasible for Ad-HGF delivery in pigs [70,72]. In a study by Rong and colleagues, skeletal myoblasts were transduced with Ad-HGF. Transplantation of HGF-engineered skeletal myoblasts results in reduced infarct size and collagen deposition, increased vessel density and improved cardiac function in a rat MI model [73]. The effect might be explained by the overexpression of HGF in skeletal myoblasts that confers resistance to ischemia in MI. Based on preclinical studies, the safe application of Ad-HGF was additionally confirmed by clinical trials. One such application is the use of two delivery approaches: an intramuscular myocardium direct injection while performing coronary artery bypass surgery (CABG) and a catheter-based intramyocardial injection guided by the NavX system. An open-label safety trial of Ad-HGF by a direct multipoint injection into the myocardium of patients, who suffer from coronary heart disease, showed that no subjective or objective adverse reactions were detected. Furthermore, there was evidence of revascularization in ischemic regions confirmed by instrumental examination and physical symptoms [71].

Additionally, some studies have reported the use of the stem cell factor (SCF), which is a ligand for the c-kit that is a receptor tyrosine kinase. SCF binding to c-kit leads to receptor dimerization and activation of multiple signaling pathways related to cell recruitment, differentiation, angiogenesis, and survival [74]. It has been shown that local overexpression of SCF post-MI induces the recruitment of c-kit+ cells at the infarct border area. Moreover, gene transfer of membrane-bound human SCF improved cardiac function in the model of ischemic cardiomyopathy in Yorkshire pigs [75].

The poor angiogenic nature of the adult heart prompted scientists to establish new therapeutic approaches for cardiac tissue regeneration. Thus, the effectiveness of gene-based therapy using pro-angiogenic factors may represent a promising therapeutic strategy to improve cardiac function. However, even though some clinical trials have shown promising results, further large-scale clinical trials need to be performed to clarify their efficacy and potential clinical application.

3.2. Gene Therapy for Scar Tissue Remodeling

Scar tissue forming after MI has a profound impact on the physiology of cardiac tissue. Since post-infarction scar is comprised of non-contractile fibrotic tissue, the scarred region of the heart can no longer contract and relax, leading to systolic and diastolic dysfunctions [76,77]. In addition, the fibrotic area becomes the source of various arrhythmias because it interferes with the conductive system of the heart [76,77]. Therefore, patients surviving MI are at a greater risk for morbidity and mortality due to cardiac dysfunction and arrhythmogenicity caused by the scar tissue.

Several gene therapy strategies have been successfully utilized for scar tissue remodeling (Figure 1). One of which is the direct reprogramming of cardiac fibroblasts into functional cardiomyocytes using a set of specific transcription factors. In vitro studies found that such a transformation required just three factors—such as Gata4, Mef2c and Tbx5, which were collectively designated as GMT [78]. Further studies identified that the use of combinations of 3–4 of the following factors—GATA4, MEF2C, TBX5, HAND2, MESP1, NKX2.5, and MYOCD—could be as efficient as or even better than the original GMT combination [79]. Cardiac reprogramming using this gene transfer was further accomplished in murine MI models using retroviral and lentiviral vectors. The in vivo reprogramming induced the formation of more mature cardiomyocytes compared to the same procedure in vitro. In addition, it led to the reduction of fibrosis and improved cardiac contractility of the injured heart. Despite the fact that early preclinical studies using cardiac reprogramming have achieved formation of new cardiomyocytes, the efficiency of this transformation was relatively low, frequently less than 30–40% [79]. Therefore, recent studies have attempted to combine cardiac reprogramming with other strategies to enhance cardiac repair [80]. In one of the studies, E-twenty-six Variant transcription factor 2 (ETV2) was introduced along with GMT genes into the genome of rat MI models using adenoviral and retroviral vectors, respectively [81]. ETV2 is crucial for the development of new blood vessels both during embryonic development and repair after injury [82]. Genetic therapy with a combination of ETV2 and GMT had positive effects on the recovery of the infarcted hearts. Namely, it induced the formation of new myocardium and blood vessels, reduced scar size and improved left ventricular function. Notably, the combinational gene transfer of GMT and ETV2 exhibited synergistic effects, i.e., the beneficial effects of the dual therapy were significantly more pronounced compared to individual gene transfers.

Direct cardiac reprogramming using conventional retroviral and lentiviral vectors possesses several limitations, such as low efficiency of fibroblast to cardiomyocyte transdifferentiation as well as potential insertional mutagenesis [83]. In order to address these obstacles, alternative gene transfer strategies for reprogramming of fibroblasts in the scar tissue have been considered. Isomi and colleagues have used Sendai virus as a vector to introduce GMT genes into cardiac fibroblasts of a mouse MI model [84]. The gene transfer induced the differentiation of fibroblasts into cardiomyocytes which in turn led to reduction of scar tissue and improved contractile function of the heart for 12 weeks. Importantly, this is the first study which observed long-term effects of the GMT-Sendai virus vector-based reprogramming. Another alternative gene carrier for direct cardiac reprogramming is nanoparticles. In a study by Chang and colleagues [83], cationic gold nanoparticles loaded with GMT transcription factor genes effectively induced transdifferentiation of mouse and human fibroblasts into cardiomyocytes in vitro, which was confirmed by their morphology and expression of cardiac-specific genes. Moreover, the direct injection of a gold nanoparticles-GMT construct into the infarct area of the murine heart generated more mature cardiomyocytes compared to in vitro experiments. In addition, the treatment resulted in a significant decrease in the scar dimensions two weeks after MI in mice. Importantly, it was shown in this study that the gold nanoparticles-carrier system did not cause integration of the delivered DNA into the host genome [83].

Besides direct cardiac reprogramming, gene therapy to attenuate fibrosis and scar tissue was also accomplished via ablation of genes regulating fibrosis. In a recent study, Adapala and colleagues [85] investigated the function of the transient receptor potential vanilloid 4 (TRPV4) mechanosensitive ion channel in cardiac remodeling after MI using TRPV4-knockout mice. The absence of TRPV4 was associated with a significant reduction in scar size and, remarkably, the presence of viable myocardium in the infarction area eight weeks after MI. Consequently, TRPV4-KO mice had a higher survival rate, and preserved systolic and diastolic function. The study also demonstrated that TRPV4 mediated differentiation of cardiac fibroblasts into myofibroblasts via a Rho/Rho kinase/MRTF-A pathway. This was evidenced by the observations that inhibition of TRPV4 caused reduced activity of Rho kinase as well as activation of TRPV4 with transforming growth factor-β1 (TGF-β1) or GSK (TRPV4 agonist) that led to the increased expression of MRTF-A. In another study, genetic ablation of Cullin-RING E3 ubiquitin ligase 7 (CRL7) was performed in order to attenuate fibrosis caused by pressure overload [86]. In particular, DNA recombination using Cre-recombinase transferred by AAV9 vector was utilized to obtain CRL7-knockout mice. Next, the mice were subjected to transverse aortic constriction, which is used to model chronic pressure overload and consequent fibrosis. CRL7 knock-out resulted in a 50% decrease in myocardial interstitial fibrosis and enhanced survival of cardiomyocytes. The observed beneficial effects of CRL7 genetic ablation were probably due to enhanced activation of a phosphatidylinositol 3-kinase (PI3K)/protein kinase B (Akt) signaling pathway, which regulates cell growth, proliferation, and apoptosis among other processes. In fact, the authors claim that the study is the first one that elucidated the role of CRL7 in PI3K/Akt regulation in cardiac tissue. Another signaling pathway that was found to be regulated by CRL7 was related to TGF-β1. Thus, the absence of CRL7 led to lower expression of TGF-β1 that confirms the pro-fibrotic role of CRL7. In another study, the role of low-density lipoprotein receptor-related protein 6 (LRP6) in cardiac repair after MI was investigated using miRNA-mediated cardiomyocyte-specific knock-down of the corresponding gene [87]. The depletion of LRP6 was found to be associated with a significantly smaller scar size and improved cardiac function in a mouse model of MI. Remarkably, it also caused enhanced proliferation of cardiomyocytes via the ING5/P21 pathway. It is important to note that the new cardiomyocytes originated from the existing cardiomyocytes rather than cardiac progenitor cells. This study is one of the first reports on the functions of LRP6 in cardiomyocyte proliferation in the adult heart. In summary, scar tissue remodeling using gene therapy can be accomplished by direct cardiac reprogramming as well as by ablation of fibrosis-related genes.

3.3. Gene Therapy to Regulate Reactive Oxygen Species

MI is one of the most common causes of death around the world. MI causes death of cardiomyocytes and irreversible damage to the heart tissue forming an infarcted zone, which is remodeled into fibrotic/scar tissue [88]. Mitochondrial damage is directly associated with cardiomyocyte death and myocardial dysfunction by inducing oxidative stress [89,90]. In addition, hypoxic stress resulting from MI leads to accumulation of reactive oxygen species (ROS), which causes apoptosis of cardiomyocytes and cardiac tissue injury. Most of the MI therapies aim to restore blood flow using CABG and percutaneous coronary intervention, which leads to an increase in the concentration of oxygen in ischemic cells [91]. This sharp increase in the amount of oxygen also induces the high production of ROS, causing cardiomyocyte injury. This process is called ischemia reperfusion injury (IRI) [92]. Moreover, regeneration of adult cardiomyocytes is limited and results in post-infarction cardiac dysfunction [91].

Most gene therapies aimed to eliminate post-infarction ROS are interested in the activation of antioxidant genes or inhibition of ROS producing genes. One of those studies investigated the effect of brahma-related gene 1 (BRG1), which regulates the chromatin structure and cardiac gene expression in Nuclear erythroid 2-related factor 2/Heme oxygenase 1 (Nrf2/HO-1) pathway [93]. Nrf2/HO-1 pathway is known to be a main regulator of cellular antioxidant responses. Translocation of Nrf2 to the nucleus activates the HO-1 and downstream oxidative stress pathway (HO1, glutathionetransferase p1 (GSTP1) or NAD (P)H:quinone oxidoreductase 1 (NQO1)) [94]. According to Liu and colleagues, overexpression of BRG1 gene by intramyocardial injection of adenoviral vectors induces the translocation of Nrf2 and attenuates the oxidative damage in cardiomyocytes. Moreover, the conducted studies demonstrated that Nrf2 expression, which was induced by BRG1, increased the activity of HO-1 promoters. On the other hand, Lenti-Brg1 shRNA injection, which inhibits the BRG1, showed adverse effects decreasing HO-1 expression [93]. Overall, BRG1 overexpression reduced the size of the infarct scar and improved the function of the cardiac tissue by decreasing oxidative damage and cell apoptosis. The HO-1 antioxidant molecule is also regulated by HIF-1α. HIF-1α overexpression with the help of a plasmid transfection encapsulated in 20-mer peptide-conjugated carboxymethylchitosan nanoparticle in MI mice upregulated the HO-1 activation and expression, which decreased ROS accumulation both in vitro and in vivo. Moreover, findings proved that downregulation of ROS-mediated oxidative stress inhibits the expression of BNIP3, which is responsible for cardiomyocyte apoptosis [95].

Another group of scientists investigated the effect of Zinc finger protein ZBTB20 in the treatment of post-infarction cardiac tissue. ZBTB20 was delivered with the help of AAV9 system [96]. Based on the results, overexpression of ZBTB20 in post-MI heart increased the superoxide dismutase (SOD) enzymatic activity, an important antioxidant that converts superoxide radicals to molecular oxygen, provided cellular defense and inhibited the activity of malondialdehyde (MDA) and NADPH oxidase [96,97].

Some of recent findings focus on the amelioration of mitochondrial dysfunction to eliminate ROS and to improve post-infarct cardiac function. Transverse aortic constriction (TAC) preconditioning of mice before the left coronary artery ligation induced MI showed a significant reduction in oxidative stress and decreased mitochondrial ROS production. TAC preconditioning can also be mimicked by the cardiac overexpression of SIRT3 in vivo with the help of AAV-SIRT3 transfection. Both TAC preconditioning and SIRT3 gene overexpression considerably increased the contractile function of heart and, in contrast, decreased the myocardial scar area and death of cardiomyocytes [98]. In addition, expression of deacetylated isocitrate dehydrogenase 2, a protein that regulates mitochondrial redox, reduced the production of mitochondrial ROS and alleviated post-MI injury [98,99].

Furthermore, there has been much research done recently to study the effects of miRNA, an important regulator in the development and pathophysiology of the cardiovascular system, on oxidative stress after MI [100]. Expression of miR-323-3p, Bax, Bcl-2, SOD1, and SOD2 genes and oxidative stress in cardiomyocytes (H9c2 cells) of miR-323-3p transfected mice were compared to mice without transfection and with miR-323-3p transfected and H_2O_2 treated group at seventh day after MI. Results indicated that miR-323-3p was downregulated in MI heart and overexpression of miR-323-3p decreased Bax and significantly increased Bcl-2, SOD1, and SOD2, consequently decreasing ROS production. Moreover, apoptosis of H9c2 cells decreased and cardiac function of mice improved considerably by targeting the TGF-β2/c-Jun N-terminal kinase (JNK) pathway [101]. Overexpression of miRNA-187 also attenuated the production of ROS by increasing the expression of SOD and reducing the intracellular MDA level, while inhibition of miRNA-187 had an inverse effect under the hypoxia/reoxygenation conditions of cardiomyocytes. This was achieved by the inhibition of DYRK2, a critical regulator of cell cycle and apoptosis, using miRNA-187, which consequently reversed the oxidative stress and apoptosis induced by cardiomyocyte hypoxia/reoxygenation conditions [102]. In addition, upregulation of miR-340-5p, which has an inhibitory effect on apoptosis, could suppress oxidative stress and apoptosis after hypoxia/reoxygenation of H9c2 cardiomyocytes by regulating the expression of Akt1 that upregulates the JNK and NF-κB signaling pathways, the main mediators of cell apoptosis [103]. Similarly, an aging-regulated long non-coding RNA (lncRNA) Sarrah demonstrated a protective effect against cardiomyocyte apoptosis by directly targeting nuclear factor erythroid 2-related factor 2 (NRF2) gene, which regulates the expression of antioxidant proteins that protect against oxidative damage. In an MI mouse model, overexpression of lncRNA Sarrah reduced cardiomyocyte apoptosis, while inducing endothelial cell proliferation and augmenting cardiac contractile function [104]. In contrast, other studies demonstrated that miRNA-124 is upregulated after MI under oxidative stress [105,106]. Application of antisense inhibitor oligodeoxyribonucleotides (AMO-124) alleviated the oxidative stress by neutralizing the miRNA-124, whose main target is STAT3, a key cellular survival factor that suppresses the apoptosis pathway [105]. Overall, gene therapy, specifically the application of various miRNAs, antioxidant gene activators and molecules, whose target is oxidative stress, is a very promising approach to eliminate ROS after MI and hypoxia/reoxygenation conditions and to improve cardiac function.

3.4. Gene Therapy to Reduce Apoptosis

Apoptosis has a significant impact on the regeneration of myocardial cells. Excessive apoptosis following hypoxia, oxidative stress and endoplasmic reticulum stress can contribute to the development of myocardial ischemia, IRI, cardiac remodeling, and atherosclerosis [107]. It also significantly increases the death of cardiomyocytes in ischemic heart disease (IHD) [108]. Therefore, targeting apoptotic agents and pathways can be an efficient strategy to promote cardiac tissue regeneration after CVDs. In the previous chapter, alleviation of oxidative stress using gene therapy to prevent apoptosis was discussed. In this section, other strategies to suppress apoptosis, namely, those that target molecular pathways involved in cellular death and survival will be covered.

Non-coding RNAs (ncRNAs) such as miRNAs, lncRNAs and circular RNAs (circRNAs), play an important role in the regulation of physiologic and pathologic signaling pathways in cardiomyocytes and can be used to regulate apoptosis in CVDs and improve cardiac tissue regeneration [109]. Yan and colleagues demonstrated that in vitro overexpression of miR-31a-5p, a leading member of the miRNA-31 family, can protect against apoptosis and increase myocardial cell survival through suppression of angiotensin II-induced apoptotic pathway and caspase-3 activity by targeting Tp53 [110]. Another miRNAs, miR378 *, demonstrated a cardioprotective effect by inhibiting endoplasmic reticulum stress-induced cell apoptosis via controlling expression of calcium-binding protein called calumenin in doxorubicin (Dox)-induced cardiomyopathic mouse hearts [111]. Moreover, another miRNA, miR-181c, was found to be suppressed in a mouse model of Dox-induced HF and its overexpression impeded cardiomyocyte apoptosis via PI3K/Akt pathway [112]. At the same time, it was reported that lncRNA UCA1 protects rat cardiomyocytes against hypoxia/reoxygenation induced apoptosis by inhibiting miR-143/MDM2/p53 signaling axis [113]. Moreover, adenovirus mediated expression of lncRNA GAS5 also decreases cardiomyocyte apoptosis through downregulation of transmembrane protein sema3a in an MI mouse model [114]. LncRNA FTX is also downregulated after IRI injury and enhanced expression of FTX attenuates cardiomyocyte apoptosis by targeting miR-29b-1-5p and Bcl2l2 [115]. Finally, a circRNA circ-Amotl1 demonstrated its ability to facilitate the cardio-protective nuclear translocation of pAkt by binding to both phosphoinositide dependent kinase-1 (PDK1) and Akt1. Injection of a circ-Amotl1 plasmid resulted in increased cardiomyocyte survival, decreased apoptosis and demonstrated a protective effect against Dox-induced cardiomyopathy in mice [116]. Zhu and colleagues also reported that AAV9 mediated cardiac overexpression of circRNA SNRK can target miR-103-3p by promoting cardiac repair via GSK3β/β-catenin pathway in rats with MI. Particularly, it reduced apoptosis, promoted cardiomyocyte proliferation, improved cardiac functions and increased angiogenesis [117]. Furthermore, Garikipati and colleagues demonstrated that AAV9 mediated delivery of circRNA CircFndc3b decreased cardiac apoptosis, enhanced neovascularization and left ventricle (LV) functions by interacting with the RNA binding protein fused in Sarcoma to regulate expression of VEGFs [118]. Therefore, these findings demonstrate that gene therapy with ncRNAs has great potential as a therapeutic strategy to protect against apoptosis in CVDs. However, further preclinical studies are needed to determine the anti-apoptotic effects of ncRNAs in large animal models.

Another strategy against cardiac apoptosis is to target genes encoding S100 family proteins. S100 proteins are highly acidic calcium-binding proteins involved in intracellular calcium homeostasis in different tissues and organs. Several members of this family are also expressed in cardiac tissue and their impairment can lead to the development of HF [119]. S100A1, a predominant member of the S100 protein family, is a crucial regulator of cardiac contractility. Due to its ability to enhance sarcoplasmic reticulum Ca^{2+} fluxes and increase SERCA2a enzyme activity, it can lead to the significant limitation of myocardial necrosis and development of HF [120]. A recent study conducted by Jungi and colleagues reported that S100A1 gene overexpression through the AAV9 vector can have cardioprotective effects in male Lewis rats after IRI. Particularly, S100A1 overexpressing hearts demonstrated improvement in LV functions, whereas the presence of necrotic markers, including troponin T (TnT), lactate dehydrogenase (LDH), and FoxO pro-apoptotic transcription factor were decreased [121]. Furthermore, a new liquid chromatography tandem mass spectrometry (LC-MS/MS)-based bioanalytical method was utilized for simultaneous quantitation of high homologous human and pig S100A1 proteins (only a single amino acid difference between the sequences). Therefore, the ability to accurately measure human S100A1 in pig hearts can enable future gene therapy studies in large animals [122]. Another member of the S100 family, S100A6, increases in the peri-infarct zone after MI in rats and protects cardiomyocytes from TNF-α-induced apoptosis by binding to p53 and decreasing its phosphorylation [123]. It was recently demonstrated that S100A6 gene delivery and its further cardiac overexpression attenuates cardiomyocyte apoptosis and reduces infarct size after myocardial IRI in both in vitro and in vivo models [124]. Thus, these results demonstrate that gene therapy mediated S100 protein expression can be an efficient method to attenuate cardiac apoptosis and protect cardiomyocytes against IRI injury.

Genetically modified stem cells can also be applied to protect against apoptosis by enhancing expression of various anti-apoptotic genes and factors. Cho and colleagues demonstrated that overexpression of lymphoid enhancer-binding factor-1 (LEF1) gene in human umbilical cord blood-derived mesenchymal stem cells (hUCB-MSCs) increased their cell proliferation and protected from hydrogen peroxide-induced apoptosis in in vitro experiments. Moreover, construction of hUCB-MSCs with overexpressed LEF1 using CRISPR/Cas9 system and further transplantation, demonstrated an enhanced cell survival rate and increased cardio-protective effects in an animal model of MI [125]. In another recent study, interleukin (IL)-10 gene was transfected into the genomic locus of amniotic mesenchymal stem cells (AMM) using transcription activator–like effector nucleases (TALEN). Further transplantation of IL-10 gene-edited AMM in an MI mouse model decreased the number of apoptotic cells and increased capillary density in an ischemic heart and as a result, increased LV functions and reduced infarct size [126]. At the same time, transplantation of bone marrow derived MSCs edited by CRISPR activation system to overexpress IL-10 also decreased apoptosis of cardiac cells, increased angiogenesis, and inhibited infiltration and production of proinflammatory factors in a diabetic MI mouse model [127]. Therefore, the therapeutic effects of genetically modified stem cells can be beneficial against apoptosis caused by cardiac ischemia. In summary, gene therapy with ncRNAs, S100 proteins or modified stem cells can be a potential and efficient strategy against cardiomyocyte death by targeting molecular pathways involved in apoptosis and cell survival in heart diseases.

4. Recent Ongoing Clinical Trials

Currently, several ongoing clinical trials are investigating the effects of various gene therapies on cardiac regeneration in CVDs. According to the ClinicalTrials.gov (accessed on 12 July 2021), there are a total of seven gene therapy trials, including six studies that are actively recruiting (EXACT, ReGenHeart, Korean trial, NAN-CS101) or planning to recruit (AFFIRM, CUPID-3) and one active clinical trial (EPICCURE).

The EXACT Trial is a phase I/II, first-in-human, multicenter, open-label, single arm dose escalation study recruiting 44 patients with refractory angina caused by CAD to evaluate the induction of therapeutic angiogenesis in ischemic myocardium by XC001 (AdVEGFXC1) [NCT04125732]. AdVEGFXC1 is an adenovirus type-5 vector expressing the hybrid VEGF with its three major isoforms (121, 165, and 189) [128]. In this trial, administration is at various doses by transthoracic epicardial procedure to 12 patients, followed by an increase in the maximum tolerated dose with 32 additional patients, and the main outcome measurements will be safety and side effect assessment.

ReGenHeart is a randomized, double-blinded, placebo-controlled multicenter phase II study enrolling patients with refractory angina pectoris to evaluate safety and efficacy of AdVEGF-D gene transfer [NCT03039751]. The trial is recruiting 180 participants in six different centers to whom revascularization cannot be performed. AdVEGF-D will be injected into myocardium through a catheter at 10 different sites and compared with a similar placebo treatment. The primary endpoint is to assess improvement in exercise tolerance and relief of angina symptoms at six months after injection.

EPICCURE is a randomized, placebo-controlled, double-blind, multicenter, phase II clinical trial testing the effect of VEGF-A165 mRNA loaded in biocompatible citrate-buffered saline (AZD8601) in patients with decreased LV function and undergoing CABG [NCT03370887] [129]. AZD8601 will be injected during the surgery as 30 epicardial injections in a 10-min extension of cardioplegia under the control of quantitative 15O-water positron emission tomography (PET) imaging. At the moment, 11 patients are enrolled in the study, and the primary endpoint is to investigate safety and tolerability of the gene therapy up to six months after the surgery. MRNA-based technology used in the EPICCURE trial is a novel revolutionizing approach for gene delivery that can be utilized for a variety of purposes such as protein replacement, gene editing, infectious diseases, cancer vaccines, and others [130]. There are multiple advantages of using an mRNA approach over conventional DNA-based strategies [131]. Firstly, unlike DNA, mRNA does not require entering the nucleus to be functional, making it more efficient. Secondly, mRNA-based strategy is much safer compared to DNA delivery with viral vectors since mRNA does not integrate into the host genome, eliminating the risk of insertional mutagenesis. Finally, as it has been proven by SARS-CoV-2 mRNA-based vaccines, mRNA production can be very rapid and capable of adapting to even high emergency conditions like pandemics. Indeed, two SARS-CoV-2 mRNA-based vaccines approved by the US Food and Drug Administration, Pfizer/BioNTech (BNT162b2), and Moderna (mRNA-1273), have been shown to be highly efficient in protecting against COVID-19 with 90–95% efficiency [132]. Moreover, a recent study published in *Nature* found that the two mRNA vaccines induced a persistent and robust germinal center response indicating that these vaccines produced strong and efficient humoral immune responses [133]. The benefits of mRNA-based technology have been utilized for cardiovascular gene therapy as well [134]. Specifically, mRNA therapeutics using genes for VEGF, IGF-1, and other proteins for heart regeneration were shown to be safe and effective in various animal models.

The AFFIRM study is a phase III clinical trial designed to determine the effect of intracoronary infusion of the human FGF-4 DNA sequence encoded in Ad5FGF-4 on ameliorating refractory angina symptoms and improving patient quality of life. This study will enroll 160 patients with refractory angina and the primary endpoint is the change of exercise tolerance test (ETT) duration over six months after intervention (NCT02928094). However, previous clinical trials with Ad5FGF-4 have demonstrated specific beneficial effects of this gene therapy only in females, while no difference was observed between placebo and treatment groups among male patients [68].

A phase II trial is ongoing in Korea to test the safety and efficacy of VM202RY, which is a DNA plasmid loaded with two isoforms of HGFs (723 and 728) (NCT03404024). A previous phase I trial demonstrated that VM202RY can improve cardiac perfusion and inhibit cardiac remodeling in patients with MI and angina [135]. Now they are recruiting 108 patients with acute MI for transendocardial injection of VM202RY through catheter to evaluate its safety, tolerability, and efficacy compared with placebo. Primary outcome measurements are to determine the maximum tolerated dose and assess LV ejection fraction after six months. Estimated study completion date was April, 2020, however, no results have yet been reported in English.

The CUPID-3 study is a randomized, double-blind, placebo-controlled phase I/II trial investigating the effect of SRD-001, an adeno-associated virus serotype 1 vector expressing the transgene for sarcoplasmic/endoplasmic reticulum Ca^{2+} ATPase 2a isoform (SERCA2a), on patients with cardiomyopathy and HF with reduced ejection fraction (NCT04703842). According to the authors, targeting SERCA2a enzyme expression by SRD-001 can correct defective intracellular Ca^{2+} hemostasis, subsequently improving cardiac contractility. In addition, SRD-001 can also correct the impaired endothelium-dependent vasodilatation, improve coronary blood flow and prevent further progression of HF. To determine this, 56 participants will be enrolled in the study until June 2021 and SRD-001 will be delivered through one-time intracoronary infusion. As a primary endpoint, change from baseline in symptomatic parameters (quality of life, HF severity), physical parameters, LV function/remodeling, and rate of recurrent and adverse events will be measured up to 12 months (NCT04703842).

NAN-CS101 is a phase I, prospective, multi-center, open-label, sequential dose escalation study to test the safety and efficacy of one dose intracoronary infusion of BNP116.sc-CMV.I1c in patients with HF (NYHA class III) (NCT04179643). BNP116.sc-CMV.I1c is an AAV vector with a chimeric AAV2/AAV8 capsid containing an active inhibitor of protein phosphatase-1 (I-1) transgene. Previous preclinical studies on large animals demonstrated that chimeric AAV vectors can selectively target cardiac muscle, while I-1 inhibits expression of protein phosphatase-1 and increases cardiac expression of SERCA2a enzyme leading to the improvement of cardiac functions [136,137]. In this small phase I study, 12 participants will be enrolled and injected intracoronary with different doses of BNP116.sc-CMV.I1. Primary endpoints will be the safety indicators measured by adverse events, mortality, and hospitalization up to 12 months after the procedure.

Lastly, another randomized, double-blind, placebo-controlled, phase II clinical trial has tested the effect of intracoronary delivery of adenovirus 5 encoding adenylyl cyclase 6 (Ad5.hAC6/RT100) in patients with HF. Adenylyl cyclase 6 (AC) is an enzyme that catalyzes the conversion of adenosine triphosphate to cyclic adenosine monophosphate and can have beneficial effects on cardiac myocytes and eventually improving LV function. Fifty-six subjects received intracoronary administration of Ad5.hAC6 or placebo and were followed up to 1 year. AC6 gene transfer safely increased LV function compared to a standard single-dose HF therapy and decreased admission rate in the treatment group [138]. A larger phase III trial (FLOURISH) with the same drug was recently registered in clinicaltrials.gov (accessed on 12 July 2021); however, the current status of the study is withdrawn due to the reevaluation of "clinical development plans and strategy for RT-100" (NCT03360448). Therefore, despite its withdrawn status, a newly designed trial with Ad5.hAC6/RT100 can be conducted in the future. Table 1 summarizes the ongoing clinical trials.

Overall, most ongoing trials related to cardiac regeneration are focused on gene therapy targeting the expression of different growth factors and their isoforms to improve angiogenesis. This might be justified by the fact that extended angiogenesis effectively improves vascular and cardiac tissue regeneration leading to the enhancement of cardiac functions. However, there are still two gene therapy studies targeting the enzyme expression (CUPID-3 and NAN-CS101). Thus, this data demonstrates that gene therapy can be an efficient and safe approach to improve cardiac regeneration and further implementation of this strategy into clinical trials has the potential to develop into novel methods of treatment for CVDs.

Table 1. Ongoing clinical trials in gene therapy for cardiac tissue regeneration in CVDs.

#	Study Title	Disease	Treatment Mechanism (Intervention)	Estimated Enrollment	Current Status and Phase	Trial Number
1	Epicardial Delivery of XC001 Gene Therapy for Refractory Angina Coronary Treatment (EXACT)	Coronary Artery Disease, Ischemia, Angina Refractory, Cardiovascular Diseases, Heart Diseases	Subjects in this study will receive one of four intramyocardial doses of XC001 that expresses human VEGF	44 participants	Recruiting, Phase I/II	NCT04125732
2	Adenovirus Vascular Endothelial Growth Factor D (AdvVEGF-D) Therapy for Treatment of Refractory Angina Pectoris (ReGenHeart)	Refractory Angina Pectoris, Coronary Artery Disease	A catheter mediated endocardial adenovirus-mediated VEGF-D (AdVEGF-D) regenerative gene transfer in patients with refractory angina to whom revascularization cannot be performed	180 participants	Recruiting, Phase II	NCT03039751
3	AZD8601 Study in CABG Patients (EPICCURE)	Heart Failure	Epicardial injections of VEGF-A165 mRNA formulated in biocompatible citrate-buffered saline (AZD8601) in a 10-min extension of cardioplegia during the CABG surgery under PET imaging	11 participants	Active, not recruiting, Phase II	NCT03370887
4	Ad5FGF-4 In Patients with Refractory Angina Due to Myocardial Ischemia (AFFIRM)	Angina, Stable	A single intracoronary infusion of an adenovirus serotype 5 virus expressing the gene for human FGF-4 (Ad5FGF-4)	160 participants	Not yet recruiting, Phase III	NCT02928094
5	Safety and Efficacy Study of Gene Therapy for Acute Myocardial Infarction	Ischemic Heart Disease, Acute Myocardial Infarction	VM202RY injected via transendocardial route using C-Cathez® catheter in subjects with acute MI	108 participants	Recruiting, Phase II	NCT03404024
6	Calcium Up-Regulation by Percutaneous Administration of Gene Therapy in Cardiac Disease (CUPID-3)	Congestive Heart Failure, Systolic Heart Failure	A single intracoronary infusion of adeno-associated virus serotype 1 vector expressing the transgene for sarco(endo)plasmic reticulum Ca^{2+} ATPase 2a isoform (SRD-001)	56 participants	Not yet recruiting, Phase I/II	NCT04703842
7	NAN-101 in Patients with Class III Heart Failure (NAN-CS101)	Heart Failure, Heart Disease, Ischemic Cardiovascular Diseases, Heart Arrhythmia	A single intracoronary infusion of chimeric adeno-associated virus serotype 2/8 with an active I-1 transgene	12 participants	Recruiting, Phase I	NCT04179643

5. Conclusions

The substantial health and economic burden of CVDs could be alleviated in the future with the advancement of novel therapeutic strategies that target pathogenetic mechanisms of the diseases. Gene therapy has been considered as a promising treatment option for a variety of CVDs. As it was reviewed in this paper, multiple preclinical studies have succeeded in reversing the outcomes of ischemic heart disease and MI by targeting angiogenesis, stimulating regeneration of cardiomyocytes, and alleviating oxidative stress and apoptosis. However, it will likely take many years for the translation of these beneficial effects from animal models to human patients. Many clinical trials of gene replacement therapy did not achieve efficiency, which might be associated with concentration of the replaced gene and its product at the target site and/or insufficient knowledge about pathogenetic mechanisms of CVDs. Moreover, safety considerations including potential off-target effects, tumorigenesis of certain viral vectors and arrhythmogenesis should be addressed before implementation of gene therapy in clinical practice. In addition, it is important to mention that individual use of the strategies described in this paper will most likely be insufficient in achieving complete heart regeneration after injury in humans. The main reason for this is that the pathways activated during cardiac injury are diverse and complex involving multiple cells and signaling cascades. Therefore, an efficient gene therapy for clinical use will likely combine several approaches described in this review—including but not limited to—strategies targeting fibrosis, oxidative stress, apoptosis, angiogenesis, and other processes.

Author Contributions: Conceptualization and editing, A.S.; Writing—original draft preparation, Y.K., Z.Z., M.S., and G.Y. All authors have read and agreed to the published version of the manuscript.

Funding: This research was funded by a Collaborative Research grant from Nazarbayev University (021220CRP0722).

Conflicts of Interest: The authors declare no conflict of interest.

References

1. Roth, G.A.; Mensah, G.A.; Johnson, C.O.; Addolorato, G.; Ammirati, E.; Baddour, L.M.; Barengo, N.C.; Beaton, A.Z.; Benjamin, E.J.; Benziger, C.P. Global burden of cardiovascular diseases and risk factors, 1990–2019: Update from the GBD 2019 study. *J. Am. Coll. Cardiol.* **2020**, *76*, 2982–3021. [CrossRef] [PubMed]
2. Wilkins, E.; Wilson, L.; Wickramasinghe, K.; Bhatnagar, P.; Leal, J.; Luengo-Fernandez, R.; Burns, R.; Rayner, M.; Townsend, N. *European Cardiovascular Disease Statistics 2017*; University of Bath: Bath, UK, 2017.
3. Giedrimiene, D.; King, R. Burden of cardiovascular disease (CVD) on economic cost. Comparison of outcomes in US and Europe. *Circ. Cardiovasc. Qual. Outcomes* **2017**, *10*, A207. [CrossRef]
4. Lähteenvuo, J.; Ylä-Herttuala, S. Advances and Challenges in Cardiovascular Gene Therapy. *Hum. Gene Ther.* **2017**, *28*, 1024–1032. [CrossRef] [PubMed]
5. Braunwald, E. Cell-based therapy in cardiac regeneration: An overview. *Circ. Res.* **2018**, *123*, 132–137. [CrossRef]
6. Liew, L.C.; Ho, B.X.; Soh, B.S. Mending a broken heart: Current strategies and limitations of cell-based therapy. *Stem. Cell. Res. Ther.* **2020**, *11*, 138. [CrossRef]
7. Smagul, S.; Kim, Y.; Smagulova, A.; Raziyeva, K.; Nurkesh, A.; Saparov, A. Biomaterials Loaded with Growth Factors/Cytokines and Stem Cells for Cardiac Tissue Regeneration. *Int. J. Mol. Sci.* **2020**, *21*, 5952. [CrossRef]
8. Raziyeva, K.; Smagulova, A.; Kim, Y.; Smagul, S.; Nurkesh, A.; Saparov, A. Preconditioned and Genetically Modified Stem Cells for Myocardial Infarction Treatment. *Int. J. Mol. Sci.* **2020**, *21*, 7301. [CrossRef]
9. Ylä-Herttuala, S.; Baker, A.H. Cardiovascular Gene Therapy: Past, Present, and Future. *Mol. Ther.* **2017**, *25*, 1095–1106. [CrossRef]
10. Ishikawa, K.; Weber, T.; Hajjar, R.J. Human Cardiac Gene Therapy. *Circ. Res.* **2018**, *123*, 601–613. [CrossRef]
11. Shimamura, M.; Nakagami, H.; Sanada, F.; Morishita, R. Progress of Gene Therapy in Cardiovascular Disease. *Hypertension* **2020**, *76*, 1038–1044. [CrossRef]
12. Kieserman, J.M.; Myers, V.D.; Dubey, P.; Cheung, J.Y.; Feldman, A.M. Current Landscape of Heart Failure Gene Therapy. *J. Am. Heart Assoc.* **2019**, *8*, e012239. [CrossRef]
13. Bezzerides, V.J.; Prondzynski, M.; Carrier, L.; Pu, W.T. Gene therapy for inherited arrhythmias. *Cardiovasc Res.* **2020**, *116*, 1635–1650. [CrossRef]
14. Lázár, E.; Sadek, H.A.; Bergmann, O. Cardiomyocyte renewal in the human heart: Insights from the fall-out. *Eur. Heart. J.* **2017**, *38*, 2333–2342. [CrossRef]

15. Bergmann, O.; Bhardwaj, R.D.; Bernard, S.; Zdunek, S.; Barnabé-Heider, F.; Walsh, S.; Zupicich, J.; Alkass, K.; Buchholz, B.A.; Druid, H.; et al. Evidence for cardiomyocyte renewal in humans. *Science* **2009**, *324*, 98–102. [CrossRef]

16. Bergmann, O.; Zdunek, S.; Felker, A.; Salehpour, M.; Alkass, K.; Bernard, S.; Sjostrom, S.L.; Szewczykowska, M.; Jackowska, T.; Dos Remedios, C.; et al. Dynamics of Cell Generation and Turnover in the Human Heart. *Cell* **2015**, *161*, 1566–1575. [CrossRef]

17. Eschenhagen, T.; Bolli, R.; Braun, T.; Field, L.J.; Fleischmann, B.K.; Frisén, J.; Giacca, M.; Hare, J.M.; Houser, S.; Lee, R.T. Cardiomyocyte regeneration: A consensus statement. *Circulation* **2017**, *136*, 680–686. [CrossRef]

18. Verjans, R.; van Bilsen, M.; Schroen, B. Reviewing the Limitations of Adult Mammalian Cardiac Regeneration: Noncoding RNAs as Regulators of Cardiomyogenesis. *Biomolecules* **2020**, *10*, 262. [CrossRef]

19. Lam, N.T.; Sadek, H.A. Neonatal Heart Regeneration: Comprehensive Literature Review. *Circulation* **2018**, *138*, 412–423. [CrossRef]

20. Ahmed, A.; Wang, T.; Delgado-Olguin, P. Ezh2 is not required for cardiac regeneration in neonatal mice. *PLoS ONE* **2018**, *13*, e0192238. [CrossRef]

21. Bassat, E.; Mutlak, Y.E.; Genzelinakh, A.; Shadrin, I.Y.; Baruch Umansky, K.; Yifa, O.; Kain, D.; Rajchman, D.; Leach, J.; Riabov Bassat, D.; et al. The extracellular matrix protein agrin promotes heart regeneration in mice. *Nature* **2017**, *547*, 179–184. [CrossRef]

22. Ingason, A.B.; Goldstone, A.B.; Paulsen, M.J.; Thakore, A.D.; Truong, V.N.; Edwards, B.B.; Eskandari, A.; Bollig, T.; Steele, A.N.; Woo, Y.J. Angiogenesis precedes cardiomyocyte migration in regenerating mammalian hearts. *J. Thorac. Cardiovasc. Surg.* **2018**, *155*, 1118–1127.e1111. [CrossRef] [PubMed]

23. Sampaio-Pinto, V.; Rodrigues, S.C.; Laundos, T.L.; Silva, E.D.; Vasques-Nóvoa, F.; Silva, A.C.; Cerqueira, R.J.; Resende, T.P.; Pianca, N.; Leite-Moreira, A.; et al. Neonatal Apex Resection Triggers Cardiomyocyte Proliferation, Neovascularization and Functional Recovery Despite Local Fibrosis. *Stem Cell Rep.* **2018**, *10*, 860–874. [CrossRef] [PubMed]

24. Zhu, W.; Zhang, E.; Zhao, M.; Chong, Z.; Fan, C.; Tang, Y.; Hunter, J.D.; Borovjagin, A.V.; Walcott, G.P.; Chen, J.Y.; et al. Regenerative Potential of Neonatal Porcine Hearts. *Circulation* **2018**, *138*, 2809–2816. [CrossRef] [PubMed]

25. Ye, L.; D'Agostino, G.; Loo, S.J.; Wang, C.X.; Su, L.P.; Tan, S.H.; Tee, G.Z.; Pua, C.J.; Pena, E.M.; Cheng, R.B.; et al. Early Regenerative Capacity in the Porcine Heart. *Circulation* **2018**, *138*, 2798–2808. [CrossRef] [PubMed]

26. Wang, H.; Paulsen, M.J.; Hironaka, C.E.; Shin, H.S.; Farry, J.M.; Thakore, A.D.; Jung, J.; Lucian, H.J.; Eskandari, A.; Anilkumar, S.; et al. Natural Heart Regeneration in a Neonatal Rat Myocardial Infarction Model. *Cells* **2020**, *9*, 229. [CrossRef] [PubMed]

27. Notari, M.; Ventura-Rubio, A.; Bedford-Guaus, S.J.; Jorba, I.; Mulero, L.; Navajas, D.; Martí, M.; Raya, Á. The local microenvironment limits the regenerative potential of the mouse neonatal heart. *Sci. Adv.* **2018**, *4*, eaao5553. [CrossRef]

28. Wang, Z.; Cui, M.; Shah, A.M.; Ye, W.; Tan, W.; Min, Y.L.; Botten, G.A.; Shelton, J.M.; Liu, N.; Bassel-Duby, R.; et al. Mechanistic basis of neonatal heart regeneration revealed by transcriptome and histone modification profiling. *Proc. Natl. Acad. Sci. USA* **2019**, *116*, 18455–18465. [CrossRef]

29. Wang, Z.; Cui, M.; Shah, A.M.; Tan, W.; Liu, N.; Bassel-Duby, R.; Olson, E.N. Cell-Type-Specific Gene Regulatory Networks Underlying Murine Neonatal Heart Regeneration at Single-Cell Resolution. *Cell Rep.* **2020**, *33*, 108472. [CrossRef]

30. Heallen, T.R.; Kadow, Z.A.; Kim, J.H.; Wang, J.; Martin, J.F. Stimulating Cardiogenesis as a Treatment for Heart Failure. *Circ. Res.* **2019**, *124*, 1647–1657. [CrossRef]

31. Zhao, M.T.; Ye, S.; Su, J.; Garg, V. Cardiomyocyte Proliferation and Maturation: Two Sides of the Same Coin for Heart Regeneration. *Front. Cell Dev. Biol.* **2020**, *8*, 594226. [CrossRef]

32. Mohamed, T.M.A.; Ang, Y.S.; Radzinsky, E.; Zhou, P.; Huang, Y.; Elfenbein, A.; Foley, A.; Magnitsky, S.; Srivastava, D. Regulation of Cell Cycle to Stimulate Adult Cardiomyocyte Proliferation and Cardiac Regeneration. *Cell* **2018**, *173*, 104–116.e112. [CrossRef] [PubMed]

33. Diez-Cuñado, M.; Wei, K.; Bushway, P.J.; Maurya, M.R.; Perera, R.; Subramaniam, S.; Ruiz-Lozano, P.; Mercola, M. miRNAs that Induce Human Cardiomyocyte Proliferation Converge on the Hippo Pathway. *Cell Rep.* **2018**, *23*, 2168–2174. [CrossRef] [PubMed]

34. Monroe, T.O.; Hill, M.C.; Morikawa, Y.; Leach, J.P.; Heallen, T.; Cao, S.; Krijger, P.H.L.; de Laat, W.; Wehrens, X.H.T.; Rodney, G.G.; et al. YAP Partially Reprograms Chromatin Accessibility to Directly Induce Adult Cardiogenesis In Vivo. *Dev. Cell* **2019**, *48*, 765–779.e767. [CrossRef] [PubMed]

35. Ali, H.; Braga, L.; Giacca, M. Cardiac regeneration and remodelling of the cardiomyocyte cytoarchitecture. *FEBS J.* **2020**, *287*, 417–438. [CrossRef]

36. Hashimoto, H.; Olson, E.N.; Bassel-Duby, R. Therapeutic approaches for cardiac regeneration and repair. *Nat. Rev. Cardiol.* **2018**, *15*, 585–600. [CrossRef] [PubMed]

37. Huntington, J.; Pachauri, M.; Ali, H.; Giacca, M. RNA interference therapeutics for cardiac regeneration. *Curr. Opin. Genet. Dev.* **2021**, *70*, 48–53. [CrossRef] [PubMed]

38. Shiba, Y. Pluripotent Stem Cells for Cardiac Regeneration-Current Status, Challenges, and Future Perspectives. *Circ. J.* **2020**, *84*, 2129–2135. [CrossRef]

39. Saparov, A.; Chen, C.W.; Beckman, S.A.; Wang, Y.; Huard, J. The role of antioxidation and immunomodulation in postnatal multipotent stem cell-mediated cardiac repair. *Int. J. Mol. Sci.* **2013**, *14*, 16258–16279. [CrossRef]

40. Madeddu, P. Cell therapy for the treatment of heart disease: Renovation work on the broken heart is still in progress. *Free Radic. Biol. Med.* **2021**, *164*, 206–222. [CrossRef]

41. Tang, J.N.; Cores, J.; Huang, K.; Cui, X.L.; Luo, L.; Zhang, J.Y.; Li, T.S.; Qian, L.; Cheng, K. Concise Review: Is Cardiac Cell Therapy Dead? Embarrassing Trial Outcomes and New Directions for the Future. *Stem Cells Transl. Med.* **2018**, *7*, 354–359. [CrossRef]

42. Chen, Z.; Chen, L.; Zeng, C.; Wang, W.E. Functionally improved mesenchymal stem cells to better treat myocardial infarction. *Stem Cells Int.* **2018**, *2018*, 7045245. [CrossRef]

43. Baldari, S.; Di Rocco, G.; Piccoli, M.; Pozzobon, M.; Muraca, M.; Toietta, G. Challenges and Strategies for Improving the Regenerative Effects of Mesenchymal Stromal Cell-Based Therapies. *Int. J. Mol. Sci.* **2017**, *18*, 2087. [CrossRef]

44. Zhao, Y.; Zhang, M.; Lu, G.L.; Huang, B.X.; Wang, D.W.; Shao, Y.; Lu, M.J. Hypoxic Preconditioning Enhances Cellular Viability and Pro-angiogenic Paracrine Activity: The Roles of VEGF-A and SDF-1a in Rat Adipose Stem Cells. *Front. Cell Dev. Biol.* **2020**, *8*, 580131. [CrossRef]

45. Ferreira, J.R.; Teixeira, G.Q.; Santos, S.G.; Barbosa, M.A.; Almeida-Porada, G.; Gonçalves, R.M. Mesenchymal Stromal Cell Secretome: Influencing Therapeutic Potential by Cellular Pre-conditioning. *Front. Immunol.* **2018**, *9*, 2837. [CrossRef]

46. Mansurov, N.; Chen, W.C.; Awada, H.; Huard, J.; Wang, Y.; Saparov, A. A controlled release system for simultaneous delivery of three human perivascular stem cell-derived factors for tissue repair and regeneration. *J. Tissue Eng. Regener. Med.* **2018**, *12*, e1164–e1172. [CrossRef]

47. Sultankulov, B.; Berillo, D.; Kauanova, S.; Mikhalovsky, S.; Mikhalovska, L.; Saparov, A. Composite Cryogel with Polyelectrolyte Complexes for Growth Factor Delivery. *Pharmaceutics* **2019**, *11*, 650. [CrossRef]

48. Jimi, S.; Jaguparov, A.; Nurkesh, A.; Sultankulov, B.; Saparov, A. Sequential Delivery of Cryogel Released Growth Factors and Cytokines Accelerates Wound Healing and Improves Tissue Regeneration. *Front. Bioeng. Biotechnol.* **2020**, *8*, 345. [CrossRef]

49. Nurkesh, A.; Jaguparov, A.; Jimi, S.; Saparov, A. Recent Advances in the Controlled Release of Growth Factors and Cytokines for Improving Cutaneous Wound Healing. *Front. Cell Dev. Biol.* **2020**, *8*, 638. [CrossRef]

50. Chingale, M.; Zhu, D.; Cheng, K.; Huang, K. Bioengineering Technologies for Cardiac Regenerative Medicine. *Front. Bioeng. Biotechnol.* **2021**, *9*, 361. [CrossRef]

51. Melhem, M.R.; Park, J.; Knapp, L.; Reinkensmeyer, L.; Cvetkovic, C.; Flewellyn, J.; Lee, M.K.; Jensen, T.W.; Bashir, R.; Kong, H.; et al. 3D Printed Stem-Cell-Laden, Microchanneled Hydrogel Patch for the Enhanced Release of Cell-Secreting Factors and Treatment of Myocardial Infarctions. *ACS Biomater. Sci. Eng.* **2017**, *3*, 1980–1987. [CrossRef]

52. Cannatà, A.; Ali, H.; Sinagra, G.; Giacca, M. Gene Therapy for the Heart Lessons Learned and Future Perspectives. *Circ. Res.* **2020**, *126*, 1394–1414. [CrossRef] [PubMed]

53. Ylä-Herttuala, S.; Bridges, C.; Katz, M.G.; Korpisalo, P. Angiogenic gene therapy in cardiovascular diseases: Dream or vision? *Eur. Heart J.* **2017**, *38*, 1365–1371. [CrossRef] [PubMed]

54. Greenberg, B. Gene therapy for heart failure. *Trends Cardiovasc. Med.* **2017**, *27*, 216–222. [CrossRef] [PubMed]

55. Hassinen, I.; Kivelä, A.; Hedman, A.; Saraste, A.; Knuuti, J.; Hartikainen, J.; Ylä-Herttuala, S. Intramyocardial Gene Therapy Directed to Hibernating Heart Muscle Using a Combination of Electromechanical Mapping and Positron Emission Tomography. *Hum. Gene Ther.* **2016**, *27*, 830–834. [CrossRef]

56. Braile, M.; Marcella, S.; Cristinziano, L.; Galdiero, M.R.; Modestino, L.; Ferrara, A.L.; Varricchi, G.; Marone, G.; Loffredo, S. VEGF-A in Cardiomyocytes and Heart Diseases. *Int. J. Mol. Sci.* **2020**, *21*, 5294. [CrossRef]

57. Korpela, H.; Järveläinen, N.; Siimes, S.; Lampela, J.; Airaksinen, J.; Valli, K.; Turunen, M.; Pajula, J.; Nurro, J.; Ylä-Herttuala, S. Gene therapy for ischaemic heart disease and heart failure. *J. Intern. Med.* **2021**, in press. [CrossRef]

58. Apte, R.S.; Chen, D.S.; Ferrara, N. VEGF in Signaling and Disease: Beyond Discovery and Development. *Cell* **2019**, *176*, 1248–1264. [CrossRef]

59. Mangialardi, G.; Madeddu, P. New vascular endothelial growth factor isoforms promise to increase myocardial perfusion without stealing. *Heart* **2016**, *102*, 1697–1698. [CrossRef]

60. Nurro, J.; Halonen, P.J.; Kuivanen, A.; Tarkia, M.; Saraste, A.; Honkonen, K.; Lähteenvuo, J.; Rissanen, T.T.; Knuuti, J.; Ylä-Herttuala, S. AdVEGF-B186 and AdVEGF-DΔNΔC induce angiogenesis and increase perfusion in porcine myocardium. *Heart* **2016**, *102*, 1716–1720. [CrossRef]

61. Huusko, J.; Lottonen, L.; Merentie, M.; Gurzeler, E.; Anisimov, A.; Miyanohara, A.; Alitalo, K.; Tavi, P.; Ylä-Herttuala, S. AAV9-mediated VEGF-B gene transfer improves systolic function in progressive left ventricular hypertrophy. *Mol. Ther.* **2012**, *20*, 2212–2221. [CrossRef]

62. Makarevich, P.I.; Dergilev, K.V.; Tsokolaeva, Z.I.; Boldyreva, M.A.; Shevchenko, E.K.; Gluhanyuk, E.V.; Gallinger, J.O.; Menshikov, M.Y.; Parfyonova, Y.V. Angiogenic and pleiotropic effects of VEGF165 and HGF combined gene therapy in a rat model of myocardial infarction. *PLoS ONE* **2018**, *13*, e0197566. [CrossRef]

63. Hassinen, I.; Hartikainen, J.; Hedman, A.; Kivelä, A.; Saraste, A.; Knuuti, J.; Husso, M.; Mussalo, H.; Hedman, M.; Toivanen, P. Adenoviral intramyocardial VEGF-D gene transfer increases myocardial perfusion in refractory angina patients. *Circulation* **2015**, *132*, A11987. [CrossRef]

64. Hartikainen, J.; Hassinen, I.; Hedman, A.; Kivelä, A.; Saraste, A.; Knuuti, J.; Husso, M.; Mussalo, H.; Hedman, M.; Rissanen, T.T. Adenoviral intramyocardial VEGF-DΔNΔC gene transfer increases myocardial perfusion reserve in refractory angina patients: A phase I/IIa study with 1-year follow-up. *Eur. Heart J.* **2017**, *38*, 2547–2555. [CrossRef]

65. Itoh, N.; Ohta, H.; Nakayama, Y.; Konishi, M. Roles of FGF signals in heart development, health, and disease. *Front. Cell Dev. Biol.* **2016**, *4*, 110. [CrossRef]

66. Khosravi, F.; Ahmadvand, N.; Bellusci, S.; Sauer, H. The Multifunctional Contribution of FGF Signaling to Cardiac Development, Homeostasis, Disease and Repair. *Front. Cell Dev. Biol.* **2021**, *9*, 1217. [CrossRef]

67. Xie, Y.; Su, N.; Yang, J.; Tan, Q.; Huang, S.; Jin, M.; Ni, Z.; Zhang, B.; Zhang, D.; Luo, F.; et al. FGF/FGFR signaling in health and disease. *Signal Transduct. Target. Ther.* **2020**, *5*, 181. [CrossRef]
68. Henry, T.D.; Grines, C.L.; Watkins, M.W.; Dib, N.; Barbeau, G.; Moreadith, R.; Andrasfay, T.; Engler, R.L. Effects of Ad5FGF-4 in patients with angina: An analysis of pooled data from the AGENT-3 and AGENT-4 trials. *J. Am. Coll. Cardiol.* **2007**, *50*, 1038–1046. [CrossRef]
69. Kaski, J.C.; Consuegra-Sanchez, L. Evaluation of ASPIRE trial: A Phase III pivotal registration trial, using intracoronary administration of Generx (Ad5FGF4) to treat patients with recurrent angina pectoris. *Expert Opin. Biol. Ther.* **2013**, *13*, 1749–1753. [CrossRef]
70. Wang, L.S.; Wang, H.; Zhang, Q.L.; Yang, Z.J.; Kong, F.X.; Wu, C.T. Hepatocyte Growth Factor Gene Therapy for Ischemic Diseases. *Hum. Gene Ther.* **2018**, *29*, 413–423. [CrossRef]
71. Yuan, B.; Zhao, Z.; Zhang, Y.R.; Wu, C.T.; Jin, W.G.; Zhao, S.; Wangl, W.; Zhang, Y.Y.; Zhu, X.L.; Wang, L.S.; et al. Short-term safety and curative effect of recombinant adenovirus carrying hepatocyte growth factor gene on ischemic cardiac disease. *In Vivo* **2008**, *22*, 629–632.
72. Chen, B.; Tao, Z.; Zhao, Y.; Chen, H.; Yong, Y.; Liu, X.; Wang, H.; Wu, Z.; Yang, Z.; Yuan, L. Catheter-based intramyocardial delivery (NavX) of adenovirus achieves safe and accurate gene transfer in pigs. *PLoS ONE* **2013**, *8*, e53007. [CrossRef]
73. Rong, S.L.; Wang, X.L.; Zhang, C.Y.; Song, Z.H.; Cui, L.H.; He, X.F.; Li, X.J.; Du, H.J.; Li, B. Transplantation of HGF gene-engineered skeletal myoblasts improve infarction recovery in a rat myocardial ischemia model. *PLoS ONE* **2017**, *12*, e0175807. [CrossRef]
74. Sun, Z.; Lee, C.J.; Mejia-Guerrero, S.; Zhang, Y.; Higuchi, K.; Li, R.K.; Medin, J.A. Neonatal transfer of membrane-bound stem cell factor improves survival and heart function in aged mice after myocardial ischemia. *Hum. Gene Ther.* **2012**, *23*, 1280–1289. [CrossRef]
75. Ishikawa, K.; Fish, K.; Aguero, J.; Yaniz-Galende, E.; Jeong, D.; Kho, C.; Tilemann, L.; Fish, L.; Liang, L.; Eltoukhy, A.A.; et al. Stem cell factor gene transfer improves cardiac function after myocardial infarction in swine. *Circ. Heart Fail.* **2015**, *8*, 167–174. [CrossRef]
76. Frangogiannis, N.G. Cardiac fibrosis: Cell biological mechanisms, molecular pathways and therapeutic opportunities. *Mol. Asp. Med.* **2019**, *65*, 70–99. [CrossRef]
77. Hinderer, S.; Schenke-Layland, K. Cardiac fibrosis—A short review of causes and therapeutic strategies. *Adv. Drug Deliv. Rev.* **2019**, *146*, 77–82. [CrossRef]
78. Bektik, E.; Fu, J.D. Ameliorating the Fibrotic Remodeling of the Heart through Direct Cardiac Reprogramming. *Cells* **2019**, *8*, 679. [CrossRef] [PubMed]
79. Adams, E.; McCloy, R.; Jordan, A.; Falconer, K.; Dykes, I.M. Direct Reprogramming of Cardiac Fibroblasts to Repair the Injured Heart. *J. Cardiovasc. Dev. Dis.* **2021**, *8*, 72. [CrossRef] [PubMed]
80. Ghiroldi, A.; Piccoli, M.; Ciconte, G.; Pappone, C.; Anastasia, L. Regenerating the human heart: Direct reprogramming strategies and their current limitations. *Basic Res. Cardiol.* **2017**, *112*, 68. [CrossRef]
81. Lee, S.; Park, B.-W.; Lee, Y.J.; Ban, K.; Park, H.-J. In vivo combinatory gene therapy synergistically promotes cardiac function and vascular regeneration following myocardial infarction. *J. Tissue Eng.* **2020**, *11*, 2041731420953413. [CrossRef]
82. Lee, D.H.; Kim, T.M.; Kim, J.K.; Park, C. ETV2/ER71 Transcription Factor as a Therapeutic Vehicle for Cardiovascular Disease. *Theranostics* **2019**, *9*, 5694–5705. [CrossRef]
83. Chang, Y.; Lee, E.; Kim, J.; Kwon, Y.W.; Kwon, Y.; Kim, J. Efficient in vivo direct conversion of fibroblasts into cardiomyocytes using a nanoparticle-based gene carrier. *Biomaterials* **2019**, *192*, 500–509. [CrossRef]
84. Isomi, M.; Sadahiro, T.; Fujita, R.; Abe, Y.; Yamada, Y.; Akiyama, T.; Mizukami, H.; Shu, T.; Fukuda, K.; Ieda, M. Direct reprogramming with Sendai virus vectors repaired infarct hearts at the chronic stage. *Biochem. Biophys. Res. Commun.* **2021**, *560*, 87–92. [CrossRef]
85. Adapala, R.K.; Kanugula, A.K.; Paruchuri, S.; Chilian, W.M.; Thodeti, C.K. TRPV4 deletion protects heart from myocardial infarction-induced adverse remodeling via modulation of cardiac fibroblast differentiation. *Basic Res. Cardiol.* **2020**, *115*, 14. [CrossRef]
86. Anger, M.; Scheufele, F.; Ramanujam, D.; Meyer, K.; Nakajima, H.; Field, L.J.; Engelhardt, S.; Sarikas, A. Genetic ablation of Cullin-RING E3 ubiquitin ligase 7 restrains pressure overload-induced myocardial fibrosis. *PLoS ONE* **2020**, *15*, e0244096. [CrossRef]
87. Wu, Y.; Zhou, L.; Liu, H.; Duan, R.; Zhou, H.; Zhang, F.; He, X.; Lu, D.; Xiong, K.; Xiong, M.; et al. LRP6 downregulation promotes cardiomyocyte proliferation and heart regeneration. *Cell Res.* **2021**, *31*, 450–462. [CrossRef]
88. Zhou, H.; Yang, J.; Xin, T.; Zhang, T.; Hu, S.; Zhou, S.; Chen, G.; Chen, Y. Exendin-4 enhances the migration of adipose-derived stem cells to neonatal rat ventricular cardiomyocyte-derived conditioned medium via the phosphoinositide 3-kinase/Akt-stromal cell-derived factor-1α/CXC chemokine receptor 4 pathway. *Mol. Med. Rep.* **2015**, *11*, 4063–4072. [CrossRef]
89. Zhu, H.; Jin, Q.; Li, Y.; Ma, Q.; Wang, J.; Li, D.; Zhou, H.; Chen, Y. Melatonin protected cardiac microvascular endothelial cells against oxidative stress injury via suppression of IP3R-[Ca 2+] c/VDAC-[Ca 2+] m axis by activation of MAPK/ERK signaling pathway. *Cell Stress Chaperones* **2018**, *23*, 101–113. [CrossRef]
90. Saparov, A.; Ogay, V.; Nurgozhin, T.; Chen, W.C.; Mansurov, N.; Issabekova, A.; Zhakupova, J. Role of the immune system in cardiac tissue damage and repair following myocardial infarction. *Inflamm. Res.* **2017**, *66*, 739–751. [CrossRef]

91. Wang, X.; Song, Q. Mst1 regulates post-infarction cardiac injury through the JNK-Drp1-mitochondrial fission pathway. *Cell. Mol. Biol. Lett.* **2018**, *23*, 21. [CrossRef]

92. Rodius, S.; de Klein, N.; Jeanty, C.; Sánchez-Iranzo, H.; Crespo, I.; Ibberson, M.; Xenarios, I.; Dittmar, G.; Mercader, N.; Niclou, S.P.; et al. Fisetin protects against cardiac cell death through reduction of ROS production and caspases activity. *Sci. Rep.* **2020**, *10*, 2896. [CrossRef]

93. Liu, X.; Yuan, X.; Liang, G.; Zhang, S.; Zhang, G.; Qin, Y.; Zhu, Q.; Xiao, Q.; Hou, N.; Luo, J.D. BRG1 protects the heart from acute myocardial infarction by reducing oxidative damage through the activation of the NRF2/HO1 signaling pathway. *Free Radic. Biol. Med.* **2020**, *160*, 820–836. [CrossRef]

94. Liu, X.; Zhu, Q.; Zhang, M.; Yin, T.; Xu, R.; Xiao, W.; Wu, J.; Deng, B.; Gao, X.; Gong, W.; et al. Isoliquiritigenin Ameliorates Acute Pancreatitis in Mice via Inhibition of Oxidative Stress and Modulation of the Nrf2/HO-1 Pathway. *Oxid. Med. Cell. Longev.* **2018**, *2018*, 7161592. [CrossRef]

95. Datta Chaudhuri, R.; Banik, A.; Mandal, B.; Sarkar, S. Cardiac-specific overexpression of HIF-1α during acute myocardial infarction ameliorates cardiomyocyte apoptosis via differential regulation of hypoxia-inducible pro-apoptotic and anti-oxidative genes. *Biochem. Biophys. Res. Commun.* **2021**, *537*, 100–108. [CrossRef]

96. Li, F.; Yang, Y.; Xue, C.; Tan, M.; Xu, L.; Gao, J.; Xu, L.; Zong, J.; Qian, W. Zinc Finger Protein ZBTB20 protects against cardiac remodelling post-myocardial infarction via ROS-TNFα/ASK1/JNK pathway regulation. *J. Cell. Mol. Med.* **2020**, *24*, 13383–13396. [CrossRef]

97. Tanwir, K.; Javed, M.T.; Shahid, M.; Akram, M.S.; Ali, Q. Antioxidant defense systems in bioremediation of organic pollutants. In *Handbook of Bioremediation*; Academic Press: Cambridge, MA, USA, 2021; pp. 505–521.

98. Ma, L.L.; Kong, F.J.; Ma, Y.J.; Guo, J.J.; Wang, S.J.; Dong, Z.; Sun, A.J.; Zou, Y.Z.; Ge, J.B. Hypertrophic preconditioning attenuates post-myocardial infarction injury through deacetylation of isocitrate dehydrogenase 2. *Acta Pharmacol. Sin.* **2021**, in press. [CrossRef]

99. Lee, J.H.; Go, Y.; Kim, D.Y.; Lee, S.H.; Kim, O.H.; Jeon, Y.H.; Kwon, T.K.; Bae, J.H.; Song, D.K.; Rhyu, I.J.; et al. Isocitrate dehydrogenase 2 protects mice from high-fat diet-induced metabolic stress by limiting oxidative damage to the mitochondria from brown adipose tissue. *Exp. Mol. Med.* **2020**, *52*, 238–252. [CrossRef]

100. Çakmak, H.A.; Demir, M. MicroRNA and Cardiovascular Diseases. *Balkan Med. J.* **2020**, *37*, 60–71. [CrossRef]

101. Shi, C.C.; Pan, L.Y.; Zhao, Y.Q.; Li, Q.; Li, J.G. MicroRNA-323-3p inhibits oxidative stress and apoptosis after myocardial infarction by targeting TGF-β2/JNK pathway. *Eur. Rev. Med. Pharmacol. Sci.* **2020**, *24*, 6961–6970. [CrossRef]

102. Zhu, F.; Yu, Z.; Li, D. *miR-187 Modulates Cardiomyocyte Apoptosis and Oxidative Stress in Myocardial Infarction Mice via Negatively Regulating DYRK2*; Research Square: Durham, NC, USA, 2021.

103. Li, D.; Zhou, J.; Yang, B.; Yu, Y. microRNA-340-5p inhibits hypoxia/reoxygenation-induced apoptosis and oxidative stress in cardiomyocytes by regulating the Act1/NF-κB pathway. *J. Cell. Biochem.* **2019**, *120*, 14618–14627. [CrossRef]

104. Trembinski, D.J.; Bink, D.I.; Theodorou, K.; Sommer, J.; Fischer, A.; van Bergen, A.; Kuo, C.C.; Costa, I.G.; Schürmann, C.; Leisegang, M.S.; et al. Aging-regulated anti-apoptotic long non-coding RNA Sarrah augments recovery from acute myocardial infarction. *Nat. Commun.* **2020**, *11*, 2039. [CrossRef] [PubMed]

105. Ellery, J.; Dickson, L.; Cheung, T.; Ciuclan, L.; Bunyard, P.; Mack, S.; Buffham, W.J.; Farnaby, W.; Mitchell, P.; Brown, D.; et al. Identification of compounds acting as negative allosteric modulators of the LPA1 receptor. *Eur. J. Pharmacol.* **2018**, *833*, 8–15. [CrossRef] [PubMed]

106. Han, F.; Chen, Q.; Su, J.; Zheng, A.; Chen, K.; Sun, S.; Wu, H.; Jiang, L.; Xu, X.; Yang, M.; et al. MicroRNA-124 regulates cardiomyocyte apoptosis and myocardial infarction through targeting Dhcr24. *J. Mol. Cell. Cardiol.* **2019**, *132*, 178–188. [CrossRef] [PubMed]

107. Dong, Y.; Chen, H.; Gao, J.; Liu, Y.; Li, J.; Wang, J. Molecular machinery and interplay of apoptosis and autophagy in coronary heart disease. *J. Mol. Cell. Cardiol.* **2019**, *136*, 27–41. [CrossRef]

108. Teringova, E.; Tousek, P. Apoptosis in ischemic heart disease. *J. Transl. Med.* **2017**, *15*, 87. [CrossRef]

109. Yuan, T.; Krishnan, J. Non-coding RNAs in Cardiac Regeneration. *Front. Physiol.* **2021**, *12*, 358. [CrossRef]

110. Yan, M.J.; Tian, Z.S.; Zhao, Z.H.; Yang, P. MiR-31a-5p protects myocardial cells against apoptosis by targeting Tp53. *Mol. Med. Rep.* **2018**, *17*, 3898–3904. [CrossRef]

111. Wang, Y.; Cui, X.; Wang, Y.; Fu, Y.; Guo, X.; Long, J.; Wei, C.; Zhao, M. Protective effect of miR378* on doxorubicin-induced cardiomyocyte injury via calumenin. *J. Cell Physiol.* **2018**, *233*, 6344–6351. [CrossRef]

112. Li, X.; Zhong, J.; Zeng, Z.; Wang, H.; Li, J.; Liu, X.; Yang, X. MiR-181c protects cardiomyocyte injury by preventing cell apoptosis through PI3K/Akt signaling pathway. *Cardiovasc. Diagn. Ther.* **2020**, *10*, 849–858. [CrossRef]

113. Wang, Q.S.; Zhou, J.; Li, X. LncRNA UCA1 protects cardiomyocytes against hypoxia/reoxygenation induced apoptosis through inhibiting miR-143/MDM2/p53 axis. *Genomics* **2020**, *112*, 574–580. [CrossRef]

114. Hao, S.; Liu, X.; Sui, X.; Pei, Y.; Liang, Z.; Zhou, N. Long non-coding RNA GAS5 reduces cardiomyocyte apoptosis induced by MI through sema3a. *Int. J. Biol. Macromol.* **2018**, *120*, 371–377. [CrossRef]

115. Long, B.; Li, N.; Xu, X.X.; Li, X.X.; Xu, X.J.; Guo, D.; Zhang, D.; Wu, Z.H.; Zhang, S.Y. Long noncoding RNA FTX regulates cardiomyocyte apoptosis by targeting miR-29b-1-5p and Bcl2l2. *Biochem. Biophys. Res. Commun.* **2018**, *495*, 312–318. [CrossRef]

116. Zeng, Y.; Du, W.W.; Wu, Y.; Yang, Z.; Awan, F.M.; Li, X.; Yang, W.; Zhang, C.; Yang, Q.; Yee, A.; et al. A Circular RNA Binds to and Activates AKT Phosphorylation and Nuclear Localization Reducing Apoptosis and Enhancing Cardiac Repair. *Theranostics* **2017**, *7*, 3842–3855. [CrossRef]

117. Zhu, Y.; Zhao, P.; Sun, L.; Lu, Y.; Zhu, W.; Zhang, J.; Xiang, C.; Mao, Y.; Chen, Q.; Zhang, F. Overexpression of circRNA SNRK targets miR-103-3p to reduce apoptosis and promote cardiac repair through GSK3β/β-catenin pathway in rats with myocardial infarction. *Cell Death Discov.* **2021**, *7*, 84. [CrossRef]

118. Garikipati, V.N.S.; Verma, S.K.; Cheng, Z.; Liang, D.; Truongcao, M.M.; Cimini, M.; Yue, Y.; Huang, G.; Wang, C.; Benedict, C.; et al. Circular RNA CircFndc3b modulates cardiac repair after myocardial infarction via FUS/VEGF-A axis. *Nat. Commun.* **2019**, *10*, 4317. [CrossRef]

119. Imbalzano, E.; Mandraffino, G.; Casciaro, M.; Quartuccio, S.; Saitta, A.; Gangemi, S. Pathophysiological mechanism and therapeutic role of S100 proteins in cardiac failure: A systematic review. *Heart Fail. Rev.* **2016**, *21*, 463–473. [CrossRef]

120. Fan, L.; Liu, B.; Guo, R.; Luo, J.; Li, H.; Li, Z.; Xu, W. Elevated plasma S100A1 level is a risk factor for ST-segment elevation myocardial infarction and associated with post-infarction cardiac function. *Int. J. Med. Sci.* **2019**, *16*, 1171–1179. [CrossRef]

121. Jungi, S.; Fu, X.; Segiser, A.; Busch, M.; Most, P.; Fiedler, M.; Carrel, T.; Tevaearai Stahel, H.; Longnus, S.L.; Most, H. Enhanced Cardiac S100A1 Expression Improves Recovery from Global Ischemia-Reperfusion Injury. *J. Cardiovasc. Transl. Res.* **2018**, *11*, 236–245. [CrossRef]

122. Sleczka, B.G.; Levesque, P.C.; Adam, L.P.; Olah, T.V.; Shipkova, P. LC/MS/MS-based quantitation of pig and human S100A1 protein in cardiac tissues: Application to gene therapy. *Anal. Biochem.* **2020**, *602*, 113766. [CrossRef]

123. Donato, R.; Sorci, G.; Giambanco, I. S100A6 protein: Functional roles. *Cell. Mol. Life Sci.* **2017**, *74*, 2749–2760. [CrossRef]

124. Mofid, A.; Newman, N.S.; Lee, P.J.; Abbasi, C.; Matkar, P.N.; Rudenko, D.; Kuliszewski, M.A.; Chen, H.H.; Afrasiabi, K.; Tsoporis, J.N.; et al. Cardiac Overexpression of S100A6 Attenuates Cardiomyocyte Apoptosis and Reduces Infarct Size After Myocardial Ischemia-Reperfusion. *J. Am. Heart Assoc.* **2017**, *6*, e004738. [CrossRef]

125. Cho, H.M.; Lee, K.H.; Shen, Y.M.; Shin, T.J.; Ryu, P.D.; Choi, M.C.; Kang, K.S.; Cho, J.Y. Transplantation of hMSCs Genome Edited with LEF1 Improves Cardio-Protective Effects in Myocardial Infarction. *Mol. Ther. Nucleic Acids* **2020**, *19*, 1186–1197. [CrossRef]

126. Meng, D.; Han, S.; Jeong, I.S.; Kim, S.W. Interleukin 10-Secreting MSCs via TALEN-Mediated Gene Editing Attenuates Left Ventricular Remodeling after Myocardial Infarction. *Cell. Physiol. Biochem.* **2019**, *52*, 728–741. [CrossRef]

127. Meng, X.; Zheng, M.; Yu, M.; Bai, W.; Zuo, L.; Bu, X.; Liu, Y.; Xia, L.; Hu, J.; Liu, L.; et al. Transplantation of CRISPRa system engineered IL10-overexpressing bone marrow-derived mesenchymal stem cells for the treatment of myocardial infarction in diabetic mice. *J. Biol. Eng.* **2019**, *13*, 49. [CrossRef] [PubMed]

128. Kaminsky, S.M.; Quach, L.; Chen, S.; Pierre-Destine, L.; Van de Graaf, B.; Monette, S.; Rosenberg, J.B.; De, B.P.; Sondhi, D.; Hackett, N.R.; et al. Safety of direct cardiac administration of AdVEGF-All6A+, a replication-deficient adenovirus vector cDNA/genomic hybrid expressing all three major isoforms of human vascular endothelial growth factor, to the ischemic myocardium of rats. *Hum. Gene Ther. Clin. Dev.* **2013**, *24*, 38–46. [CrossRef]

129. Anttila, V.; Saraste, A.; Knuuti, J.; Jaakkola, P.; Hedman, M.; Svedlund, S.; Lagerström-Fermér, M.; Kjaer, M.; Jeppsson, A.; Gan, L.M. Synthetic mRNA Encoding VEGF-A in Patients Undergoing Coronary Artery Bypass Grafting: Design of a Phase 2a Clinical Trial. *Mol. Ther. Methods Clin. Dev.* **2020**, *18*, 464–472. [CrossRef]

130. Weng, Y.; Li, C.; Yang, T.; Hu, B.; Zhang, M.; Guo, S.; Xiao, H.; Liang, X.J.; Huang, Y. The challenge and prospect of mRNA therapeutics landscape. *Biotechnol. Adv.* **2020**, *40*, 107534. [CrossRef] [PubMed]

131. Meng, Z.; O'Keeffe-Ahern, J.; Lyu, J.; Pierucci, L.; Zhou, D.; Wang, W. A new developing class of gene delivery: Messenger RNA-based therapeutics. *Biomater. Sci.* **2017**, *5*, 2381–2392. [CrossRef] [PubMed]

132. Teijaro, J.R.; Farber, D.L. COVID-19 vaccines: Modes of immune activation and future challenges. *Nat. Rev. Immunol.* **2021**, *21*, 195–197. [CrossRef] [PubMed]

133. Turner, J.S.; O'Halloran, J.A.; Kalaidina, E.; Kim, W.; Schmitz, A.J.; Zhou, J.Q.; Lei, T.; Thapa, M.; Chen, R.E.; Case, J.B.; et al. SARS-CoV-2 mRNA vaccines induce persistent human germinal centre responses. *Nature* **2021**, *596*, 109–113. [CrossRef] [PubMed]

134. Magadum, A.; Kaur, K.; Zangi, L. mRNA-Based Protein Replacement Therapy for the Heart. *Mol. Ther.* **2019**, *27*, 785–793. [CrossRef]

135. Kim, J.S.; Hwang, H.Y.; Cho, K.R.; Park, E.A.; Lee, W.; Paeng, J.C.; Lee, D.S.; Kim, H.K.; Sohn, D.W.; Kim, K.B. Intramyocardial transfer of hepatocyte growth factor as an adjunct to CABG: Phase I clinical study. *Gene Ther.* **2013**, *20*, 717–722. [CrossRef]

136. Ishikawa, K.; Fish, K.M.; Tilemann, L.; Rapti, K.; Aguero, J.; Santos-Gallego, C.G.; Lee, A.; Karakikes, I.; Xie, C.; Akar, F.G.; et al. Cardiac I-1c overexpression with reengineered AAV improves cardiac function in swine ischemic heart failure. *Mol. Ther.* **2014**, *22*, 2038–2045. [CrossRef]

137. Watanabe, S.; Ishikawa, K.; Fish, K.; Oh, J.G.; Motloch, L.J.; Kohlbrenner, E.; Lee, P.; Xie, C.; Lee, A.; Liang, L.; et al. Protein Phosphatase Inhibitor-1 Gene Therapy in a Swine Model of Nonischemic Heart Failure. *J. Am. Coll. Cardiol.* **2017**, *70*, 1744–1756. [CrossRef]

138. Hammond, H.K.; Penny, W.F.; Traverse, J.H.; Henry, T.D.; Watkins, M.W.; Yancy, C.W.; Sweis, R.N.; Adler, E.D.; Patel, A.N.; Murray, D.R.; et al. Intracoronary Gene Transfer of Adenylyl Cyclase 6 in Patients with Heart Failure: A Randomized Clinical Trial. *JAMA Cardiol.* **2016**, *1*, 163–171. [CrossRef]

International Journal of
Molecular Sciences

Review

Engineering and Assessing Cardiac Tissue Complexity

Karine Tadevosyan [1,2,†], Olalla Iglesias-García [1,2,3,4,*,†] , Manuel M. Mazo [3,4,5] , Felipe Prósper [3,4,5,6] and Angel Raya [1,2,7,*]

1   Regenerative Medicine Program, Bellvitge Institute for Biomedical Research (IDIBELL) and Program for Clinical Translation of Regenerative Medicine in Catalonia (P-CMRC), 08908 L'Hospitalet del Llobregat, Spain; ktadevosyan@idibell.cat
2   Center for Networked Biomedical Research on Bioengineering, Biomaterials and Nanomedicine (CIBER-BBN), 28029 Madrid, Spain
3   Regenerative Medicine Program, Cima Universidad de Navarra, Foundation for Applied Medical Research, 31008 Pamplona, Spain; mmazoveg@unav.es (M.M.M.); fprosper@unav.es (F.P.)
4   IdiSNA, Navarra Institute for Health Research, 31008 Pamplona, Spain
5   Hematology and Cell Therapy Area, Clínica Universidad de Navarra, 31008 Pamplona, Spain
6   Center for Networked Biomedical Research on Cancer (CIBERONC), 28029 Madrid, Spain
7   Catalan Institution for Research and Advanced Studies (ICREA), 08010 Barcelona, Spain
*   Correspondence: oiglesias@unav.es (O.I.-G.); araya@idibell.cat (A.R.)
†   These authors contributed equally to this work.

Abstract: Cardiac tissue engineering is very much in a current focus of regenerative medicine research as it represents a promising strategy for cardiac disease modelling, cardiotoxicity testing and cardiovascular repair. Advances in this field over the last two decades have enabled the generation of human engineered cardiac tissue constructs with progressively increased functional capabilities. However, reproducing tissue-like properties is still a pending issue, as constructs generated to date remain immature relative to native adult heart. Moreover, there is a high degree of heterogeneity in the methodologies used to assess the functionality and cardiac maturation state of engineered cardiac tissue constructs, which further complicates the comparison of constructs generated in different ways. Here, we present an overview of the general approaches developed to generate functional cardiac tissues, discussing the different cell sources, biomaterials, and types of engineering strategies utilized to date. Moreover, we discuss the main functional assays used to evaluate the cardiac maturation state of the constructs, both at the cellular and the tissue levels. We trust that researchers interested in developing engineered cardiac tissue constructs will find the information reviewed here useful. Furthermore, we believe that providing a unified framework for comparison will further the development of human engineered cardiac tissue constructs displaying the specific properties best suited for each particular application.

Keywords: pluripotent stem cells; cardiomyocytes; cardiac tissue engineering; human heart; tissue maturation

Citation: Tadevosyan, K.; Iglesias-García, O.; Mazo, M.M.; Prósper, F.; Raya, A. Engineering and Assessing Cardiac Tissue Complexity. *Int. J. Mol. Sci.* **2021**, 22, 1479. https://doi.org/10.3390/ijms22031479

Academic Editor: Nicole Wagner

Received: 12 January 2021

Accepted: 28 January 2021

Published: 2 February 2021

Publisher's Note: MDPI stays neutral with regard to jurisdictional claims in published maps and institutional affiliations.

1. Introduction

Cardiovascular diseases (CVD) remain the main cause of morbidity and mortality worldwide despite many advances in preventive cardiology. According to 2017 estimates, CVD accounted for over 420 million cases every year and resulted in close to 18 million deaths [1]. Acute myocardial infarction (MI) and subsequent heart failure continue to have high prevalence worldwide [2] and are among the major health issues. As the human adult heart has minimal regenerative capacity [3], cardiomyocytes (CMs) that are lost during ischemic injuries are usually replaced with fibrotic scar tissue leading to the partial or total cardiac dysfunction [4]. Today the only curative option for patients with end-stage heart failure is heart transplantation, which is usually limited due to the scarcity of donor hearts [5]. Preclinical research studies have demonstrated that cell therapy can attenuate myocardial damage and reduce the progression of cardiac remodeling to heart failure [6]. However, clinical studies have failed to show significant improvements and preliminary data indicate that stem cells have the potential to enhance tissue perfusion and contractile performance [7]. Numerous studies have demonstrated that the therapeutic benefits exerted by cells are mainly attributable to the release of complementary paracrine factors and the efficacy is limited as only a small percentage of transplanted cells engrafted in the infarcted tissue [8]. Studies on animal models showed that combining cell therapy with tissue engineering techniques for the creation of cell sheets and patches, can increase stem cell survival and boost therapeutic action [9]. Therefore, tissue engineering has been considered as a potential approach for cardiac regeneration after MI. Current research in the field is, to a large extent, aimed towards the development of functional heart tissue for application in cell-based regenerative therapies [10,11], cardiac disease modeling [12–15] and drug screening [16]. In particular, the application of cardiac tissue engineering (CTE) systems for investigating mechanisms of disease has been recently reviewed [17] and will not be specifically covered here. Despite the evident progress realized to date, engineered cardiac tissues (ECT) remain immature and still differ from the adult human heart. One of the major challenges in CTE is the organization of stem-cell derived CMs into a functional tissue of large dimensions suitable for the intended therapeutic use, and that recapitulates the main physiological properties of the real heart. In this review, we describe the main approaches undertaken toward the development of ECT constructs, along with the methodologies utilized for assessing their functionality, and discuss the current state and future directions.

2. Cell Sources

The heart muscle is composed of electrically and mechanically connected CMs closely surrounded by endothelial cells (ECs), fibroblasts (FBs), smooth muscle cells (SMCs), pericytes, and heart resident immune cells [18]. Even though CMs constitute approximately 70–85% of the heart by volume [19], non-myocyte cells are more abundant by cell number and bear critical roles in heart homeostasis, supporting the extracellular matrix, intercellular communication, and vascular supply essential for CM survival and contraction [19,20]. Thus, the most advanced current CTE approaches aim at generating 3-dimensional (3D) constructs that incorporate CMs as well as non-myocyte cardiac cells to reproduce the myocardial niche and allow the creation of a functional mature structure (Figure 1 and Table 1).

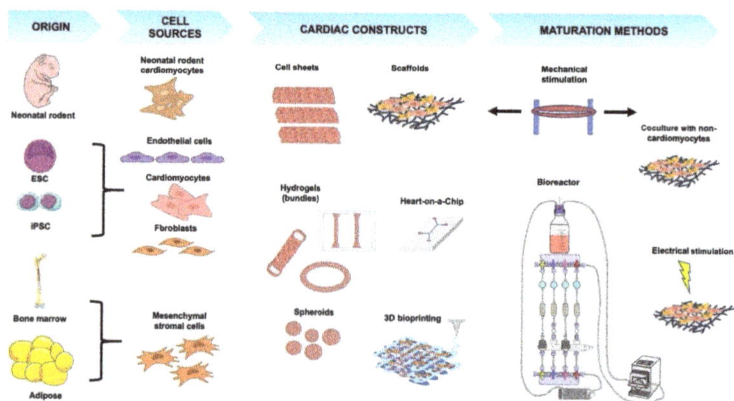

Figure 1. Main types of cell sources, cardiac constructs and maturation methods used in CTE. The cell types most frequently used as source for CTE include neonatal rodent cardiomyocytes, ESC-or iPSC-derived endothelial cells, cardiomyocytes and fibroblasts, and mesenchymal stromal cells derived from bone marrow or adipose tissue. Different systems have been developed to generate ECT including cell sheets, scaffolds, hydrogels, bundles, spheroids, heart-on-a-chip, and 3D-bioprinted constructs. The main methods used to increase maturation of cardiac constructs comprise mechanical and/or electrical stimulation, co-culturing with non-myocyte cells and continuous perfusion within bioreactors. See main text for details. Figure created with BioRender.com and smart.servier.com.

Table 1. Representative studies on CTE.

CTE Construct			Cell Source		Culture Conditions			Functional Analyses		Ref.
Type	Area (mm²)	Thickness (mm)	Species	Cell Type	Stimulation	Perfusion	Time (d)	CM Maturation	Tissue Maturation	
Scaffold	70	2	Rat	Neonatal CM	No	Yes	10	Protein expression	No functional analyses	[21]
Scaffold	95	1.5	Rat	Neonatal CM	No	Yes	7	Protein expression	Contractile activity	[22]
Scaffold	48	1.5	Rat	Neonatal CM	Electrical stimulation	No	8	Protein expression, sarcomere structure	Contractile activity	[23]
Scaffold	95	1.5	Rat	Neonatal CM	No	Yes	14	No functional analyses	Contractile activity	[24]
Scaffold	48	1.5	Rat	Neonatal CM	Electrical stimulation	No	8	Protein expression, sarcomere structure	Contractile activity	[25]
Scaffold	20	2	Rat	Neonatal CM	Electrical stimulation	Yes	4	Protein expression	No functional analyses	[26]
Scaffold	50	1	Rat	Neonatal CM	Electrical stimulation	Yes	8	No functional analyses	Contractile activity	[27]
Scaffold	20	2	Rat	Neonatal CM	Mechanical stimulation	Yes	4	Protein expression	No functional analyses	[28]
Scaffold	80	5	Rat	Neonatal CM	No	No	7	No functional analyses	Calcium imaging	[29]
Scaffold	n/a	n/a	Human	ESC-CM	Mechanical stimulation	No	14	Protein expression	Force of contraction	[30]
Scaffold	56	3.5	Human	ESC-CM	Mechanical	No	5	Gene/Protein expression	Calcium imaging	[31]
Scaffold	80	1	Rat/human	Neonatal CM/iPSC-CM	Electrical stimulation	Yes	14	Gene/protein expression, sarcomere structure	Force of contraction, contractile activity, whole construct electrical activity	[32]
Hydrogel	n/a	n/a	Rat	Neonatal CM	Mechanical stimulation	No	14	Protein expression, sarcomere structure, electrical signal propagation	Force of contraction	[33]
Hydrogel	n/a	0.9	Rat	Neonatal CM	Mechanical stimulation	Yes	14	Gene/protein expression	Force of contraction	[34]
Hydrogel	n/a	n/a	Rat	Neonatal CM	Electro-mechanical stimulation	No	13	Protein expression	Force of contraction, calcium imaging	[35]
Hydrogel	60	n/a	Human	ESC-CM/iPSC-CM+HUVEC+MSC	Mechanical stimulation	No	4	Gene expression, sarcomere structure	Force of contraction	[36]
Hydrogel	0.4	n/a	Human	ESC-CM	No	No	14	Gene/protein expression, patch clamp	Force of contraction	[37]
Hydrogel	n/a	0.1	Rat/human	Neonatal CM/ESC-CM	Electro-mechanical stimulation	No	7	Gene/protein expression	Contractile activity, optical mapping	[38]
Hydrogel	49	n/a	Human	ESC-CM	No	No	14	Gene/protein expression, sarcomere structure	Force of contraction, optical mapping	[39]
Hydrogel	3	0.3	Human	iPSC-CM	Electrical stimulation	No	14	Protein expression, sarcomere structure, patch clamp	Contractile activity, optical mapping, calcium imaging	[40]
Hydrogel	15	n/a	Rat/human	Neonatal CM/ESC-CM	Electrical stimulation	Yes	14	Protein expression	Force of contraction, contractile activity	[41]
Hydrogel	5	n/a	Human	ESC-CM	No	No	24	Gene/protein expression, sarcomere structure	Force of contraction, optical mapping	[42]
Hydrogel	n/a	n/a	Human	iPSC-CM	No	No	7	Protein expression, electrical signal propagation	No functional analyses	[43]
Hydrogel	0.125	n/a	Human	iPSC-CM	No	No	15	Gene/protein expression	No functional analyses	[44]

Table 1. *Cont.*

CTE Construct			Cell Source	Culture Conditions			Functional Analyses		Ref.
Hydrogel	27	0.2	Human	iPSC-CM+iPSC-EC/HUVEC	No	No	15	Gene/Protein expression	Optical mapping	[45]
Hydrogel	4	n/a	Human	iPSC-CM	No	No	40	Protein expression, sarcomere structure	Force of contraction	[46]
Hydrogel	20	0.3	Human	iPSC-CM	Electro-mechanical stimulation	No	14	Protein expression, sarcomere structure	Force of contraction, calcium imaging	[47]
Hydrogel	14	0.2	Rat/human	Neonatal CM/iPSC-CM	No	No	14	Gene/protein expression	Force of contraction, optical mapping	[48]
Hydrogel	200	n/a	Human	iPSC-CM	No	No	60	Protein expression, electrical signal propagation	No functional analyses	[49]
Hydrogel	900	n/a	Human	iPSC-CM	No	No	28	Protein expression	Force of contraction, contractile activity	[50]
Hydrogel	1190	0.5	Human	ESC-CM/iPSC-CM	Mechanical stimulation	No	45	Gene expression, sarcomere structure, patch clamp	Force of contraction	[51]
Hydrogel	1296	0.1	Human	iPSC-CM	No	No	21	Gene/protein expression, sarcomere structure	Force of contraction, optical mapping	[11]
Hydrogel	100	0.1	Rat/human	Neonatal CM/ESC-CM	Electrical stimulation	No	7	No functional analyses	Force of contraction, contractile activity	[52]
Hydrogel	800	n/a	Human	iPSC-CM+iPSC-EC+iPS-SMC	No	No	7	Protein expression	Force of contraction, optical mapping	[53]
Hydrogel	5	0.3	Human	ESC-CM/iPSC-CM+hcFB	Electrical stimulation	No	42	Gene/Protein expression, sarcomere structure, patch clamp	Force of contraction, contractile activity, calcium imaging	[54]
Hydrogel	11	n/a	Human	iPSC-CM	Electrical stimulation	No	30	Gene/protein expression, sarcomere structure, patch clamp	Force of contraction, calcium imaging	[55]
Hydrogel	14	2	Human/Pig/Rat	iPSC-CM+iPSC-EC/Neonatal CM+HUVEC+FB	No	No	7	Gene/Protein expression	Optical mapping, calcium imaging	[56]
Cell sheets	116	0.045	Rat	Neonatal CM	No	No	4	Protein expression, sarcomere structure, electrical signal propagation	Force of contraction	[57]
Cell sheets	960	0.1	Rat/human	Neonatal CM+EC	No	Yes	10	Protein expression	No functional analyses	[58]
Cell sheets	70	0.1	Human	iPSC-CM+MSC	No	No	4	Protein expression, electrical signal propagation	Contractile activity	[59]

2.1. Neonatal Rodent Myocytes

The first studies on cardiac engineering in the late 1990s used cells derived from neonatal rat hearts since human CMs were not readily available and could not be expanded from cardiac biopsies [60–62]. Moreover, while human embryonic stem cells (hESC) were first derived in 1998, robust protocols to guide cardiac differentiation took another decade to be perfected and yield human CMs with the desired purity and reproducibility [63–65]. On the other hand, the use of adult ventricular myocytes from both rats and mice showed low viability and proliferation rate in long-term cultures which made them useless for CTE [66,67]. Thus, by this period of time neonatal rat cardiac myocytes (NRCMs) became the preferred cell source for CTE as they could be isolated with high yield and quality and presented comparably long-term viability in culture. Therefore, NRCMs were used to generate 3D functional cardiac tissue constructs in which cells underwent progressive electromechanical maturation over time in static conditions [22,35,38,41,48,57,60,68] while culturing these 3D constructs in dynamic systems further enhanced their functionality [24–26,28,34,41,69–72].

2.2. Pluripotent Stem Cells

Pluripotent stem cells (PSCs), including embryonic and induced pluripotent stem cells (ESC and iPSCs), show promise for CTE as they can be expanded indefinitely in vitro and differentiated into multiple cell types including CMs. Robust protocols have been developed for cardiac differentiation of PSCs allowing to obtain high yield of CMs [63,73,74]. The therapeutic potential of both human ESC- and iPSCs-derived CMs has been previously demonstrated in different animal models of MI including mice [75], rats [76–78], guinea pigs [79], pigs [80], and non-human primates [81]. Remarkably, Shiba et al. demonstrated in 2016 that allogeneic transplantation of iPSC-CMs can regenerate the damaged heart in a non-human primate MI model. The authors generated iPSCs from an animal homozygous for a specific major histocompatibility complex haplotype (HT4), which were then differentiated into CMs. A total of 4×10^8 iPSC-CMs were injected in the infarcted heart of HT4-heterozygous primates. Cells survived and engrafted for at least 12 weeks with no evidence of immune rejection. However, some grafts remained isolated from the host tissue and the incidence of arrhythmias increased after cell transplantation, although the incidence was transient and decreased gradually over time [82].

Transplantation of ECT constructs comprising human PSC-derived CMs has been shown to enhance cell survival and engraftment, promote tissue revascularization and improve functional properties of the injured heart. The Murry laboratory generated ECT constructs using hESC-CMs or hiPSC-CMs together with human umbilical vein ECs, and mesenchymal stromal cells (MSCs) in a type I collagen hydrogel and subjected them to uniaxial mechanical loading [36]. One week after transplantation onto injured hearts of athymic rats, the ECT constructs survived and formed grafts containing a microvascular network that was perfused by the host coronary circulation. Kawamura et al. used the cell sheet method to deliver hiPSC-CMs onto infarcted swine hearts. This procedure significantly improved contractile function, promoted vascularization and attenuated left ventricular remodeling. However, poor level of engraftment was detectable 8 weeks post-transplantation [83]. Transplantation of ECT constructs from hESC-CMs in a chronic rodent MI model has been shown to enhance engraftment rate, leading to increased long-term survival, and progressive maturation of the human CMs [30]. Engineered tissues containing hiPSC-derived vascular cells (ECs and SMCs) without CMs have also been associated with significant functional improvement, reduction of infarct size, increase in neovascularization, and recruitment of endogenous cardiac progenitor cells in a MI swine model [84]. Cardiac repair has also been reported after implantation of ECT constructs comprising hiPSC-derived cells in guinea pigs [45] and pigs [85].

Representative publications on CTE, indicating the sources and types of cells used, the type and dimension of the constructs generated, the culture conditions, and the functional analyses performed. CM, cardiomyocytes; iPSC-CM, cardiomyocytes derived from induced pluripotent stem cells; ESC-CM, cardiomyocytes derived from embryonic stem cells; EC, endothelial cells; HUVEC, human umbilical vein endothelial cells; FB, fibroblasts; hcFB, human cardiac fibroblasts; SMC, smooth muscle cells; MSC, mesenchymal stromal cells; n/a, information not available.

Notably, Menasché et al. reported the first clinical use of hESC-derived cardiovascular progenitor cells in a fibrin patch of 20 cm^2, which was delivered to the heart of a patient suffering from severe ischemic left ventricular dysfunction. The study demonstrated the feasibility of generating a clinical-grade population of hESC-derived cardiovascular progenitors and improved the patient's functional outcome after a 3-month follow-up [86]. In a subsequent study, the group tested the safety of this approach on 6 patients with up to one year follow-up, during which the patients did not present any clinically significant arrhythmias and showed an increase in systolic function. One patient died after surgery from treatment-unrelated comorbidities and another one at 22 months due to heart failure [87]. No tumors were detected during the follow-up period, demonstrating the safety of this approach at 1 year.

2.3. Mesenchymal Stromal Cells

Mesenchymal stromal cells (MSCs) are multipotent cells that have the potential to differentiate into a variety of cell types like adipocytes, chondrocytes and osteocytes, limited self-renewal capacity, and low immunogenicity [88]. Despite the adipocyte tissue and bone marrow being the most common sources for MSCs [89], they can be also isolated from synovial tissue, umbilical cord and peripheral blood [90]. MSCs were considered in autologous and allogeneic therapies for cardiac injuries due to their high expansion rate and immunomodulatory properties. In particular, numerous preclinical trials have shown that MSCs have the potential to promote cardiac repair in heart injury models through paracrine effects [91–95]. Moreover, engineered constructs containing MSCs have also been evaluated in preclinical models of MI. Patches consisting of autologous porcine MSCs in a fibrin hydrogel were transplanted onto infarcted swine hearts and resulted in improvement in contractile function and increase in neovascularization in the patch covered area [96]. Later, human MSCs cultured in a scaffold made of decellularized human myocardium with fibrin hydrogel were transplanted in nude rat models of acute and chronic MI. The treatments showed cardiac functional improvement 4 weeks after transplantation that was associated with the release of proangiogenic factors by MSCs [97]. Moreover, researchers have found an important structural role of MSCs for the organization of cardiac tissues. In particular, co-culture of human MSCs with hPSC-derived CMs and vascular cells within ECT constructs markedly improved their self-organization, vascular network formation and stabilization [36,59]. Finally, cell heterogeneity in ECT constructs due to the presence of MSCs has also been reported to influence successful engraftment [36] and to aid in the generation of human in vitro disease models of high pathophysiological relevance [59].

2.4. Non-Myocytes

Within the native heart, CMs are never alone but rather organize in a complex 3D tissue with intimately coupled non-myocyte cells. Moreover, heart homeostasis and response to disease strongly depends on the interaction between non-myocytes and cardiomyocytes. It therefore follows that recapitulating the normal heart tissue organization in the laboratory would necessitate including CMs along with non-myocytes [3,98]. Liau et al. generated two types of patches containing mouse ESC-derived CMs alone, or in combination with different ranges of cardiac FBs (3%, 6% and 12%). Based on their results, patches containing only ESC-CMs failed to form a functional syncytium, while the presence of cardiac FBs allowed the generation of synchronously contractile constructs with functional properties close to the native heart [99]. Building on those findings, Zhang et al. generated cardiac patches using human ESC-derived CMs with varying purity (49–90%) together with cardiac FBs, endothelial cells, and vascular SMCs, which inadvertently arose during cardiac differentiation. Interestingly, the conduction velocity in patches linearly increased with CMs purity, and when compared to matched 2D monolayers, 3D tissues showed higher conduction velocities but not different action potential duration or maximum capture rates. Moreover, the authors found that inclusion of non-myocytes increased the survival and the maturation level of CMs [100]. Tiburcy et al. also demonstrated the critical importance of defining the non-myocyte fraction for engineering force-generating cardiac tissues. Specifically, these authors found that the ratio of 70% hPSC-derived CMs and 30% human foreskin FBs was optimal for promoting maturation at the cellular and tissue levels [51]. Many subsequent studies have confirmed the significance of including non-myocytes in engineered cardiac constructs [32,36,51,53,55,101–104].

3. Cardiac Tissue Engineering Systems

A wide variety of 3D ECT constructs with different shapes and sizes/thicknesses have been fabricated with the purpose of being used as model systems for drug/toxicity testing or for application in regenerative medicine strategies [45,105]. The main categories of ECT constructs are depicted in Figure 1, with representative publications listed in Table 1 and described below.

3.1. Cell Sheets

The cell sheet technique was first reported by Shimizu et al. in 2002 for creating a transplantable 3D cell patch [57]. Cell sheet technology, also referred to as scaffold-free system or "Cell Sheet Engineering" [57,106,107], is based on stacking monolayers (or sheets) of CMs cultured to confluency to form 3D tissue-like structures. By using cell culture surfaces coated with the temperature-responsive polymer poly(*N*-isopropylacrylamide) (PIPAAm), it is possible to readily detach intact cellular monolayers of CMs as cell sheets by lowering the temperature, without any enzymatic treatments. Overlaying these thin 2D monolayers then results in 3D cardiac constructs [108]. The benefits of this system have been analyzed in vivo in murine animal models of MI showing improvements in cell survival, cardiac function and tissue remodeling [109]. The use of cell sheets created new opportunities for in vitro tissue engineering and helped exploring new therapies and drugs for heart diseases [58,59]. Interestingly, Sekine et al. produced in vitro vascularized cardiac tissues with perfusable blood vessels by overlaying additional triple-layer cell sheets made by NRCMs cocultured with endothelial cells. Such sheets were then transplanted under the neck skin of nude rats and connected to the local vasculature. Constructs with vessel anastomoses survived and maintained their vascular structure up to two weeks after transplantation. However, the thickness of the constructs decreased over time indicating that uniform perfusion was insufficient for whole tissue survival. Moreover, no functional analyses were performed in the study to evaluate maturation at tissue level [58]. More recently, Kawatou et al. developed an in vitro drug-induced Torsade de Pointes arrhythmia model using 3D cardiac tissue sheets. Importantly, the authors showed the importance of using multi-layered 3D structures containing a hiPSC-derived heterogeneous cell mixture (CMs and non-myocytes) in order to recapitulate disease-related phenotypes in vitro [59]. In addition, phase II clinical trials have been performed by Japanese scientists to evaluate the efficacy and safety of autologous skeletal myoblast sheet transplantation in patients with advanced heart failure. They demonstrated that the transplantation of engineered tissue promoted left ventricular remodeling, improved the heart failure symptoms and prevented cardiac death with a 2 year follow-up period [110]. Also, the potential use of cell-sheets that contain allogeneic hiPSC-CMs for clinical transplantation is under investigation [111].

A clear advantage of cell sheet technology for therapeutic applications is the absence of biomaterials, which reduces the risk of immune rejection that could arise from xenobiotic or non-autologous materials, and that no suture is needed to ligate the construct to the injured heart. Moreover, sheet technology enables direct cell-to-cell communications between cells in the transplanted sheets and the host tissue, facilitating electrical communication and vascular network formation within the cell sheet structure. On the downside, it has been argued that the fragility of these sheets makes them difficult to manipulate during implantation onto the heart [112]. Although cardiac tissue sheets have many advantages over other tissue engineering methods, these structures are not thick enough to reproduce the high complexity of the native myocardial tissue.

3.2. Scaffolds

ECT constructs made by repopulating cell-free scaffolds with suitable cells are usually referred to as cardiac patches. Scaffolds for CTE usually consist on a 3D polymeric porous structure that contributes to cell attachment and leads to the desirable cell interaction for further tissue formation [11,23,25–27,69,113–115]. Many different materials have been tested for the fabrication of scaffolds suitable for CTE, including natural and synthetic biomaterials. A commonly used synthetic material is polylactic acid (PLA), which is easily degradable forming lactic acid. PLA scaffolds were tested in some cardiovascular studies [116,117]. Another example is polyglycolic acid (PGA) and its copolymer with PLA poly(D,L-lactic-*co*-glycolic acid) (PLGA), an FDA approved biomaterial among the first tested for CTE due to its high porosity, biodegradability and processability [69,70,118]. However, it has been noted that the high stiffness of PLGA may limit the capacity of CMs to remodel the scaffold and ultimately impinge on their maturation [112]. Collagen, being the most abundant protein in the cardiac extracellular matrix (ECM), has a fibrillar structure that facilitates CM scaffolding. In addition to good biocompatibility, biodegradability and permeability, collagen also elicits low immunogenicity and can be engineered in various formats including high porosity scaffolds, all of which make of collagen the most commonly used biomaterial in scaffold-based CTE [21–25,28,119]. Other natural biomaterials used in this context include alginate, a polysaccharide derived from algae used in ECT constructs [26,120] due its high biocompatibility and appropriate chemical and mechanical properties [121], and albumin fibers, which have been used to create biocompatible scaffolds of high porosity and elasticity [29,122].

The combination of different approaches has enabled the development of scaffold systems of increasing complexity, thus bringing them morphologically closer to heart tissue. For example, researchers have generated fibrous scaffold with spatially distributed cues that enabled CM alignment within the patch [29,123,124]. However, major limitations of CTE still remain the generation of thick constructs (over ~100 μm in thickness) and the lack of electromechanical coupling between the cardiac patch and the host myocardium [100,112]. The generation of ECT constructs with a clinically relevant size requires ensuring that appropriate levels of oxygen and nutrients are maintained within the construct to satisfy the metabolic demand of CMs. Perfusion bioreactor systems pioneered by the Vunjak-Novakovic laboratory have proven to be of great value for the generation of thick ECT constructs full of viable cells with aerobic metabolism. In this case, cells were seeded and cultured in porous collagen scaffolds (11 mm in diameter, 1.5 mm in thickness) under continuous perfusion for 7 to 14 days, which led to the formation of contractile thick cardiac tissues [113,125]. More complex bioreactor systems designed to perfuse ECT constructs while also delivering electrical signals mimicking those in the native heart have also been developed [26,27,32]. Maidhof et al. used NRCMs seeded under perfusion into porous poly(glycerol sebacate) (PGS) scaffolds (8 mm in diameter, 1 mm in thickness), which were maintained under continuous perfusion at a flow rate of 18 μL/min and electrically stimulated at a frequency of 3 Hz. After 8 days, the combination of perfusion and electrical stimulation resulted in cell elongation, structural organization and improved contractility of ECT constructs [27]. Recently, our laboratories have reported the generation of 3D engineered thick human cardiac macrotissues (CardioSlices). Human iPSC-CMs were seeded together with human FBs into large 3D porous collagen/elastin scaffolds and cultured under perfusion and electrical stimulation in a custom-made bioreactor. Two weeks after culture, stimulated ECT constructs exhibited contractile and electrophysiological properties close to those of working human myocardium [32].

In addition to scaffolds made from synthetic or natural biomaterials, the use of matrices obtained by decellularizing native tissues has gained popularity for CTE. The process of decellularization allows obtaining natural ECM that can be used to mimic the native tissue structure. In essence, decellularized ECM could be recellularized with CMs or mixtures of CMs and other cell types, or with PSCs that would be differentiated in situ toward the desired cell types [126]. Tissues from a wide variety of sources including human, animals and plants have been used for this purpose [112]. The porcine heart is a prime example of tissue source for animal-derived decellularized scaffolds, due to its large size and to it being a preferred experimental model for cardiovascular research. In this case, it has been reported that the decellularization procedure allows obtaining a cardiac scaffold with preserved vasculature, mechanical integrity and biocompatibility [127]. Nevertheless, limitations noted with this approach include the extent of preservation of the ECM composition (which can be altered by the decellularization process), the difficulty in recellularizing the scaffold with clinically relevant numbers of CMs (in the order of billions), and the risk of eliciting immune rejection [112]. The issue of immune intolerance of animal-derived decellularized scaffolds has prompted research on plant-derived biomaterials as a source for ECT constructs [128]. Even though promising results have been reported using biomaterials derived from decellularized apple [129,130], spinach and parsley leaves [131], along with other cellulose-based scaffolds [132,133], further evaluation will be necessary to assess the usefulness of this type of materials for in vitro bioengineering and in vivo therapeutic applications.

Alternatively, human tissue might be a more appropriate source for decellularized ECM for therapeutic purposes, as it would address some of the limitations of animal- and/or plant-sourced materials described above [134–137]. In this respect, studies by Sanchez et al. demonstrated that the human acellular heart matrix can serve as a biocompatible scaffold for recellularization with parenchymal and vascular cells [138]. Moreover, Guyete et al. also used human decellularized heart tissue, which was in this case recellularized with iPSC-CMs and maintained in a custom-made bioreactor that provided coronary perfusion and mechanical stimulation. After 14 days in culture, the recellularized cardiac segment presented high metabolic activity and contractile function but exhibited low maturation state [139].

3.3. Hydrogels

Embedding suitable cells in hydrogels provide important 3D information cues and, in the context of CTE, the constructs generated in manner are typically known as cardiac grafts. Hydrogels are among the most widely studied types of biomaterials in CTE (see Table 1). In particular, hydrogel-based materials have been shown to provide structural/mechanical support to cells [140], promote vascularization [141] and cell migration, differentiation and proliferation [142], and to improve cardiac function after implantation in murine and porcine models of MI [30,45,103]. Hydrogels can be made from different biomaterials that are usually classified into three types: natural (type I collagen, fibrin, gelatin, alginate, keratin, among other), synthetic [polycaprolactone (PCL), polyethylene-glycol (PEG), PLA, PGA and their co-polymer PLGA], and hybrid hydrogels, which are made by combining natural and synthetic polymers [143,144]. Natural-based hydrogels are often preferred for generating ECT constructs because of their high bioactivity, biocompatibility and biodegradability [142].

Cardiac "bundles" are the most common structures generated when using hydrogel-based systems and are cylindrical ECT constructs in the form of cables, ribbons or rings [112]. These structures are usually formed by embedding CMs from various sources within hydrogels made up of fibrin, type I collagen or other biomaterials, and maintaining them in culture until constructs become spontaneously contractile. The formation of these bundles results in self-alignment and anisotropic organization of CMs, which is a hallmark of cell maturation. Moreover, these constructs provide an easy way to analyze the electrical and mechanical properties of CMs, thus enabling the readily evaluation of their maturation state and facilitating their use in drug screening and toxicity assessment platforms [16,36,40,48,50,51,55,145–147].

In a pioneering approach, Eschenhagen and Zimmermann generated cardiac bundles (which they termed engineered heart tissues, or EHTs) by casting a mixture of NRCMs and a blend of type I collagen type I and Matrigel into cylindrical molds. Under conditions of mechanical stretching, the resulting ring-shaped constructs exhibited improved contractile function and a high degree of cardiac myocyte differentiation [33]. In subsequent work, five of such rings were stacked on a custom-made structure creating multiloop tissue constructs that survived after implantation and improved the cardiac function of infarcted rats [103]. Using the same principle, Kensah et al. produced cardiac bundles by seeding NRCMs with FBs in a collagen/Matrigel hydrogel into a Teflon casting mold between two titanium rods and subjected to mechanical and/or chemical stimulation [34]. Similarly, human ECT constructs have also been generated by casting a cell/hydrogel suspension in different types of molds between or around flexible posts. Schaaf et al. used hESC-CMs in a fibrin hydrogel seeded into an agarose casting mold between 2 elastic silicone posts for 5 weeks [37]. Controlling the 3D microenvironment has been further reported to induce spatial organization and promote CM maturation in hydrogel-based systems. In a study by Thavandiran et al., hESC-CMs and hESC-derived cardiac FBs were seeded in a collagen/Matrigel hydrogel into polydimethylsiloxane (PDMS) microwells with integrated posts. The authors compared two-well designs side by side: an elongated microwell containing posts in both extremes (capable of inducing uniaxial mechanical stress) and a square well containing posts around the edges (thus effecting biaxial mechanical conditioning). These studies demonstrated that constructs on elongated microwells showed comparatively better aligned sarcomeres and more elongated and longitudinally oriented CMs [38]. In turn, the Bursac laboratory created thin (~70 µm in thickness) 3D sheet-like constructs of large surface dimensions (7×7 mm) by casting hESC-derived CMs in fibrin hydrogels into PDMS molds with hexagonal posts, resulting in improved maturation at the functional (conduction velocities of up to 25 cm/s and contractile forces and stresses of 3.0 mN and 11.8 mN/mm^2, respectively) and structural (increased sarcomeric organization and expression of cardiac genes) level [100]. Similarly, Turnbull et al. generated human ECT constructs with hESC-derived cells mixed in a collagen/Matrigel hydrogel in rectangular PDMS casting molds with integrated posts at each end and removable inserts. Forty-eight hours after casting, the inserts were removed from the mold, allowing the self-assembly of the tissues between the two flexible posts, which were used as force sensors. The resulting tissues exhibited typical features of human newborn myocardium tissue including contractile, structural and molecular characteristics [42].

Similar to hESC-CMs, iPSC-CMs also have been successfully cultured in hydrogel-based structures [11,40,46,47,49–51,53,55] (see Table 1). The Radisic laboratory pioneered the use of hiPSC-CMs to generate human ECT constructs by developing a platform in which cells in a collagen hydrogel organized around a surgical suture in a PDMS channel. The resulting 3D microstructures (3 mm^2) were termed Biowires and contained aligned CMs that exhibited well-developed striations and showed improved cardiac tissue function after electrical stimulation [40]. A further improvement to this platform was the use of a polytetrafluoroethylene tube that allowed perfusion of the ECT microstructures and facilitated their use for drug toxicity testing [41]. More importantly, three independent studies reported in 2017 on the generation of clinical-size cardiac tissues by using hydrogel-based systems and hiPSC-CMs. Shadrin et al. generated human cardiac tissues of 36×36 mm that showed electromechanical properties close to those of working myocardium (conduction velocity of 30 cm/s and specific forces of 20 mN/mm^2) by seeding hiPSC-CMs in a fibrin hydrogel into square PDMS molds [11]. Using a mixture of type I collagen and Matrigel with hiPSC-derived CMs and endothelial and vascular cells (in a 3:1:1 ratio), Nakane et al. generated rectangular ECT constructs with different shapes (bundles and mesh junctions, parallel bundles, plain sheets) and sizes (from 15×15 mm to 30×30 mm). They analyzed the association of CM orientation and survival with construct architecture and found that bundles and mesh junctions resulted in the highest myofiber alignment and lowest percentage of dead cells. Moreover, functional integration was observed after

4 weeks of transplantation onto rat uninjured hearts [50]. Large ECT constructs (35 × 34 mm) were also generated in the Zimmermann laboratory by seeding hiPSC-CMs and FBs in a collagen hydrogel on a 3D-printed construct holder with flexible poles in a hexagonal casting mold [51]. In a further refinement of this approach, Gao et al. generated human ECT constructs of 4 × 2 cm comprising hiPSC-derived CMs, SMCs, and ECs (3:1:1 ratio) in a fibrin hydrogel and maintained them in culture on a rocking platform. After 7 days of culture, the constructs showed improved electromechanical coupling, calcium-handling, and force generation [53].

Synthetic hydrogels have received comparatively less attention for CTE than those of natural origin. Ma et al. used PEG to create cardiac microchambers (100 to 300 μm in height) that induced spatial organization of hESCs and hiPSCs and directed their cardiac differentiation [43,44]. In addition, hybrid hydrogels have a great potential for CTE as they can mimic biological properties of the ECM and, at the same time, be tuned to suit the mechanical properties expected or desired for cardiac constructs [148]. Despite their great potential, much research is still needed to ascertain the specific advantages that synthetic and hybrid hydrogels may have over commonly used natural hydrogels in the context of CTE. At any rate, irrespective of the type of hydrogel used, current 3D cardiac grafts are constrained in maximum thickness by the ~300 μm limit of oxygen diffusion in passive culture systems and, therefore, while ideal for miniature structures with some tissue-like functionalities, they may not be suited for applications that require fully capturing the high complexity of the native heart tissue structure.

3.4. Cardiac Spheroids (and 'Organoids')

Spheroids are scaffold-free 3D cell constructs that rely on cell aggregation or self-organization and simulate aspects of the native cell microenvironment [149]. Cardiac spheroids can be constructed with CMs [150,151] and also include other cardiac resident cells such as FBs [152] or ECs [104]. Different percentages of various cell types have been tested for the generation of cardiac spheroids [153,154]. Spheroid-based systems are attractive to scientists for studying heterocellular interactions and drug effects because they only need low cell numbers to be formed. On the other hand, the absence of functional architecture limits the physiological analyses of the cells, like force generation and electrical conduction. Nevertheless, using the spheroid-based systems to deliver CMs into the damaged region of the heart has been reported. For instance, intramyocardial injection of cardiac spheroids in mice resulted in higher engraftment rates and improved electrical coupling with host myocardium, compared to single cell injection, which reveals potential for future clinical applications [155,156]. Moreover, several research groups are working on the generation of thicker functional structures using multicellular spheroids for further clinical testing. Noguchi et al. created a scaffold-free 3D tissue construct based on self-organization of 1×10^4 spheroids. The individual spheroids were, in turn, obtained by combining 3 cell sources: NRCMs, hECs and hFBs in a 7:1.5: 1.5 ratio. ECT constructs generated in this way remained adherent and presented signs of vascularization seven days after transplantation onto the heart of nude rats [157]. In a related approach, scientists created a biomaterial-free cardiac tissue by 3D printing multicellular cardiac spheroids that displayed spontaneous beating and ventricular-like action potentials, which were engrafted as well into the rat heart tissue [158]. A final example of this approach is the study by Kim et al., who generated elongated 3D heterocellular microtissues by fusing together multicellular cardiac spheroids containing CMs and cardiac FBs. The authors demonstrated that such microtissues formed an electrical syncytium after seven hours in culture [159].

Similar to spheroids, organoids are scaffold-free 3D cell constructs that simulate aspects of the native environmental conditions. However, a critically important characteristic of organoids that sets them apart from spheroids, is that organoids contain organ-specific cell types that self-organize in a way that is architecturally similar to that of the native organ [160]. While some researchers may use the terms organoids and spheroids interchangeably, there are important differences between them. Both spheroids and organoids can be generated from PSCs, PSC derivatives, or tissue-specific stem/progenitor cells. When using PSCs, the technique of organoid formation is inspired in the embryoid body system [160], 3D aggregates of PSCs where cells undergo specification and differentiation into cell derivatives of the three main embryo germ layers [161]. In contrast, spheroids are much simpler than organoids in terms of the cell types that conform them, do not self-organize into organ-like patterns or structures, and depend to a lesser extent on ECM properties and composition [162]. Even though several published reports describe the generation of human "cardiac organoids", these rely on direct cardiac differentiation of hiPSC-derived embryoid bodies [163] or aggregation of cardiac cell types (CMs, cardiac FBs and cardiac ECs) [164–168], which actually represent typical examples of spheroids [160,161]. We believe that the use of the term "cardiac organoids" in this context is misleading since the structures generated in those studies lacked the organ-like complexity characteristic of true organoids. Very recently, Lee et al. have described the generation of *bona fide* cardiac organoids in the mouse system that showed atrium- and ventricular-like structures highly reminiscent of the native embryonic heart. For this purpose, the authors induced sequential morphological changes in PSC-derived cells by including a laminin-entactin complex in the ECM and FGF-4 in serum-free medium [169]. Perhaps the application of similar approaches to hiPSC derivatives could lead to the generation of true human cardiac organoids containing relevant organ-specific cell types with the capacity to self-organize in organ-like structures.

3.5. Heart-on-a-Chip

Microfluidic cell culture technologies enable researchers to create in vitro cell microenvironments that mimic organ-level physiology [170]. The term 'organ-on-a-chip' is generally applied to a microphysiological system, including the slides or plates that are connected to microfluidic devices to control perfusion of culture medium and exposure of defined stimuli [171]. Heart-on-a-chip technology refers specifically to microphysiological systems mimicking the function of heart tissue. In vivo-like cardiac cell culture systems could lead to a better understanding of (1) cardiac cell physiology; (2) cardiotoxicity of drugs intended for human used; (3) personalized treatments for CVD patients; and (3) mechanisms of heart regeneration [39,172,173]. In physiological conditions, the heart tissue is in direct contact with body fluids such as blood and lymph that exert physical forces (shear stress) on the cells. This continuous flow stimulation determines the cardiac cells structure, phenotype, intra- and extracellular interactions [174]. In vivo-like cardiac cell culture systems try to replicate these conditions in vitro. Thus, the heart-on-a-chip system provides suitable conditions to imitate biochemical, mechanical, and physical signals characteristic of heart tissue [175–177]. For example, it was shown that continuous perfusion enhances cell proliferation and parallel alignment of cells compared to static conditions [178]. In addition to perfusion, integrating mechanical and electrical stimulation into heart-on-a-chip devices also improves the maturation state of CMs [54,179–181], one of the key features for successful modeling of cardiac diseases [182]. Moreover, heart-on-a-chip systems are amenable to parallelization and thus to be used in high-throughput assays for drug screening and cardiotoxicity testing [173,183]. Particularly, the possibility of using hPSC-CMs brings an additional level of personalization to heart-on-a-chip systems [46,54,104,184–187]. For example, the Radisic laboratory developed a powerful platform, dubbed AngioChip, that integrated tissue engineering and organ-on-a-chip technologies to produce vascularized polymer-based microfluidic cardiac scaffolds. Such a platform can be used to generate both in vitro heart tissue models and in vivo implants for potential clinical application [186].

3.6. 3D Bioprinting

3D bioprinting is one of the latest additions to the tissue engineering toolbox, and one that could be used to create complex and large vascularized tissues [188,189]. Several methods of 3D bioprinting have been used in the context of CTE, including cell-laden hydrogel 3D structures [190], inkjet bioprinting [191], laser-assisted bioprinting [192], and extrusion-based bioprinting [193]. Biomaterials used in 3D bioprinting are based on piezo-resistive, high-conductance, and biocompatible soft materials. Gaetani et al. bioprinted a 2 × 2 cm ECT construct using human cardiac progenitor cells and alginate matrices, which was maintained for 2 weeks in vitro [120] or transplanted onto rat infarcted hearts where it led to cell engraftment [194]. Generation of 3D-bioprinted vascularized heart tissues using mouse iPSC-CMs with human ECs in a PEG/fibrin hydrogel has been recently reported showing improved connectivity to the host vasculature after subcutaneous transplantation in mice [195]. Despite the early stage of development, 3D bioprinting is a very promising technology for recapitulating the complex structure of heart tissue and already shows enormous potential in CTE. In a recent study, Noor et al. succeeded in bioprinting thick (2 mm) 3D vascularized and perfused ECT constructs that had high cell viability using an extrusion-based bioprinter. As bioinks, the authors used an ECM-based hydrogel derived from human decellularized omentum containing hiPSC-CMs, and gelatin containing iPSC-derived ECs and FBs. Computerized tomography of a patient's heart was used to reproduce in vitro the structure and orientation of the blood vessels into the tissue. Bioprinted ECT constructs were transplanted into the omentum of rats and analyzed after seven days, when elongated and well-aligned CMs with massive striations were observed [56]. This study demonstrated that the possibility of using fully personalized materials makes 3D bioprinting technology very promising for clinical application by reducing the risk of immune rejection after transplantation. However, the system is still limited and further analyses should be performed to evaluate if heart tissue bioprinted in this manner could sustain normal blood pressure levels after transplantation [56].

3D bioprinting can also be combined with microfluidic systems to provide superior organ-level response with greater prediction of drug-induced capability [196]. On the other hand, recent advances in nanomaterial technology present an attractive platform for the creation of ECT constructs for biomedical applications. Electrospinning technology allows creating nanofibers with controlled dimensions and further development of 3D structures [197]. In addition, the nanofibrous structure provides appropriate conditions for pluripotent cells to anchor, migrate and differentiate [198]. Increasing attention is being given to these types of structures due to their distinct mechanical properties, high porosity and potential to induce formation of aligned tissues that can be successfully implanted to the heart [199].

3.7. Other Structures

So far, we have described the variety of approaches undertaken for the generation of ECT constructs that reproduce increasingly complex features of the human myocardial tissue. However, several groups have pursued approaches intended to model whole heart chambers. In an early attempt in 2007, scientists created a "pouch-like" single ventricle using a mixture of hydrogel composed of type I collagen and Matrigel and NRCMs, which was cultured in an agarose casting mold with the dimensions of a rat ventricle. After formation, it was transplanted onto the rat heart showing limited functional integration [200]. One year later, Lee et al. created a "cardiac organoid chamber" by seeding NRCMs mixed with a collagen/Matrigel hydrogel around a balloon-like shaped mold with a thickness of 0.65 mm and sizes of 4.5 to 9 mm for the unloaded outer diameter, equivalent to the sizes of embryonic rat hearts at 9.5–11 days of development. The authors succeeded in the creation of a heart ventricle but with a moderate contractile capacity [201]. In subsequent work, the same group used a similar method to produce a functional human cardiac chamber using hESC-CMs embedded in collagen-based hydrogel. They proved that such technology could facilitate drug discovery as it provides the capacity to measure clinically meaningful parameters of the heart like ejection fraction and developed pressure, as well as electrophysiological properties [202]. The Parker laboratory has also recently generated tissue-engineered ventricles by using nanofibrous scaffolds composed of PCL and gelatin seeded with either NRCMs or hiPSC-CMs on a rotating ellipsoidal collector. The authors showed that cells aligned following the fiber orientation, thus recapitulating the orientation of the native myocardium, although the contractile function was much smaller than those of the native rat/human ventricles most likely owing to the small thickness of the chamber [203]. Notably, the Feinberg group printed human cardiac ventricles in 2019 using the freeform embedding of suspended hydrogels (FRESH) method. In this case, type I collagen was used as a support material to create an ellipsoidal shell that was filled with hESC-CMs in a fibrinogen suspension. These engineered tissues displayed synchronized functional activity [204]. Finally, Kupfer et al. have very recently generated a 3D bioprinted chambered muscle pump using an optimized bioink formulation of ECM proteins that allowed hiPSC proliferation prior to differentiation. In this way, 3D bioprinted hiPSCs underwent cell expansion and differentiated into CMs in situ, yielding contiguous muscle walls of up to 500 μm in thickness. The resulting human chambered organoids showed macroscale contractions and continuous action potential propagation and were responsive to drugs and to external pacing [205].

4. Maturation of Engineered Cardiac Tissues

Regardless of the cell source or type chosen, biomaterial used, or method employed to generate ECT constructs, a critical aspect common to all of them is the requirement to promote or maintain the maturation state of CMs in culture. The application of biophysical stimuli mimicking those experienced by CMs in the native heart has been investigated in greatest detail and found to be among the most efficient ways to obtain highly mature ECT constructs. Thus, a variety of existing CTE technologies have incorporated the means to apply relevant biophysical stimuli during culture to promote cellular growth, orientation and/or maturation, and are reviewed below and summarized in Figures 1 and 2.

4.1. Mechanical Stimulation

Mechanical cues are among the most important stimuli during heart development, ultimately determining the overall size and architecture of the organ, as well as regulating physiological postnatal changes in the heart resulting from the growth process, pregnancy, sports or injury [206]. Building on this notion, early studies investigated the importance of mechanical stimulation for CM maturation using embryonic chick cells [207] and NRCMs [60] embedded in collagen gels that were subjected to cyclic stretching. In later work, Zimmermann et al. used linear cyclic stretching to generate ECT constructs that exhibited contractile and electrophysiological properties similar to those of the rat native tissue. Specifically, the authors used circular molds to cast collagen-based constructs that, after 7 days in culture, were transferred to a device applying circular cyclic stretching. Transplantation of the constructs generated in this way to the healthy rat heart showed its fast integration and the high percentage of terminal cardiac differentiation achieved [33]. The positive role of mechanical stimulation for CM maturation was also reported by Kensah et al. [34]. In this case, cyclic mechanical stretching together with electrical stimulation of NRCMs resulted in improved maturation at the cell and tissue levels, which was proved by the increased cardiac gene expression and improved functional properties of the construct [34]. Similar results were obtained in another study where ECT constructs were subjected to mechanical compression and to the shear stress induced by medium perfusion in a bioreactor system. Additionally, elevated levels of bFGF and TGF-β and higher expression of Cx43 were observed under intermittent compression, which suggested improvement in cell-cell communication [28]. Moreover, combining mechanical stimulation with other strategies such as co-culturing CMs with ECs [36] or using highly defined media based on specific growth factors (FGF-2, IGF-1, TGF-β1, VEGF) [51] further increased the structural and functional maturation of ECT constructs derived from hESC and hiPSC [31,36,51]. Overall, the application of mechanical stimulation improves sarcomere organization, cardiomyocyte contractility, increases tensile stiffness and cell size.

On the other hand, the application of mechanical stimulation alone may not be enough for obtaining mature cardiac tissues. During heart development, CMs experience mechanical forces and electrical fields resulting from the synchronized contraction of the organ itself, which greatly influence their further differentiation and maturation [208,209]. For this reason, recent CTE studies used the combination of these two stimuli [35,38]. For example, Ruan et al. compared the influence of mechanical stimulation on the maturation of cardiac constructs alone or in combination with electrical stimulation. The authors created hiPSC-derived collagen-based 3D scaffolds that either underwent static stress conditioning for 2 weeks or were subjected to mechanical stimulation for one week followed by a second week with additional electrical stimulation. They demonstrated that, in response to mechanical loading alone, the engineered tissue showed increase in CM alignment and size, and improved contractility, force generation and passive stiffness of the constructs. The addition of electrical stimulation further improved force generation and increased expression of Ca-handling proteins but without further changes in cell size and alignment [47].

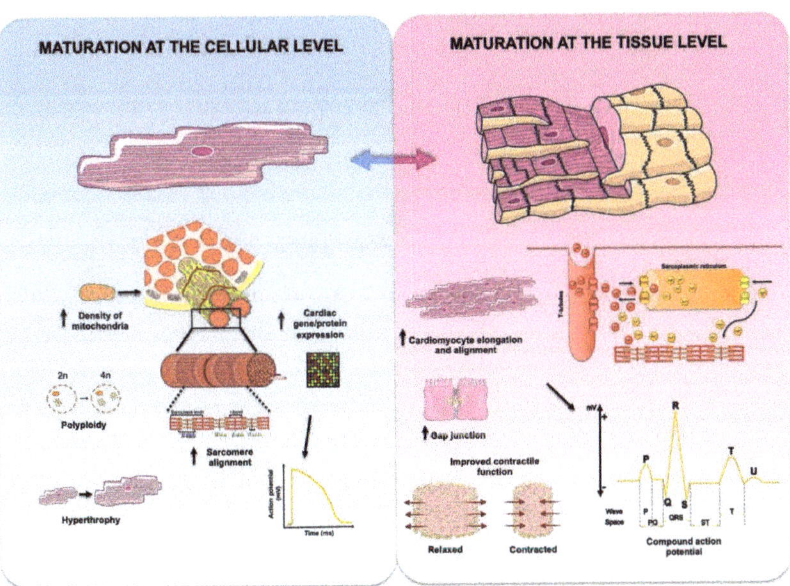

Figure 2. General features of cardiac maturation at the cellular and tissue levels. Cardiac maturation at the level of cardiomyocytes (blue-shaded background) can be experimentally ascertained by a more adult-like cardiac gene expression profile, changes in sarcomeric ultrastructure resulting in sarcomere elongation and alignment, increased mitochondrial density to accommodate for higher energy demand, and massive growth in cardiomyocyte size (hypertrophy) with increased percentage of polyploid cells. Electrophysiologically, cardiomyocyte maturation results in a more negative resting membrane potential and increased duration and amplitude of action potential. Cardiac maturation at the tissue level (pink-shaded background) is characterized by elongated cardiomyocytes preferentially aligned along a specific axis, increased expression of Cx43 and distinct localization of gap junction proteins, and improvements in impulse propagation and synchronization of cardiac contraction, leading to cardiomyocytes performing as a functional syncytium. Maturation at the tissue level results in increased contractility and force generation and, electrophysiologically can be registered as compound action potentials reminiscent of surface ECG recordings. Figure created with BioRender.com and smart.servier.com.

4.2. Electrical Stimulation

Electromechanical coupling is one of the major characteristics of mature CMs resulting in their synchronous response to the electrical pacing signals following with contractile function and pumping. The process of cardiac muscle contraction is generated by the depolarization of the CM membrane, which in the end leads to calcium release from the sarcoplasmic reticulum and subsequent contraction of the myofibrils [210]. Thus, researchers began to test the effects of electrical stimulation to train cardiac constructs. The Vunjak-Novakovic laboratory pioneered the application of electrical stimulation to ECT constructs by culturing NRCMs on thick collagen sponges and subjecting them to electrical pulses at 1 Hz for up to 8 days. They observed that electrical field stimulation induced cell alignment and improved ultrastructure organization along with an increase in the amplitude of synchronous contractions, whereas non-stimulated constructs displayed poorly developed ultrastructural features [23]. For the purpose of electrical stimulation, the group initially developed a simple electrical stimulation chamber with two carbon rod electrodes placed lengthwise along the bottom of a Petri dish that also allowed to evaluate the functional properties of the constructs [25]. In further refinements, more complex systems were developed that better mimicked conditions in the native heart, including a bioreactor system that allowed simultaneous perfusion and electrical stimulation of the ECT constructs [27]. The benefits of this combined system were highlighted by the presence of elongated CMs with marked striations, a well-defined intracellular structure, high levels of expression and proper localization Cx43, and increased conduction velocity and enhanced contractility of the construct. The combination of perfusion and continuous electrical stimulation has been extensively used by other research groups to promote functional maturity. For instance, Barash et al. integrated custom-made stimulating electrodes in a bioreactor system to electrically stimulate thick rat ECT constructs under perfusion. After 4 days in culture in this system, the constructs showed enhanced levels of Cx43 and improved cell striation and elongation [26]. In another study, Xiao et al. designed a microfabricated bioreactor with integrated electrodes to engineer mature rat microtissues for assessing in vitro pharmacological effects [41].

The positive effect of electric stimulation on CM maturation has been validated for human cells using hESC- and hiPSC-derived ECT constructs. Nunes et al. were the first to use a daily step-up protocol to test the effect of increasing stimulation frequency on human 3D self-assembled cardiac bundles (Biowires). Specifically, ramping up stimulation frequency over a 1-week period enhanced maturation, with a 1 to 6 Hz paradigm resulting in better results than a 1 to 3 Hz, as judged by improved sarcomeric organization, increased number of desmosomes and subsequent higher conduction velocities, and overall more mature Ca-handling and electrophysiological properties [40]. More recently, Ronaldson-Bouchard et al. generated ECT constructs of early-stage hiPSC-CMs (immediately after the initiation of spontaneous contraction) in fibrin hydrogels and maintained them in culture for one month while increasing the frequency of electric stimulation from 2 Hz to 6 Hz at a rate of 0.33 Hz per day. Tissues cultured under these conditions displayed a remarkably advanced state of maturation including adult-like cardiac gene expression and tissue ultrastructure, a positive force-frequency relationship, and functional calcium handling [55].

While the effects of electric stimulation have been analyzed in detail on the maturation of human ECT microconstructs and of perfused rat macroconstructs, perfusion bioreactors incorporating electrical stimulation have been more rarely employed to generate human ECT macroconstructs [43,51,52]. Indeed, only recently have hiPSC-derived cells and 3D thick scaffolds been combined with continuous electrical stimulation to develop macrotissues of human CMs. For this purpose, the group of one the authors developed a parallelized perfusion bioreactor with custom-made culture chambers endowed with electrostimulation capabilities. We showed that hiPSC-CMs readily survive and mature after long-term culture within large (10 mm in diameter by 2 mm in thickness) 3D porous scaffolds, giving rise to macroscopically contractile tissues, which we termed 'CardioSlices'. More importantly, we found that continuous electrical stimulation of these large ECT constructs for 2 weeks promoted the alignment and synchronization of CMs, and the emergence of complex cardiac tissue-like behaviors, including the spontaneous generation of electrophysiological signals strikingly similar to human electrocardiograms, which could be readily registered from the construct surface, and a response to proarrhythmic drugs that was predictive of their effect in humans [32].

4.3. Coculture with Non-CMs

Since cardiac cells other than CMs play important roles in heart homeostasis, several groups have explored whether their incorporation into ECT constructs improved cardiac maturation. Indeed, epicardial cells closely interact with CMs and regulate myocardial wall development [211] and, through paracrine mechanisms, influence heart electrophysiology [212,213]. Cardiac endocardial cells are found in direct contact with CMs, supporting their metabolic activity and regulating their contractility via paracrine signals [214]. Studies on neonatal and fetal CMs have shown that inclusion of cardiac FBs and/or ECs into ECT constructs improved the morphological and functional maturation of CMs [113,215,216]. Thavandiran et al. directly tested whether the cell composition of ECT constructs affected their functionality. For this purpose, the authors incorporated ESC-derived CMs and CFs in different ratios into collagen based microtissues, and found that the 3:1 (CMs:CFs) ratio provided the best microenvironment based on increased expression of cardiac maturation genes [38]. Similarly, Gao et al. generated a clinical size human ECT construct using hiPSC-CMs, -ECs and -SMC (2:1:1) in a fibrin hydrogel and reported a significant functional maturation of the construct in vitro and high engraftment rates in vivo [53]. Thus, the combination of more than one cell type within the ECT constructs can also aid in promoting their morphological and functional maturation [36,51,58,59].

5. Functional Assessment of Engineered Cardiac Constructs

The development of CTE approaches to increase the cardiac maturity of ECT constructs is necessarily mirrored by advances in the methodologies used to evaluate their degree of maturation. As described in the previous sections, the vast majority of human ECT constructs developed to date are hydrogel based and do achieve measurable increases in CM maturation when analyzed at the cellular level, both in miniature constructs and thin sheets. Specific improvements in CM maturation were most frequently analyzed in terms of gene/protein expression profile, sarcomeric structure and/or cell electrophysiological properties being closer to those of adult CMs. Some studies also analyzed functional tissue-like properties of the engineered constructs, such as force generation and contractility, and a few studies also measured propagation of the electrical signals by analyzing calcium transients or voltage-sensitive dyes (see Table 1). The advantages and limitations of the technologies used to assess the functional maturation of ECT constructs are summarized in Table 2 and discussed below.

Table 2. Main approaches to functional assessment of ECT constructs.

Approach	Advantages	Disadvantages	Ref.
Patch-clamp	- Direct recording of action potentials - Ideal for recording isolated CMs	- Requires specialized equipment/personnel - Requires direct cell visualization - Not well suited for ECT constructs	[217]
Multielectrode arrays	- Allows assessment of single CMs and ECT constructs - Noninvasive method. - Allows long-term recording	- Measurements restricted to small area of the ECT construct. - Incomplete information of action potential propagation	[59,108,217–219]
Optical mapping	- Allows assessment of CM monolayers and ECT constructs - Noninvasive method - Unsophisticated equipment	- Indirect measurement of electrophysiological parameters - Cytotoxicity of dyes may limit measurement duration and preclude sample reassessment	[29,35,39,48,55,118,220,221]
Force transducers	- Allows assessment of single CMs and ECT constructs - Allows measuring direct and indirect contractile force - Allows long-term measurements	- Unsuitable for assessing contractile stress from single CMs within ECT constructs	[36,39,42,47,48,50,57,217]

5.1. Patch-Clamp and Microelectrodes

Patch-clamp is the gold standard technique used to study the electrophysiological properties of excitable single cells such as PSC-CMs [37,40,49,51,55]. This technique allows measuring transmembrane voltage and currents using pulled glass micropipettes [222]). In addition, electrical coupling between two cells can be measured by applying patch electrodes to each of them separately [223]. Changes in membrane potential like action potentials can be measured by a whole-cell current clamp or using sharp microelectrodes with narrow bores [33,49,224], which allow precise registering of action potentials due to preservation of the true intracellular milieu [217]. While ideal for recording isolated CMs, patch-clamp technology presents important limitations for the study of ECT constructs. In addition to the inherent low throughput of the technique and the need for specialized equipment and highly trained personnel, patch-clamp requires direct visualization of the cell(s) being analyzed, which is not straightforward when cells are inside an ECT construct. Moreover, patch-clamp measures the electrophysiological properties of CMs themselves, rather than those of the construct as a tissue.

Functional assays used to evaluate the cardiac maturation state of the constructs, both at the cellular and the tissue levels, indicating the main advantages and disadvantages of each approach, as well as references to publications where those approaches were used in the context of CTE. CM, cardiomyocyte; ECT, engineered cardiac tissue.

5.2. Multielectrode Arrays

Multielectrode arrays (MEAs) are devices that contain arrays of electrodes distributed over a small surface on which cells can be cultured directly. MEAs are generally used for millimeter-scale analysis of extracellular potential within beating clusters of CMs [43,225], where they can measure extracellular field potential, spontaneous beating rate, repolarization characteristics, and conduction velocity and trajectories [226]. Unlike patch-clamp, MEA technology can be used to assess the electrophysiological properties of CMs at both the single-cell and tissue levels in cell monolayers or ECT constructs [59,108,218], and allows long-term recording of electrophysiological parameters noninvasively [219]. The main limitations of MEA technology for the functional evaluation of ECT constructs are that measurements are restricted to the small areas detected by the electrodes, and that it provides incomplete information about the shape of the propagated action potentials [217].

5.3. Optical Mapping

Optical mapping is a non-invasive method for the assessment of electrical activity that is well established for the study of cell monolayers [227] and has also been applied to ECT constructs [29,35,40,48,55,99,118,220]. This technique relies on the use of voltage-sensitive (di-4-ANEPPS, di-8-ANEPPS and RH237) [228] or Ca2+-sensitive fluorescent dyes (rhod-2 and fluo-4). The main advantage of this technique is the ability to monitor the electrical activity and shape of action potentials from many areas of the construct. On the other hand, these dyes have cytotoxic effects that limit the maximum duration of experiments [221]. Moreover, the dyes are light-sensitive and subject to photobleaching, further limiting the time of analysis and reassessment of the sample. Optical mapping has been combined with MEA technology for the study of conduction velocity, direction of the signal and visualization of arrhythmias [229].

5.4. Force Transducers

The mechanical force exerted by contracting CMs can be measured in single cells as well as in whole ECT constructs [36,42,47,48,50,57,100]. For single-cell measurement, one end of the CM is attached to a rigid steel needle and the other end to a piezo-electric motor for its stretching. Force measurements are then taken by a sensitive force transducer connected to the needle [230]. For tissue-level properties, the whole construct is connected to a force transducer [36,47,48,100]. Here, the main disadvantage is the inability to measure the contractile stress from each CM. Advantages, on the other hand, include the larger magnitude of the signals obtained and the possibility to measure the therapeutic potential of ECT constructs before implantation [217].

To date, those are the most complex multicellular (tissue-level) behaviors analyzed in the context of CTE. New methods with greater sensitivity will be needed for evaluating tissue-like functional properties. Recently, the group of one the authors has explored the use of surface electrodes to record the electrical activity of thick ECT constructs. The bioelectrical signal detected in this way was extremely informative of the activity of electrically-active cells within the construct and, most importantly, of the intercellular coupling at the macro (whole construct) scale, resulting in complex waves very reminiscent of surface ECG recording in humans. In our study, we showed that electrical stimulation of the constructs resulted in improvements in tissue-like properties (shape of ECG-like signals and response to drugs) that could not be predicted by the small increases in maturation achieved at the CM level [32].

6. Summary and Future Directions

One of the most pressing current challenges in CTE is how to achieve a level of cardiac maturation that resembles that of the adult human myocardium. This is even more important for experimental approaches based on PSC-CMs that, while clearly taking the center stage of CTE, are notoriously difficult to mature ex vivo [182,231]. The use of immature cells/constructs could lead to the improper interpretation of in vitro drug testing results and inaccurate prediction of their effect in vivo upon transplantation. We have reviewed here the variety of imaginative approaches researchers have employed to obtain mature CET constructs using mechanical and/or electrical stimulation and coculture with non-myocyte cells, as well as biochemical interventions. These approaches have proven successful in promoting CM maturation from an embryonic-like state to cells displaying late fetal phenotypes. We have also reviewed how transplantation of such CET constructs had positive impact in cardiac function in experimental animal models and even patients. However, it should be noted that, to our knowledge, no approach has yet succeeded in obtaining CMs with maturity features beyond those of early postnatal CMs. Moreover, much of the earlier work attempting to obtain adult-like cardiac constructs concentrated on increasing the degree of maturation at the CM cell level, and had limited success in achieving functional tissue-like properties [40,47,51,55]. Taking into account that the myocardial tissue is a complex system that requires the synchronized operation of CMs as a functional syncytium [232], we contend that improving high-order behaviors at the macroscale (tissue) level could lead to more adult-like phenotypes, even when the maturation state of individual CMs were only marginally improved [32,47]. Future research will be necessary to explore whether the synergistic exploitation of approaches improving cardiac maturation at both the CM and the tissue levels will result in ECT constructs with the features of adult human myocardium. Moreover, pending issues such as improving the functional integration of ECT constructs within the host myocardium, increasing their durability and scalability, along with making them amenable to cryopreservation, surely warrant further research. Considering the multidisciplinary nature of CTE, further developments in the field will require the coordinated efforts from researchers with diverse backgrounds.

Author Contributions: Conceptualization, O.I.-G. and A.R.; writing-original draft preparation, K.T. and O.I.-G.; writing-review and editing, K.T., O.I.-G., M.M.M., F.P., and A.R.; supervision, O.I.-G. and A.R.; project administration, A.R. All authors have read and agreed to the published version of the manuscript.

Funding: Research in the authors' laboratories on this topic is supported by the Spanish Ministry of Economy and Competitiveness-MINECO (RTI2018-095377-B-100), Instituto de Salud Carlos III-ISCIII/FEDER (Red de Terapia Celular—TerCel RD16/0011/0024 and RD16/0011/0005; PI 19/01350), Gobierno de Navarra Departamento de Salud (GNª8/2019), co-funded by FEDER funds, European Union's H2020 Program under grant agreement No 874827 (BRAV∃), AGAUR (2017-SGR-899), and CERCA Programme/Generalitat de Catalunya.

Institutional Review Board Statement: Not applicable.

Informed Consent Statement: Not applicable.

Data Availability Statement: Not applicable.

Acknowledgments: The authors are grateful to all members of the P-CMR[C] for productive discussions.

Conflicts of Interest: The authors declare no conflict of interest. The funders had no role in the design of the study; in the collection, analyses, or interpretation of data; in the writing of the manuscript, or in the decision to publish the results.

References

1. Roth, G.A.; Johnson, C.; Abajobir, A.; Abd-Allah, F.; Abera, S.F.; Abyu, G.; Ahmed, M.; Aksut, B.; Alam, T.; Alam, K.; et al. Global, Regional, and National Burden of Cardiovascular Diseases for 10 Causes, 1990 to 2015. *J. Am. Coll. Cardiol.* **2017**, *70*, 1–25. [CrossRef] [PubMed]
2. Mozaffarian, D.; Benjamin, E.J.; Go, A.S.; Arnett, D.K.; Blaha, M.J.; Cushman, M.; Das, S.R.; de Ferranti, S.; Després, J.-P.; Fullerton, H.J.; et al. Heart Disease and Stroke Statistics-2016 Update: A Report from the American Heart Association. *Circulation* **2016**, *133*, e38–e360. [CrossRef] [PubMed]
3. Bergmann, O.; Zdunek, S.; Felker, A.; Salehpour, M.; Alkass, K.; Bernard, S.; Sjostrom, S.L.; Szewczykowska, M.; Jackowska, T.; dos Remedios, C.; et al. Dynamics of Cell Generation and Turnover in the Human Heart. *Cell* **2015**, *161*, 1566–1575. [CrossRef] [PubMed]
4. See, F.; Kompa, A.; Martin, J.; Lewis, D.; Krum, H. Fibrosis as a Therapeutic Target Post-Myocardial Infarction. *Curr. Pharm. Design* **2005**, *11*, 477–487. [CrossRef]
5. Yacoub, M. Cardiac Donation after Circulatory Death: A Time to Reflect. *Lancet* **2015**, *385*, 2554–2556. [CrossRef]
6. Müller, P.; Lemcke, H.; David, R. Stem Cell Therapy in Heart Diseases–Cell Types, Mechanisms and Improvement Strategies. *Cell. Physiol. Biochem.* **2018**, *48*, 2607–2655. [CrossRef]
7. Menasché, P. Cell Therapy Trials for Heart Regeneration-Lessons Learned and Future Directions. *Nat. Rev. Cardiol.* **2018**, *15*, 659–671. [CrossRef]
8. Mirotsou, M.; Jayawardena, T.M.; Schmeckpeper, J.; Gnecchi, M.; Dzau, V.J. Paracrine Mechanisms of Stem Cell Reparative and Regenerative Actions in The Heart. *J. Mol. Cell. Cardiol.* **2011**, *50*, 280–289. [CrossRef]
9. Nguyen, P.K.; Rhee, J.-W.; Wu, J.C. Adult Stem Cell Therapy and Heart Failure, 2000 to 2016: A Systematic Review. *JAMA Cardiol* **2016**, *1*, 831–841. [CrossRef]
10. Jackman, C.; Li, H.; Bursac, N. Long-Term Contractile Activity and Thyroid Hormone Supplementation Produce Engineered Rat Myocardium with Adult-Like Structure and Function. *Acta Biomater.* **2018**, *78*, 98–110. [CrossRef]
11. Shadrin, I.Y.; Allen, B.W.; Qian, Y.; Jackman, C.P.; Carlson, A.L.; Juhas, M.E.; Bursac, N. Cardiopatch Platform Enables Maturation and Scale-Up of Human Pluripotent Stem Cell-Derived Engineered Heart Tissues. *Nat. Commun.* **2017**, *8*, 1825. [CrossRef] [PubMed]
12. Benam, K.H.; Dauth, S.; Hassell, B.; Herland, A.; Jain, A.; Jang, K.-J.; Karalis, K.; Kim, H.J.; MacQueen, L.; Mahmoodian, R.; et al. Engineered In Vitro Disease Models. *Annu. Rev. Pathol. Mech. Dis.* **2015**, *10*, 195–262. [CrossRef] [PubMed]
13. Sadeghi, A.H.; Shin, S.R.; Deddens, J.C.; Fratta, G.; Mandla, S.; Yazdi, I.K.; Prakash, G.; Antona, S.; Demarchi, D.; Buijsrogge, M.P.; et al. Engineered 3D Cardiac Fibrotic Tissue to Study Fibrotic Remodeling. *Adv. Healthc. Mater.* **2017**, *6*. [CrossRef] [PubMed]
14. Turnbull, I.C.; Mayourian, J.; Murphy, J.F.; Stillitano, F.; Ceholski, D.K.; Costa, K.D. Cardiac Tissue Engineering Models of Inherited and Acquired Cardiomyopathies. *Methods Mol. Biol.* **2018**, *1816*, 145–159. [CrossRef] [PubMed]
15. Wang, G.; McCain, M.L.; Yang, L.; He, A.; Pasqualini, F.S.; Agarwal, A.; Yuan, H.; Jiang, D.; Zhang, D.; Zangi, L.; et al. Modeling the Mitochondrial Cardiomyopathy of Barth Syndrome with Induced Pluripotent Stem Cell and Heart-On-Chip Technologies. *Nat. Med.* **2014**, *20*, 616–623. [CrossRef]
16. Hansen, A.; Eder, A.; Bönstrup, M.; Flato, M.; Mewe, M.; Schaaf, S.; Aksehirlioglu, B.; Schwörer, A.; Uebeler, J.; Eschenhagen, T. Development of a Drug Screening Platform Based on Engineered Heart Tissue. *Circ. Res.* **2010**, *107*, 35–44. [CrossRef]
17. Veldhuizen, J.; Migrino, R.Q.; Nikkhah, M. Three-Dimensional Microengineered Models of Human Cardiac Diseases. *J. Biol. Eng.* **2019**, *13*, 29. [CrossRef]
18. Hulsmans, M.; Clauss, S.; Xiao, L.; Aguirre, A.D.; King, K.R.; Hanley, A.; Hucker, W.J.; Wülfers, E.M.; Seemann, G.; Courties, G.; et al. Macrophages Facilitate Electrical Conduction in the Heart. *Cell* **2017**, *169*, 510–522.e20. [CrossRef]
19. Pinto, A.R.; Ilinykh, A.; Ivey, M.J.; Kuwabara, J.T.; D'Antoni, M.L.; Debuque, R.; Chandran, A.; Wang, L.; Arora, K.; Rosenthal, N.A.; et al. Revisiting Cardiac Cellular Composition. *Circ. Res.* **2016**, *118*, 400–409. [CrossRef]
20. Zhou, P.; Pu, W.T. Recounting Cardiac Cellular Composition. *Circ. Res.* **2016**, *118*, 368–370. [CrossRef]
21. Carrier, R.L.; Rupnick, M.; Langer, R.; Schoen, F.J.; Freed, L.E.; Vunjak-Novakovic, G. Perfusion Improves Tissue Architecture of Engineered Cardiac Muscle. *Tissue Eng.* **2002**, *8*, 175–188. [CrossRef] [PubMed]
22. Radisic, M.; Yang, L.; Boublik, J.; Cohen, R.J.; Langer, R.; Freed, L.E.; Vunjak-Novakovic, G. Medium Perfusion Enables Engineering of Compact and Contractile Cardiac Tissue. *Am. J. Physiol. Heart Circ. Physiol.* **2004**, *286*, H507–H516. [CrossRef] [PubMed]

23. Radisic, M.; Park, H.; Shing, H.; Consi, T.; Schoen, F.J.; Langer, R.; Freed, L.E.; Vunjak-Novakovic, G. Functional Assembly of Engineered Myocardium by Electrical Stimulation of Cardiac Myocytes Cultured on Scaffolds. *Proc. Natl. Acad. Sci. USA* **2004**, *101*, 18129–18134. [CrossRef] [PubMed]

24. Radisic, M.; Marsano, A.; Maidhof, R.; Wang, Y.; Vunjak-Novakovic, G. Cardiac Tissue Engineering Using Perfusion Bioreactor Systems. *Nat. Protoc.* **2008**, *3*, 719–738. [CrossRef] [PubMed]

25. Tandon, N.; Cannizzaro, C.; Chao, P.-H.G.; Maidhof, R.; Marsano, A.; Au, H.T.H.; Radisic, M.; Vunjak-Novakovic, G. Electrical Stimulation Systems for Cardiac Tissue Engineering. *Nat. Protoc.* **2009**, *4*, 155–173. [CrossRef] [PubMed]

26. Barash, Y.; Dvir, T.; Tandeitnik, P.; Ruvinov, E.; Guterman, H.; Cohen, S. Electric Field Stimulation Integrated into Perfusion Bioreactor for Cardiac Tissue Engineering. *Tissue Eng. Part C Methods* **2010**, *16*, 1417–1426. [CrossRef] [PubMed]

27. Maidhof, R.; Tandon, N.; Lee, E.J.; Luo, J.; Duan, Y.; Yeager, K.; Konofagou, E.; Vunjak-Novakovic, G. Biomimetic Perfusion and Electrical Stimulation Applied in Concert Improved the Assembly of Engineered Cardiac Tissue. *J. Tissue Eng. Regen. Med.* **2012**, *6*, e12–e23. [CrossRef] [PubMed]

28. Shachar, M.; Benishti, N.; Cohen, S. Effects of Mechanical Stimulation Induced by Compression and Medium Perfusion on Cardiac Tissue Engineering. *Biotechnol. Prog.* **2012**, *28*, 1551–1559. [CrossRef]

29. Fleischer, S.; Shapira, A.; Feiner, R.; Dvir, T. Modular Assembly of Thick Multifunctional Cardiac Patches. *Proc. Natl. Acad. Sci. USA* **2017**, *114*, 1898–1903. [CrossRef]

30. Riegler, J.; Tiburcy, M.; Ebert, A.; Tzatzalos, E.; Raaz, U.; Abilez, O.J.; Shen, Q.; Kooreman, N.G.; Neofytou, E.; Chen, V.C.; et al. Human Engineered Heart Muscles Engraft and Survive Long Term in a Rodent Myocardial Infarction Model. *Circ. Res.* **2015**, *117*, 720–730. [CrossRef]

31. Mihic, A.; Li, J.; Miyagi, Y.; Gagliardi, M.; Li, S.-H.; Zu, J.; Weisel, R.D.; Keller, G.; Li, R.-K. The Effect of Cyclic Stretch on Maturation and 3D Tissue Formation of Human Embryonic Stem Cell-Derived Cardiomyocytes. *Biomaterials* **2014**, *35*, 2798–2808. [CrossRef] [PubMed]

32. Valls-Margarit, M.; Iglesias-García, O.; Di Guglielmo, C.; Sarlabous, L.; Tadevosyan, K.; Paoli, R.; Comelles, J.; Blanco-Almazán, D.; Jiménez-Delgado, S.; Castillo-Fernández, O.; et al. Engineered Macroscale Cardiac Constructs Elicit Human Myocardial Tissue-like Functionality. *Stem Cell Rep.* **2019**, *13*, 207–220. [CrossRef] [PubMed]

33. Zimmermann, W.-H.; Didié, M.; Wasmeier, G.H.; Nixdorff, U.; Hess, A.; Melnychenko, I.; Boy, O.; Neuhuber, W.L.; Weyand, M.; Eschenhagen, T. Cardiac Grafting of Engineered Heart Tissue in Syngenic Rats. *Circulation* **2002**, *106*, I151–I157. [PubMed]

34. Kensah, G.; Gruh, I.; Viering, J.; Schumann, H.; Dahlmann, J.; Meyer, H.; Skvorc, D.; Bär, A.; Akhyari, P.; Heisterkamp, A.; et al. A Novel Miniaturized Multimodal Bioreactor for Continuous In Situ Assessment of Bioartificial Cardiac Tissue During Stimulation and Maturation. *Tissue Eng. Part C Methods* **2011**, *17*, 463–473. [CrossRef] [PubMed]

35. Godier-Furnémont, A.F.G.; Tiburcy, M.; Wagner, E.; Dewenter, M.; Lämmle, S.; El-Armouche, A.; Lehnart, S.E.; Vunjak-Novakovic, G.; Zimmermann, W.-H. Physiologic Force-Frequency Response in Engineered Heart Muscle by Electromechanical Stimulation. *Biomaterials* **2015**, *60*, 82–91. [CrossRef] [PubMed]

36. Tulloch, N.L.; Muskheli, V.; Razumova, M.V.; Korte, F.S.; Regnier, M.; Hauch, K.D.; Pabon, L.; Reinecke, H.; Murry, C.E. Growth of Engineered Human Myocardium with Mechanical Loading and Vascular Coculture. *Circ. Res.* **2011**, *109*, 47–59. [CrossRef] [PubMed]

37. Schaaf, S.; Shibamiya, A.; Mewe, M.; Eder, A.; Stöhr, A.; Hirt, M.N.; Rau, T.; Zimmermann, W.-H.; Conradi, L.; Eschenhagen, T.; et al. Human Engineered Heart Tissue as a Versatile Tool in Basic Research and Preclinical Toxicology. *PLoS ONE* **2011**, *6*, e26397. [CrossRef]

38. Thavandiran, N.; Dubois, N.; Mikryukov, A.; Massé, S.; Beca, B.; Simmons, C.A.; Deshpande, V.S.; McGarry, J.P.; Chen, C.S.; Nanthakumar, K.; et al. Design and Formulation of Functional Pluripotent Stem Cell-Derived Cardiac Microtissues. *Proc. Natl. Acad. Sci. USA* **2013**, *110*, E4698–E4707. [CrossRef]

39. Zhang, X.; Wang, Q.; Gablaski, B.; Zhang, X.; Lucchesi, P.; Zhao, Y. A Microdevice for Studying Intercellular Electromechanical Transduction in Adult Cardiac Myocytes. *Lab Chip* **2013**, *13*, 3090–3097. [CrossRef]

40. Nunes, S.S.; Miklas, J.W.; Liu, J.; Aschar-Sobbi, R.; Xiao, Y.; Zhang, B.; Jiang, J.; Massé, S.; Gagliardi, M.; Hsieh, A.; et al. Biowire: A Platform for Maturation of Human Pluripotent Stem Cell–Derived Cardiomyocytes. *Nat. Methods* **2013**, *10*, 781–787. [CrossRef]

41. Xiao, Y.; Zhang, B.; Liu, H.; Miklas, J.W.; Gagliardi, M.; Pahnke, A.; Thavandiran, N.; Sun, Y.; Simmons, C.; Keller, G.; et al. Microfabricated Perfusable Cardiac Biowire: A Platform that Mimics Native Cardiac Bundle. *Lab Chip* **2014**, *14*, 869–882. [CrossRef] [PubMed]

42. Turnbull, I.C.; Karakikes, I.; Serrao, G.W.; Backeris, P.; Lee, J.-J.; Xie, C.; Senyei, G.; Gordon, R.E.; Li, R.A.; Akar, F.G.; et al. Advancing Functional Engineered Cardiac Tissues Toward a Preclinical Model of Human Myocardium. *FASEB J.* **2014**, *28*, 644–654. [CrossRef] [PubMed]

43. Ma, Z.; Koo, S.; Finnegan, M.A.; Loskill, P.; Huebsch, N.; Marks, N.C.; Conklin, B.R.; Grigoropoulos, C.P.; Healy, K.E. Three-Dimensional Filamentous Human Diseased Cardiac Tissue Model. *Biomaterials* **2014**, *35*, 1367–1377. [CrossRef] [PubMed]

44. Ma, Z.; Wang, J.; Loskill, P.; Huebsch, N.; Koo, S.; Svedlund, F.L.; Marks, N.C.; Hua, E.W.; Grigoropoulos, C.P.; Conklin, B.R.; et al. Self-Organizing Human Cardiac Microchambers Mediated by Geometric Confinement. *Nat. Commun.* **2015**, *6*, 7413. [CrossRef] [PubMed]

45. Weinberger, F.; Breckwoldt, K.; Pecha, S.; Kelly, A.; Geertz, B.; Starbatty, J.; Yorgan, T.; Cheng, K.-H.; Lessmann, K.; Stolen, T.; et al. Cardiac Repair in Guinea Pigs with Human Engineered Heart Tissue from Induced Pluripotent Stem Cells. *Sci. Transl. Med.* **2016**, *8*, 363ra148. [CrossRef]

46. Mannhardt, I.; Breckwoldt, K.; Letuffe-Brenière, D.; Schaaf, S.; Schulz, H.; Neuber, C.; Benzin, A.; Werner, T.; Eder, A.; Schulze, T.; et al. Human Engineered Heart Tissue: Analysis of Contractile Force. *Stem Cell Rep.* **2016**, *7*, 29–42. [CrossRef]

47. Ruan, J.-L.; Tulloch, N.L.; Razumova, M.V.; Saiget, M.; Muskheli, V.; Pabon, L.; Reinecke, H.; Regnier, M.; Murry, C.E. Mechanical Stress Conditioning and Electrical Stimulation Promote Contractility and Force Maturation of Induced Pluripotent Stem Cell-Derived Human Cardiac Tissue. *Circulation* **2016**, *134*, 1557–1567. [CrossRef]

48. Jackman, C.P.; Carlson, A.L.; Bursac, N. Dynamic Culture Yields Engineered Myocardium with Near-Adult Functional Output. *Biomaterials* **2016**, *111*, 66–79. [CrossRef]

49. Lemoine, M.D.; Mannhardt, I.; Breckwoldt, K.; Prondzynski, M.; Flenner, F.; Ulmer, B.; Hirt, M.N.; Neuber, C.; Horváth, A.; Kloth, B.; et al. Human iPSC-Derived Cardiomyocytes Cultured in 3D Engineered Heart Tissue Show Physiological Upstroke Velocity and Sodium Current Density. *Sci. Rep.* **2017**, *7*, 5464. [CrossRef]

50. Nakane, T.; Masumoto, H.; Tinney, J.P.; Yuan, F.; Kowalski, W.J.; Ye, F.; LeBlanc, A.J.; Sakata, R.; Yamashita, J.K.; Keller, B.B. Impact of Cell Composition and Geometry on Human Induced Pluripotent Stem Cells-Derived Engineered Cardiac Tissue. *Sci. Rep.* **2017**, *7*, 45641. [CrossRef]

51. Tiburcy, M.; Hudson, J.E.; Balfanz, P.; Schlick, S.; Meyer, T.; Chang Liao, M.-L.; Levent, E.; Raad, F.; Zeidler, S.; Wingender, E.; et al. Defined Engineered Human Myocardium with Advanced Maturation for Applications in Heart Failure Modeling and Repair. *Circulation* **2017**, *135*, 1832–1847. [CrossRef] [PubMed]

52. Montgomery, M.; Ahadian, S.; Davenport Huyer, L.; Lo Rito, M.; Civitarese, R.A.; Vanderlaan, R.D.; Wu, J.; Reis, L.A.; Momen, A.; Akbari, S.; et al. Flexible Shape-Memory Scaffold for Minimally Invasive Delivery of Functional Tissues. *Nat. Mater.* **2017**, *16*, 1038–1046. [CrossRef] [PubMed]

53. Gao, L.; Gregorich, Z.R.; Zhu, W.; Mattapally, S.; Oduk, Y.; Lou, X.; Kannappan, R.; Borovjagin, A.V.; Walcott, G.P.; Pollard, A.E.; et al. Large Cardiac Muscle Patches Engineered from Human Induced-Pluripotent Stem Cell-Derived Cardiac Cells Improve Recovery from Myocardial Infarction in Swine. *Circulation* **2018**, *137*, 1712–1730. [CrossRef] [PubMed]

54. Zhao, Y.; Rafatian, N.; Feric, N.T.; Cox, B.J.; Aschar-Sobbi, R.; Wang, E.Y.; Aggarwal, P.; Zhang, B.; Conant, G.; Ronaldson-Bouchard, K.; et al. A Platform for Generation of Chamber-Specific Cardiac Tissues and Disease Modeling. *Cell* **2019**, *176*, 913–927.e18. [CrossRef] [PubMed]

55. Ronaldson-Bouchard, K.; Ma, S.P.; Yeager, K.; Chen, T.; Song, L.; Sirabella, D.; Morikawa, K.; Teles, D.; Yazawa, M.; Vunjak-Novakovic, G. Author Correction: Advanced Maturation of Human Cardiac Tissue Grown from Pluripotent Stem Cells. *Nature* **2019**, *572*, E16–E17. [CrossRef] [PubMed]

56. Noor, N.; Shapira, A.; Edri, R.; Gal, I.; Wertheim, L.; Dvir, T. 3D Printing of Personalized Thick and Perfusable Cardiac Patches and Hearts. *Adv. Sci.* **2019**, *6*, 1900344. [CrossRef]

57. Shimizu, T.; Yamato, M.; Isoi, Y.; Akutsu, T.; Setomaru, T.; Abe, K.; Kikuchi, A.; Umezu, M.; Okano, T. Fabrication of Pulsatile Cardiac Tissue Grafts Using a Novel 3-Dimensional Cell Sheet Manipulation Technique and Temperature-Responsive Cell Culture Surfaces. *Circ. Res.* **2002**, *90*, e40. [CrossRef]

58. Sekine, H.; Shimizu, T.; Sakaguchi, K.; Dobashi, I.; Wada, M.; Yamato, M.; Kobayashi, E.; Umezu, M.; Okano, T. In Vitro Fabrication of Functional Three-Dimensional Tissues with Perfusable Blood Vessels. *Nat. Commun.* **2013**, *4*, 1399. [CrossRef]

59. Kawatou, M.; Masumoto, H.; Fukushima, H.; Morinaga, G.; Sakata, R.; Ashihara, T.; Yamashita, J.K. Modelling Torsade de Pointes Arrhythmias In Vitro in 3D Human iPS Cell-Engineered Heart Tissue. *Nat. Commun.* **2017**, *8*, 1078. [CrossRef]

60. Zimmermann, W.H.; Fink, C.; Kralisch, D.; Remmers, U.; Weil, J.; Eschenhagen, T. Three-Dimensional Engineered Heart Tissue from Neonatal Rat Cardiac Myocytes. *Biotechnol. Bioeng.* **2000**, *68*, 106–114. [CrossRef]

61. Akins, R.E.; Boyce, R.A.; Madonna, M.L.; Schroedl, N.A.; Gonda, S.R.; McLaughlin, T.A.; Hartzell, C.R. Cardiac Organogenesis In Vitro: Reestablishment of Three-Dimensional Tissue Architecture by Dissociated Neonatal Rat Ventricular Cells. *Tissue Eng.* **1999**, *5*, 103–118. [CrossRef] [PubMed]

62. Carrier, R.L.; Papadaki, M.; Rupnick, M.; Schoen, F.J.; Bursac, N.; Langer, R.; Freed, L.E.; Vunjak-Novakovic, G. Cardiac Tissue Engineering: Cell Seeding, Cultivation Parameters, and Tissue Construct Characterization. *Biotechnol. Bioeng.* **1999**, *64*, 580–589. [CrossRef]

63. Yang, L.; Soonpaa, M.H.; Adler, E.D.; Roepke, T.K.; Kattman, S.J.; Kennedy, M.; Henckaerts, E.; Bonham, K.; Abbott, G.W.; Linden, R.M.; et al. Human Cardiovascular Progenitor Cells Develop from A KDR+ Embryonic-Stem-Cell-Derived Population. *Nature* **2008**, *453*, 524–528. [CrossRef] [PubMed]

64. Lian, X.; Hsiao, C.; Wilson, G.; Zhu, K.; Hazeltine, L.B.; Azarin, S.M.; Raval, K.K.; Zhang, J.; Kamp, T.J.; Palecek, S.P. Robust Cardiomyocyte Differentiation from Human Pluripotent Stem Cells Via Temporal Modulation of Canonical Wnt Signaling. *Proc. Natl. Acad. Sci. USA* **2012**, *109*, E1848–E1857. [CrossRef] [PubMed]

65. Burridge, P.W.; Holmström, A.; Wu, J.C. Chemically Defined Culture and Cardiomyocyte Differentiation of Human Pluripotent Stem Cells. *Curr. Protoc. Hum. Genet.* **2015**, *87*, 21.3.1–21.3.15. [CrossRef] [PubMed]

66. Louch, W.E.; Sheehan, K.A.; Wolska, B.M. Methods in Cardiomyocyte Isolation, Culture, and Gene Transfer. *J. Mol. Cell. Cardiol.* **2011**, *51*, 288–298. [CrossRef] [PubMed]

67. Wolska, B.M.; Solaro, R.J. Method for Isolation of Adult Mouse Cardiac Myocytes for Studies of Contraction and microfluorimetry. *Am. J. Physiol.* **1996**, *271*, H1250–H1255. [CrossRef]

68. Zimmermann, W.-H.; Zimmermann, W.; Schneiderbanger, K.; Schubert, P.; Didié, M.; Münzel, F.; Heubach, J.F.; Kostin, S.; Neuhuber, W.L.; Eschenhagen, T. Tissue Engineering of a Differentiated Cardiac Muscle Construct. *Circ. Res.* **2002**, *90*, 223–230. [CrossRef]

69. Bursac, N.; Papadaki, M.; Cohen, R.J.; Schoen, F.J.; Eisenberg, S.R.; Carrier, R.; Vunjak-Novakovic, G.; Freed, L.E. Cardiac Muscle Tissue Engineering: Toward an In Vitro Model for Electrophysiological Studies. *Am. J. Physiol. Heart Circul. Physiol.* **1999**, *277*, H433–H444. [CrossRef]

70. Papadaki, M.; Bursac, N.; Langer, R.; Merok, J.; Vunjak-Novakovic, G.; Freed, L.E. Tissue Engineering of Functional Cardiac Muscle: Molecular, Structural, and Electrophysiological Studies. *Am. J. Physiol. Heart Circul. Physiol.* **2001**, *280*, H168–H178. [CrossRef]

71. Bursac, N.; Papadaki, M.; White, J.A.; Eisenberg, S.R.; Vunjak-Novakovic, G.; Freed, L.E. Cultivation in Rotating Bioreactors Promotes Maintenance of Cardiac Myocyte Electrophysiology and Molecular Properties. *Tissue Eng.* **2003**, *9*, 1243–1253. [CrossRef] [PubMed]

72. Radisic, M.; Park, H.; Chen, F.; Salazar-Lazzaro, J.E.; Wang, Y.; Dennis, R.; Langer, R.; Freed, L.E.; Vunjak-Novakovic, G. Biomimetic Approach to Cardiac Tissue Engineering: Oxygen Carriers and Channeled Scaffolds. *Tissue Eng.* **2006**, *12*, 2077–2091. [CrossRef] [PubMed]

73. Lian, X.; Zhang, J.; Azarin, S.M.; Zhu, K.; Hazeltine, L.B.; Bao, X.; Hsiao, C.; Kamp, T.J.; Palecek, S.P. Directed Cardiomyocyte Differentiation from Human Pluripotent Stem Cells by Modulating Wnt/β-catenin Signaling Under Fully Defined Conditions. *Nat. Protoc.* **2013**, *8*, 162–175. [CrossRef] [PubMed]

74. Burridge, P.W.; Thompson, S.; Millrod, M.A.; Weinberg, S.; Yuan, X.; Peters, A.; Mahairaki, V.; Koliatsos, V.E.; Tung, L.; Zambidis, E.T. A Universal System for Highly Efficient Cardiac Differentiation of Human Induced Pluripotent Stem Cells that Eliminates Interline Variability. *PLoS ONE* **2011**, *6*, e18293. [CrossRef]

75. Van Laake, L.W.; Passier, R.; Monshouwer-Kloots, J.; Verkleij, A.J.; Lips, D.J.; Freund, C.; den Ouden, K.; Ward-van Oostwaard, D.; Korving, J.; Tertoolen, L.G.; et al. Human Embryonic Stem Cell-Derived Cardiomyocytes Survive and Mature in the Mouse Heart and Transiently Improve Function after Myocardial Infarction. *Stem Cell Res.* **2007**, *1*, 9–24. [CrossRef]

76. Laflamme, M.A.; Chen, K.Y.; Naumova, A.V.; Muskheli, V.; Fugate, J.A.; Dupras, S.K.; Reinecke, H.; Xu, C.; Hassanipour, M.; Police, S.; et al. Cardiomyocytes Derived from Human Embryonic Stem Cells in Pro-Survival Factors Enhance Function of Infarcted Rat Hearts. *Nat. Biotechnol.* **2007**, *25*, 1015–1024. [CrossRef]

77. Caspi, O.; Huber, I.; Kehat, I.; Habib, M.; Arbel, G.; Gepstein, A.; Yankelson, L.; Aronson, D.; Beyar, R.; Gepstein, L. Transplantation of Human Embryonic Stem Cell-Derived Cardiomyocytes Improves Myocardial Performance in Infarcted Rat Hearts. *J. Am. Coll. Cardiol.* **2007**, *50*, 1884–1893. [CrossRef]

78. Carpenter, L.; Carr, C.; Yang, C.T.; Stuckey, D.J.; Clarke, K.; Watt, S.M. Efficient Differentiation of Human Induced Pluripotent Stem Cells Generates Cardiac Cells that Provide Protection Following Myocardial Infarction in the Rat. *Stem Cells Dev.* **2012**, *21*, 977–986. [CrossRef]

79. Shiba, Y.; Fernandes, S.; Zhu, W.-Z.; Filice, D.; Muskheli, V.; Kim, J.; Palpant, N.J.; Gantz, J.; Moyes, K.W.; Reinecke, H.; et al. Human ES-Cell-Derived Cardiomyocytes Electrically Couple and Suppress Arrhythmias in Injured Hearts. *Nature* **2012**, *489*, 322–325. [CrossRef]

80. Templin, C.; Zweigerdt, R.; Schwanke, K.; Olmer, R.; Ghadri, J.-R.; Emmert, M.Y.; Müller, E.; Küest, S.M.; Cohrs, S.; Schibli, R.; et al. Transplantation and Tracking of Human-Induced Pluripotent Stem Cells in a Pig Model of Myocardial Infarction: Assessment of Cell Survival, Engraftment, and Distribution by Hybrid Single Photon Emission Computed Tomography/Computed Tomography of Sodium Iodide Symporter Transgene Expression. *Circulation* **2012**, *126*, 430–439.

81. Chong, J.J.H.; Murry, C.E. Cardiac Regeneration Using Pluripotent Stem Cells–Progression to Large Animal Models. *Stem Cell Res.* **2014**, *13*, 654–665. [CrossRef] [PubMed]

82. Shiba, Y.; Gomibuchi, T.; Seto, T.; Wada, Y.; Ichimura, H.; Tanaka, Y.; Ogasawara, T.; Okada, K.; Shiba, N.; Sakamoto, K.; et al. Allogeneic transplantation of iPS Cell-Derived Cardiomyocytes Regenerates Primate Hearts. *Nature* **2016**, *538*, 388–391. [CrossRef] [PubMed]

83. Kawamura, M.; Miyagawa, S.; Miki, K.; Saito, A.; Fukushima, S.; Higuchi, T.; Kawamura, T.; Kuratani, T.; Daimon, T.; Shimizu, T.; et al. Feasibility, Safety, and Therapeutic Efficacy of Human Induced Pluripotent Stem Cell-Derived Cardiomyocyte Sheets in a porcine Ischemic Cardiomyopathy Model. *Circulation* **2012**, *126*, S29–S37. [CrossRef]

84. Xiong, Q.; Ye, L.; Zhang, P.; Lepley, M.; Tian, J.; Li, J.; Zhang, L.; Swingen, C.; Vaughan, J.T.; Kaufman, D.S.; et al. Functional Consequences of human Induced Pluripotent Stem Cell Therapy: Myocardial ATP turnover Rate in the In Vivo Swine Heart with Postinfarction Remodeling. *Circulation* **2013**, *127*, 997–1008. [CrossRef] [PubMed]

85. Ye, L.; Chang, Y.-H.; Xiong, Q.; Zhang, P.; Zhang, L.; Somasundaram, P.; Lepley, M.; Swingen, C.; Su, L.; Wendel, J.S.; et al. Cardiac Repair in a Porcine Model of Acute Myocardial Infarction with Human Induced Pluripotent Stem Cell-Derived Cardiovascular Cells. *Cell Stem Cell* **2014**, *15*, 750–761. [CrossRef] [PubMed]

86. Menasché, P.; Vanneaux, V.; Fabreguettes, J.-R.; Bel, A.; Tosca, L.; Garcia, S.; Bellamy, V.; Farouz, Y.; Pouly, J.; Damour, O.; et al. Towards Clinical Use of Human Embryonic Stem Cell-Derived Cardiac Progenitors: A Translational Experience. *Eur. Heart J.* **2015**, *36*, 743–750. [PubMed]

87. Menasché, P.; Vanneaux, V.; Hagège, A.; Bel, A.; Cholley, B.; Parouchev, A.; Cacciapuoti, I.; Al-Daccak, R.; Benhamouda, N.; Blons, H.; et al. Transplantation of Human Embryonic Stem Cell-Derived Cardiovascular Progenitors for Severe Ischemic Left Ventricular Dysfunction. *J. Am. Coll. Cardiol.* **2018**, *71*, 429–438. [CrossRef] [PubMed]

88. Ankrum, J.A.; Ong, J.F.; Karp, J.M. Mesenchymal Stem Cells: Immune Evasive, Not Immune privileged. *Nat. Biotechnol.* **2014**, *32*, 252–260. [CrossRef]

89. Golpanian, S.; Wolf, A.; Hatzistergos, K.E.; Hare, J.M. Rebuilding the Damaged Heart: Mesenchymal Stem Cells, Cell-Based Therapy, and Engineered Heart Tissue. *Physiol. Rev.* **2016**, *96*, 1127–1168. [CrossRef]

90. Williams, A.R.; Hare, J.M. Mesenchymal Stem Cells: Biology, Pathophysiology, Translational Findings, and Therapeutic Implications for Cardiac Disease. *Circ. Res.* **2011**, *109*, 923–940. [CrossRef]

91. Valina, C.; Pinkernell, K.; Song, Y.-H.; Bai, X.; Sadat, S.; Campeau, R.J.; Le Jemtel, T.H.; Alt, E. Intracoronary Administration of Autologous Adipose Tissue-Derived Stem Cells Improves Left Ventricular Function, Perfusion, and Remodelling after Acute Myocardial Infarction. *Eur. Heart J.* **2007**, *28*, 2667–2677. [CrossRef] [PubMed]

92. Schuleri, K.H.; Feigenbaum, G.S.; Centola, M.; Weiss, E.S.; Zimmet, J.M.; Turney, J.; Kellner, J.; Zviman, M.M.; Hatzistergos, K.E.; Detrick, B.; et al. Autologous Mesenchymal Stem Cells Produce Reverse Remodelling in Chronic Ischaemic Cardiomyopathy. *Eur. Heart J.* **2009**, *30*, 2722–2732. [CrossRef] [PubMed]

93. Mazo, M.; Hernández, S.; Gavira, J.J.; Abizanda, G.; Araña, M.; López-Martínez, T.; Moreno, C.; Merino, J.; Martino-Rodríguez, A.; Uixeira, A.; et al. Treatment of Reperfused Ischemia with Adipose-Derived Stem Cells in a Preclinical Swine Model of Myocardial Infarction. *Cell Transplant.* **2012**, *21*, 2723–2733. [CrossRef] [PubMed]

94. Hatzistergos, K.E.; Quevedo, H.; Oskouei, B.N.; Hu, Q.; Feigenbaum, G.S.; Margitich, I.S.; Mazhari, R.; Boyle, A.J.; Zambrano, J.P.; Rodriguez, J.E.; et al. Bone Marrow Mesenchymal Stem Cells Stimulate Cardiac Stem Cell Proliferation and Differentiation. *Circ. Res.* **2010**, *107*, 913–922. [CrossRef]

95. Williams, A.R.; Suncion, V.Y.; McCall, F.; Guerra, D.; Mather, J.; Zambrano, J.P.; Heldman, A.W.; Hare, J.M. Durable Scar Size Reduction due to Allogeneic Mesenchymal Stem Cell Therapy Regulates Whole-Chamber Remodeling. *J. Am. Heart Assoc.* **2013**, *2*, e000140. [CrossRef]

96. Liu, J.; Hu, Q.; Wang, Z.; Xu, C.; Wang, X.; Gong, G.; Mansoor, A.; Lee, J.; Hou, M.; Zeng, L.; et al. Autologous Stem Cell Transplantation for Myocardial Repair. *Am. J. Physiol. Heart Circ. Physiol.* **2004**, *287*, H501–H511. [CrossRef]

97. Godier-Furnemont, A.F.G.; Martens, T.P.; Koeckert, M.S.; Wan, L.; Parks, J.; Arai, K.; Zhang, G.; Hudson, B.; Homma, S.; Vunjak-Novakovic, G. Composite Scaffold Provides a Cell Delivery Platform for Cardiovascular Repair. *Proc. Natl. Acad. Sci. USA* **2011**, *108*, 7974–7979. [CrossRef]

98. Gilsbach, R.; Preissl, S.; Grüning, B.A.; Schnick, T.; Burger, L.; Benes, V.; Würch, A.; Bönisch, U.; Günther, S.; Backofen, R.; et al. Dynamic DNA Methylation Orchestrates Cardiomyocyte Development, Maturation and Disease. *Nat. Commun.* **2014**, *5*, 5288. [CrossRef]

99. Liau, B.; Christoforou, N.; Leong, K.W.; Bursac, N. Pluripotent Stem Cell-Derived Cardiac Tissue Patch with Advanced Structure and Function. *Biomaterials* **2011**, *32*, 9180–9187. [CrossRef]

100. Zhang, D.; Shadrin, I.Y.; Lam, J.; Xian, H.-Q.; Snodgrass, H.R.; Bursac, N. Tissue-engineered Cardiac Patch for Advanced Functional Maturation of Human ESC-Derived Cardiomyocytes. *Biomaterials* **2013**, *34*, 5813–5820. [CrossRef]

101. Liau, B.; Jackman, C.P.; Li, Y.; Bursac, N. Developmental Stage-Dependent Effects of Cardiac Fibroblasts on Function of Stem Cell-Derived Engineered Cardiac Tissues. *Sci. Rep.* **2017**, *7*, 42290. [CrossRef] [PubMed]

102. Lin, Y.; Liu, H.; Klein, M.; Ostrominski, J.; Hong, S.G.; Yada, R.C.; Chen, G.; Navarengom, K.; Schwartzbeck, R.; San, H.; et al. Efficient Differentiation of Cardiomyocytes and Generation of Calcium-Sensor Reporter Lines from Nonhuman Primate iPSCs. *Sci. Rep.* **2018**, *8*, 5907. [CrossRef] [PubMed]

103. Zimmermann, W.-H.; Melnychenko, I.; Wasmeier, G.; Didié, M.; Naito, H.; Nixdorff, U.; Hess, A.; Budinsky, L.; Brune, K.; Michaelis, B.; et al. Engineered Heart Tissue Grafts Improve Systolic and Diastolic Function in Infarcted Rat Hearts. *Nat. Med.* **2006**, *12*, 452–458. [CrossRef] [PubMed]

104. Giacomelli, E.; Bellin, M.; Sala, L.; van Meer, B.J.; Tertoolen, L.G.J.; Orlova, V.V.; Mummery, C.L. Three-Dimensional Cardiac Microtissues Composed of Cardiomyocytes and Endothelial Cells Co-Differentiated from Human Pluripotent Stem Cells. *Development* **2017**, *144*, 1008–1017. [CrossRef] [PubMed]

105. Jackman, C.P.; Shadrin, I.Y.; Carlson, A.L.; Bursac, N. Human Cardiac Tissue Engineering: From Pluripotent Stem Cells to Heart Repair. *Curr. Opin. Chem. Eng.* **2015**, *7*, 57–64. [CrossRef]

106. Yamato, M.; Okano, T. Cell sheet engineering. *Mater. Today* **2004**, *7*, 42–47. [CrossRef]

107. Yang, J.; Yamato, M.; Kohno, C.; Nishimoto, A.; Sekine, H.; Fukai, F.; Okano, T. Cell Sheet Engineering: Recreating Tissues without Biodegradable Scaffolds. *Biomaterials* **2005**, *26*, 6415–6422. [CrossRef]

108. Shimizu, T.; Yamato, M.; Kikuchi, A.; Okano, T. Cell Sheet Engineering for Myocardial Tissue Reconstruction. *Biomaterials* **2003**, *24*, 2309–2316. [CrossRef]

109. Sekine, H.; Shimizu, T.; Dobashi, I.; Matsuura, K.; Hagiwara, N.; Takahashi, M.; Kobayashi, E.; Yamato, M.; Okano, T. Cardiac Cell Sheet Transplantation Improves Damaged Heart Function via Superior Cell Survival in Comparison with Dissociated Cell Injection. *Tissue Eng. Part A* **2011**, *17*, 2973–2980. [CrossRef]

110. Sawa, Y.; Yoshikawa, Y.; Toda, K.; Fukushima, S.; Yamazaki, K.; Ono, M.; Sakata, Y.; Hagiwara, N.; Kinugawa, K.; Miyagawa, S. Safety and Efficacy of Autologous Skeletal Myoblast Sheets (TCD-51073) for the Treatment of Severe Chronic Heart Failure Due to Ischemic Heart Disease. *Circ. J.* **2015**, *79*, 991–999. [CrossRef]

111. Yui, Y. Concerns on a New Therapy for Severe Heart Failure Using Cell Sheets with Skeletal Muscle or Myocardial Cells from iPS Cells in Japan. *NPJ Regen. Med.* **2018**, *3*, 7. [CrossRef] [PubMed]

112. Pomeroy, J.E.; Helfer, A.; Bursac, N. Biomaterializing the Promise of Cardiac Tissue Engineering. *Biotechnol. Adv.* **2019**. [CrossRef] [PubMed]

113. Radisic, M.; Park, H.; Martens, T.P.; Salazar-Lazaro, J.E.; Geng, W.; Wang, Y.; Langer, R.; Freed, L.E.; Vunjak-Novakovic, G. Pre-treatment of Synthetic Elastomeric Scaffolds by Cardiac Fibroblasts Improves Engineered Heart Tissue. *J. Biomed. Mater. Res. A* **2008**, *86*, 713–724. [CrossRef]

114. Zandonella, C. Tissue Engineering: The Beat Goes On. *Nature* **2003**, *421*, 884–886. [CrossRef] [PubMed]

115. Sultana, N.; Hassan, M.I.; Lim, M.M. Scaffolding Biomaterials. In *Composite Synthetic Scaffolds for Tissue Engineering and Regenerative Medicine*; Sultana, N., Hassan, M.I., Lim, M.M., Eds.; Springer International Publishing: Cham, Switzerland, 2015; pp. 1–11. ISBN 9783319097558.

116. Badrossamay, M.R.; McIlwee, H.A.; Goss, J.A.; Parker, K.K. Nanofiber Assembly by Rotary Jet-Spinning. *Nano Lett.* **2010**, *10*, 2257–2261. [CrossRef]

117. Kenar, H.; Kose, G.T.; Toner, M.; Kaplan, D.L.; Hasirci, V. A 3D Aligned Microfibrous Myocardial Tissue Construct Cultured Under Transient Perfusion. *Biomaterials* **2011**, *32*, 5320–5329. [CrossRef]

118. Bursac, N.; Loo, Y.; Leong, K.; Tung, L. Novel Anisotropic Engineered Cardiac Tissues: Studies of Electrical Propagation. *Biochem. Biophys. Res. Commun.* **2007**, *361*, 847–853. [CrossRef]

119. O'Brien, F.J. Biomaterials & Scaffolds for Tissue Engineering. *Mater. Today* **2011**, *14*, 88–95. [CrossRef]

120. Gaetani, R.; Doevendans, P.A.; Metz, C.H.G.; Alblas, J.; Messina, E.; Giacomello, A.; Sluijter, J.P.G. Cardiac Tissue Engineering Using Tissue Printing Technology and Human Cardiac Progenitor Cells. *Biomaterials* **2012**, *33*, 1782–1790. [CrossRef]

121. Liberski, A.; Latif, N.; Raynaud, C.; Bollensdorff, C.; Yacoub, M. Alginate for Cardiac Regeneration: From Seaweed to Clinical Trials. *Glob. Cardiol. Sci. Pract.* **2016**, *2016*, e201604. [CrossRef]

122. Fleischer, S.; Shapira, A.; Regev, O.; Nseir, N.; Zussman, E.; Dvir, T. Albumin Fiber Scaffolds for Engineering Functional Cardiac Tissues. *Biotechnol. Bioeng.* **2014**, *111*, 1246–1257. [CrossRef] [PubMed]

123. Bian, W.; Jackman, C.P.; Bursac, N. Controlling the Structural and Functional Anisotropy of Engineered Cardiac Tissues. *Biofabrication* **2014**, *6*, 024109. [CrossRef] [PubMed]

124. Liu, Y.; Wang, L.; Kikuiri, T.; Akiyama, K.; Chen, C.; Xu, X.; Yang, R.; Chen, W.; Wang, S.; Shi, S. Mesenchymal Stem Cell-Based Tissue Regeneration is Governed by Recipient T Lymphocytes via IFN-γ and TNF-α. *Nat. Med.* **2011**, *17*, 1594–1601. [CrossRef]

125. Radisic, M.; Euloth, M.; Yang, L.; Langer, R.; Freed, L.E.; Vunjak-Novakovic, G. High-Density Seeding of Myocyte Cells for Cardiac Tissue Engineering. *Biotechnol. Bioeng.* **2003**, *82*, 403–414. [CrossRef] [PubMed]

126. Yu, Y.; Alkhawaji, A.; Ding, Y.; Mei, J. Decellularized Scaffolds in Regenerative Medicine. *Oncotarget* **2016**, *7*, 58671–58683. [CrossRef]

127. Hodgson, M.J.; Knutson, C.C.; Momtahan, N.; Cook, A.D. Extracellular Matrix from Whole Porcine Heart Decellularization for Cardiac Tissue Engineering. *Methods Mol. Biol.* **2017**, 95–102. [CrossRef]

128. Fontana, G.; Gershlak, J.; Adamski, M.; Lee, J.-S.; Matsumoto, S.; Le, H.D.; Binder, B.; Wirth, J.; Gaudette, G.; Murphy, W.L. Biofunctionalized Plants as Diverse Biomaterials for Human Cell Culture. *Adv. Healthc. Mater.* **2017**, *6*. [CrossRef]

129. Modulevsky, D.J.; Lefebvre, C.; Haase, K.; Al-Rekabi, Z.; Pelling, A.E. Apple Derived Cellulose Scaffolds for 3D Mammalian Cell Culture. *PLoS ONE* **2014**, *9*, e97835. [CrossRef]

130. Modulevsky, D.J.; Cuerrier, C.M.; Pelling, A.E. Biocompatibility of Subcutaneously Implanted Plant-Derived Cellulose Biomaterials. *PLoS ONE* **2016**, *11*, e0157894. [CrossRef]

131. Gershlak, J.R.; Hernandez, S.; Fontana, G.; Perreault, L.R.; Hansen, K.J.; Larson, S.A.; Binder, B.Y.K.; Dolivo, D.M.; Yang, T.; Dominko, T.; et al. Crossing Kingdoms: Using Decellularized Plants as Perfusable Tissue Engineering Scaffolds. *Biomaterials* **2017**, *125*, 13–22. [CrossRef]

132. Entcheva, E.; Bien, H.; Yin, L.; Chung, C.-Y.; Farrell, M.; Kostov, Y. Functional Cardiac Cell Constructs on Cellulose-Based Scaffolding. *Biomaterials* **2004**, *25*, 5753–5762. [CrossRef] [PubMed]

133. Wippermann, J.; Schumann, D.; Klemm, D.; Albes, J.M.; Wittwer, T.; Strauch, J.; Wahlers, T. Preliminary Results of Small Arterial Substitutes Performed with a New Cylindrical Biomaterial Composed of Bacterial Cellulose. *Thorac. Cardiovasc. Surg.* **2008**, *56*, 592–596. [CrossRef]

134. Oberwallner, B.; Brodarac, A.; Anić, P.; Šarić, T.; Wassilew, K.; Neef, K.; Choi, Y.-H.; Stamm, C. Human Cardiac Extracellular Matrix Supports Myocardial Lineage Commitment of Pluripotent Stem Cells. *Eur. J. Cardiothorac. Surg.* **2015**, *47*, 416–425, discussion 425. [CrossRef] [PubMed]

135. Kc, P.; Hong, Y.; Zhang, G. Cardiac Tissue-Derived Extracellular Matrix Scaffolds for Myocardial Repair: Advantages and Challenges. *Regen Biomater* **2019**, *6*, 185–199. [CrossRef] [PubMed]

136. Oberwallner, B.; Brodarac, A.; Choi, Y.-H.; Saric, T.; Anić, P.; Morawietz, L.; Stamm, C. Preparation of Cardiac Extracellular Matrix Scaffolds by Decellularization of Human Myocardium. *J. Biomed. Mater. Res. A* **2014**, *102*, 3263–3272. [CrossRef] [PubMed]

137. Garreta, E.; de Oñate, L.; Fernández-Santos, M.E.; Oria, R.; Tarantino, C.; Climent, A.M.; Marco, A.; Samitier, M.; Martínez, E.; Valls-Margarit, M.; et al. Myocardial Commitment From Human Pluripotent Stem Cells: Rapid Production Of Human Heart Grafts. *Biomaterials* **2016**, *98*, 64–78. [CrossRef] [PubMed]
138. Sánchez, P.L.; Fernández-Santos, M.E.; Costanza, S.; Climent, A.M.; Moscoso, I.; Gonzalez-Nicolas, M.A.; Sanz-Ruiz, R.; Rodríguez, H.; Kren, S.M.; Garrido, G.; et al. Acellular Human Heart Matrix: A Critical Step Toward Whole Heart Grafts. *Biomaterials* **2015**, *61*, 279–289. [CrossRef] [PubMed]
139. Guyette, J.P.; Charest, J.M.; Mills, R.W.; Jank, B.J.; Moser, P.T.; Gilpin, S.E.; Gershlak, J.R.; Okamoto, T.; Gonzalez, G.; Milan, D.J.; et al. Bioengineering Human Myocardium on Native Extracellular Matrix. *Circ. Res.* **2016**, *118*, 56–72. [CrossRef]
140. Pedron, S.; Van Lierop, S.; Horstman, P.; Penterman, R.; Broer, D.J.; Peeters, E. Stimuli Responsive Delivery Vehicles for Cardiac Microtissue Transplantation. *Adv. Funct. Mater.* **2011**, *21*, 1624–1630. [CrossRef]
141. Birla, R.K.; Borschel, G.H.; Dennis, R.G.; Brown, D.L. Myocardial Engineering in Vivo: Formation and Characterization of Contractile, Vascularized Three-Dimensional Cardiac Tissue. *Tissue Eng.* **2005**, *11*, 803–813. [CrossRef]
142. Sepantafar, M.; Maheronnaghsh, R.; Mohammadi, H.; Rajabi-Zeleti, S.; Annabi, N.; Aghdami, N.; Baharvand, H. Stem Cells and Injectable Hydrogels: Synergistic Therapeutics in Myocardial Repair. *Biotechnol. Adv.* **2016**, *34*, 362–379. [CrossRef] [PubMed]
143. Kopeček, J.; Yang, J. Smart Self-Assembled Hybrid Hydrogel Biomaterials. *Angew. Chem. Int. Ed.* **2012**, *51*, 7396–7417. [CrossRef] [PubMed]
144. Camci-Unal, G.; Annabi, N.; Dokmeci, M.R.; Liao, R.; Khademhosseini, A. Hydrogels for Cardiac Tissue Engineering. *NPG Asia Mater.* **2014**, *6*, e99. [CrossRef]
145. Black, L.D., 3rd; Meyers, J.D.; Weinbaum, J.S.; Shvelidze, Y.A.; Tranquillo, R.T. Cell-Induced Alignment Augments Twitch Force in Fibrin Gel-Based Engineered Myocardium via Gap Junction Modification. *Tissue Eng. Part A* **2009**, *15*, 3099–3108. [CrossRef] [PubMed]
146. Hansen, K.J.; Laflamme, M.A.; Gaudette, G.R. Development of a Contractile Cardiac Fiber from Pluripotent Stem Cell Derived Cardiomyocytes. *Front. Cardiovasc. Med.* **2018**, *5*, 52. [CrossRef] [PubMed]
147. Naito, H.; Melnychenko, I.; Didié, M.; Schneiderbanger, K.; Schubert, P.; Rosenkranz, S.; Eschenhagen, T.; Zimmermann, W.-H. Optimizing Engineered Heart Tissue for Therapeutic Applications as Surrogate Heart Muscle. *Circulation* **2006**, *114*, I72–I78. [CrossRef]
148. Lau, H.K.; Kiick, K.L. Opportunities for Multicomponent Hybrid Hydrogels in Biomedical Applications. *Biomacromolecules* **2015**, *16*, 28–42. [CrossRef]
149. Fennema, E.; Rivron, N.; Rouwkema, J.; van Blitterswijk, C.; de Boer, J. Spheroid Culture as a Tool for Creating 3D Complex Tissues. *Trends Biotechnol.* **2013**, *31*, 108–115. [CrossRef]
150. Beauchamp, P.; Moritz, W.; Kelm, J.M.; Ullrich, N.D.; Agarkova, I.; Anson, B.D.; Suter, T.M.; Zuppinger, C. Development and Characterization of a Scaffold-Free 3D Spheroid Model of Induced Pluripotent Stem Cell-Derived Human Cardiomyocytes. *Tissue Eng. Part C Methods* **2015**, *21*, 852–861. [CrossRef]
151. Nguyen, D.C.; Hookway, T.A.; Wu, Q.; Jha, R.; Preininger, M.K.; Chen, X.; Easley, C.A.; Spearman, P.; Deshpande, S.R.; Maher, K.; et al. Microscale Generation of Cardiospheres Promotes Robust Enrichment of Cardiomyocytes Derived from Human Pluripotent Stem Cells. *Stem Cell Rep.* **2014**, *3*, 260–268. [CrossRef]
152. Kofron, C.M.; Kim, T.Y.; King, M.E.; Xie, A.; Feng, F.; Park, E.; Qu, Z.; Choi, B.; Mende, U. Gq-Activated Fibroblasts Induce Cardiomyocyte Action Potential Prolongation and Automaticity in a Three-Dimensional Microtissue Environment. *Am. J. Physiol. Heart Circ. Physiol.* **2017**, *313*, H810–H827. [CrossRef] [PubMed]
153. Yan, Y.; Bejoy, J.; Xia, J.; Griffin, K.; Guan, J.; Li, Y. Cell Population Balance of Cardiovascular Spheroids Derived from Human Induced Pluripotent Stem Cells. *Sci. Rep.* **2019**, *9*, 1295. [CrossRef] [PubMed]
154. Polonchuk, L.; Chabria, M.; Badi, L.; Hoflack, J.-C.; Figtree, G.; Davies, M.J.; Gentile, C. Cardiac Spheroids as Promising In Vitro Models to Study the Human Heart Microenvironment. *Sci. Rep.* **2017**, *7*, 7005. [CrossRef] [PubMed]
155. Oltolina, F.; Zamperone, A.; Colangelo, D.; Gregoletto, L.; Reano, S.; Pietronave, S.; Merlin, S.; Talmon, M.; Novelli, E.; Diena, M.; et al. Human Cardiac Progenitor Spheroids Exhibit Enhanced Engraftment Potential. *PLoS ONE* **2015**, *10*, e0137999. [CrossRef]
156. Mattapally, S.; Zhu, W.; Fast, V.G.; Gao, L.; Worley, C.; Kannappan, R.; Borovjagin, A.V.; Zhang, J. Spheroids of Cardiomyocytes Derived from Human-Induced Pluripotent Stem Cells Improve Recovery from Myocardial Injury in Mice. *Am. J. Physiol. Heart Circul. Physiol.* **2018**, *315*, H327–H339. [CrossRef]
157. Noguchi, R.; Nakayama, K.; Itoh, M.; Kamohara, K.; Furukawa, K.; Oyama, J.-I.; Node, K.; Morita, S. Development of a Three-Dimensional Pre-Vascularized Scaffold-Free Contractile Cardiac Patch for Treating Heart Disease. *J. Heart Lung Transplant.* **2016**, *35*, 137–145. [CrossRef]
158. Ong, C.S.; Fukunishi, T.; Zhang, H.; Huang, C.Y.; Nashed, A.; Blazeski, A.; DiSilvestre, D.; Vricella, L.; Conte, J.; Tung, L.; et al. Biomaterial-Free Three-Dimensional Bioprinting of Cardiac Tissue using Human Induced Pluripotent Stem Cell Derived Cardiomyocytes. *Sci. Rep.* **2017**, *7*, 4566. [CrossRef]
159. Kim, T.Y.; Kofron, C.M.; King, M.E.; Markes, A.R.; Okundaye, A.O.; Qu, Z.; Mende, U.; Choi, B.-R. Directed Fusion of Cardiac Spheroids into Larger Heterocellular Microtissues Enables Investigation of Cardiac Action Potential Propagation via Cardiac Fibroblasts. *PLoS ONE* **2018**, *13*, e0196714. [CrossRef]
160. Lancaster, M.A.; Knoblich, J.A. Organogenesis in a Dish: Modeling Development and Disease Using Organoid Technologies. *Science* **2014**, *345*, 1247125. [CrossRef]

161. Evans, M. Discovering Pluripotency: 30 Years of Mouse Embryonic Stem Cells. *Nat. Rev. Mol. Cell Biol.* **2011**, *12*, 680–686. [CrossRef]

162. Knight, E.; Przyborski, S. Advances in 3D Cell Culture Technologies Enabling Tissue-Like Structures to Be Created In Vitro. *J. Anat.* **2015**, *227*, 746–756. [CrossRef] [PubMed]

163. Shkumatov, A.; Baek, K.; Kong, H. Matrix Rigidity-Modulated Cardiovascular Organoid Formation from Embryoid Bodies. *PLoS ONE* **2014**, *9*, e94764. [CrossRef] [PubMed]

164. Mills, R.J.; Parker, B.L.; Quaife-Ryan, G.A.; Voges, H.K.; Needham, E.J.; Bornot, A.; Ding, M.; Andersson, H.; Polla, M.; Elliott, D.A.; et al. Drug Screening in Human PSC-Cardiac Organoids Identifies Pro-Proliferative Compounds Acting via the Mevalonate Pathway. *Cell Stem Cell* **2019**, *24*, 895–907.e6. [CrossRef] [PubMed]

165. Hoang, P.; Wang, J.; Conklin, B.R.; Healy, K.E.; Ma, Z. Generation of Spatial-Patterned Early-Developing Cardiac Organoids Using Human Pluripotent Stem Cells. *Nat. Protoc.* **2018**, *13*, 723–737. [CrossRef] [PubMed]

166. Varzideh, F.; Pahlavan, S.; Ansari, H.; Halvaei, M.; Kostin, S.; Feiz, M.-S.; Latifi, H.; Aghdami, N.; Braun, T.; Baharvand, H. Human Cardiomyocytes Undergo Enhanced Maturation in Embryonic Stem Cell-Derived Organoid Transplants. *Biomaterials* **2019**, *192*, 537–550. [CrossRef]

167. Richards, D.J.; Li, Y.; Kerr, C.M.; Yao, J.; Beeson, G.C.; Coyle, R.C.; Chen, X.; Jia, J.; Damon, B.; Wilson, R.; et al. Human Cardiac Organoids for the Modelling of Myocardial Infarction and Drug Cardiotoxicity. *Nat. Biomed. Eng.* **2020**, *4*, 446–462. [CrossRef]

168. Richards, D.J.; Coyle, R.C.; Tan, Y.; Jia, J.; Wong, K.; Toomer, K.; Menick, D.R.; Mei, Y. Inspiration from Heart Development: Biomimetic Development of Functional Human Cardiac Organoids. *Biomaterials* **2017**, *142*, 112–123. [CrossRef]

169. Lee, J.; Sutani, A.; Kaneko, R.; Takeuchi, J.; Sasano, T.; Kohda, T.; Ihara, K.; Takahashi, K.; Yamazoe, M.; Morio, T.; et al. In Vitro Generation of Functional Murine Heart Organoids via FGF4 and Extracellular Matrix. *Nat. Commun.* **2020**, *11*, 4283. [CrossRef]

170. Huh, D.; Torisawa, Y.-S.; Hamilton, G.A.; Kim, H.J.; Ingber, D.E. Microengineered Physiological Biomimicry: Organs-on-Chips. *Lab Chip* **2012**, *12*, 2156–2164. [CrossRef]

171. Ewart, L.; Dehne, E.-M.; Fabre, K.; Gibbs, S.; Hickman, J.; Hornberg, E.; Ingelman-Sundberg, M.; Jang, K.-J.; Jones, D.R.; Lauschke, V.M.; et al. Application of Microphysiological Systems to Enhance Safety Assessment in Drug Discovery. *Annu. Rev. Pharmacol. Toxicol.* **2018**, *58*, 65–82. [CrossRef]

172. Jastrzebska, E.; Tomecka, E.; Jesion, I. Heart-on-a-Chip Based on Stem Cell Biology. *Biosens. Bioelectron.* **2016**, *75*, 67–81. [CrossRef] [PubMed]

173. Ribas, J.; Sadeghi, H.; Manbachi, A.; Leijten, J.; Brinegar, K.; Zhang, Y.S.; Ferreira, L.; Khademhosseini, A. Cardiovascular Organ-on-a-Chip Platforms for Drug Discovery and Development. *Appl. In Vitro Toxicol.* **2016**, *2*, 82–96. [CrossRef] [PubMed]

174. Kujala, V.J.; Pasqualini, F.S.; Goss, J.A.; Nawroth, J.C.; Parker, K.K. Laminar Ventricular Myocardium on a Microelectrode Array-Based Chip. *J. Mater. Chem. B Mater. Biol. Med.* **2016**, *4*, 3534–3543. [CrossRef] [PubMed]

175. Chen, Y.; Chan, H.N.; Michael, S.A.; Shen, Y.; Chen, Y.; Tian, Q.; Huang, L.; Wu, H. A Microfluidic Circulatory System Integrated with Capillary-Assisted Pressure Sensors. *Lab Chip* **2017**, *17*, 653–662. [CrossRef] [PubMed]

176. Ghiaseddin, A.; Pouri, H.; Soleimani, M.; Vasheghani-Farahani, E.; Ahmadi Tafti, H.; Hashemi-Najafabadi, S. Cell Laden Hydrogel Construct on-a-Chip for Mimicry of Cardiac Tissue in-Vitro Study. *Biochem. Biophys. Res. Commun.* **2017**, *484*, 225–230. [CrossRef] [PubMed]

177. Simmons, C.S.; Petzold, B.C.; Pruitt, B.L. Microsystems for Biomimetic Stimulation of Cardiac Cells. *Lab Chip* **2012**, *12*, 3235–3248. [CrossRef] [PubMed]

178. Kobuszewska, A.; Tomecka, E.; Zukowski, K.; Jastrzebska, E.; Chudy, M.; Dybko, A.; Renaud, P.; Brzozka, Z. Heart-on-a-Chip: An Investigation of the Influence of Static and Perfusion Conditions on Cardiac (H9C2) Cell Proliferation, Morphology, and Alignment. *SLAS Technol.* **2017**, *22*, 536–546. [CrossRef] [PubMed]

179. Nguyen, M.-D.; Tinney, J.P.; Ye, F.; Elnakib, A.A.; Yuan, F.; El-Baz, A.; Sethu, P.; Keller, B.B.; Giridharan, G.A. Effects of Physiologic Mechanical Stimulation on Embryonic Chick Cardiomyocytes Using a Microfluidic Cardiac Cell Culture Model. *Anal. Chem.* **2015**, *87*, 2107–2113. [CrossRef]

180. Marsano, A.; Conficconi, C.; Lemme, M.; Occhetta, P.; Gaudiello, E.; Votta, E.; Cerino, G.; Redaelli, A.; Rasponi, M. Beating Heart on a Chip: A Novel Microfluidic Platform to Generate Functional 3D Cardiac Microtissues. *Lab Chip* **2016**, *16*, 599–610. [CrossRef]

181. Grosberg, A.; Alford, P.W.; McCain, M.L.; Parker, K.K. Ensembles of Engineered Cardiac Tissues for Physiological and Pharmacological Study: Heart on a Chip. *Lab Chip* **2011**, *11*, 4165–4173. [CrossRef]

182. Tu, C.; Chao, B.S.; Wu, J.C. Strategies for Improving the Maturity of Human Induced Pluripotent Stem Cell-Derived Cardiomyocytes. *Circ. Res.* **2018**, *123*, 512–514. [CrossRef] [PubMed]

183. Zhang, Y.S.; Aleman, J.; Arneri, A.; Bersini, S.; Piraino, F.; Shin, S.R.; Dokmeci, M.R.; Khademhosseini, A. From Cardiac Tissue Engineering to Heart-on-a-Chip: Beating Challenges. *Biomed. Mater.* **2015**, *10*, 034006. [CrossRef] [PubMed]

184. Chu, X.; Bleasby, K.; Evers, R. Species Differences in Drug Transporters and Implications for Translating Preclinical Findings to Humans. *Expert Opin. Drug Metab. Toxicol.* **2013**, *9*, 237–252. [CrossRef] [PubMed]

185. Hirt, M.N.; Boeddinghaus, J.; Mitchell, A.; Schaaf, S.; Börnchen, C.; Müller, C.; Schulz, H.; Hubner, N.; Stenzig, J.; Stoehr, A.; et al. Functional Improvement and Maturation of Rat and Human Engineered Heart Tissue by Chronic Electrical Stimulation. *J. Mol. Cell. Cardiol.* **2014**, *74*, 151–161. [CrossRef]

186. Zhang, B.; Montgomery, M.; Chamberlain, M.D.; Ogawa, S.; Korolj, A.; Pahnke, A.; Wells, L.A.; Massé, S.; Kim, J.; Reis, L.; et al. Biodegradable Scaffold with Built-In Vasculature for Organ-on-a-Chip Engineering and Direct Surgical Anastomosis. *Nat. Mater.* **2016**, *15*, 669–678. [CrossRef] [PubMed]

187. Mathur, A.; Loskill, P.; Shao, K.; Huebsch, N.; Hong, S.; Marcus, S.G.; Marks, N.; Mandegar, M.; Conklin, B.R.; Lee, L.P.; et al. Human iPSC-Based Cardiac Microphysiological System for Drug Screening Applications. *Sci. Rep.* **2015**, *5*, 8883. [CrossRef]

188. Homan, K.A.; Kolesky, D.B.; Skylar-Scott, M.A.; Herrmann, J.; Obuobi, H.; Moisan, A.; Lewis, J.A. Bioprinting of 3D Convoluted Renal Proximal Tubules on Perfusable Chips. *Sci. Rep.* **2016**, *6*, 34845. [CrossRef]

189. Kolesky, D.B.; Homan, K.A.; Skylar-Scott, M.A.; Lewis, J.A. Three-Dimensional Bioprinting of Thick Vascularized Tissues. *Proc. Natl. Acad. Sci. USA* **2016**, *113*, 3179–3184. [CrossRef]

190. Dhariwala, B.; Hunt, E.; Boland, T. Rapid Prototyping of Tissue-Engineering Constructs, Using Photopolymerizable Hydrogels and Stereolithography. *Tissue Eng.* **2004**, *10*, 1316–1322. [CrossRef]

191. Boland, T.; Xu, T.; Damon, B.; Cui, X. Application of Inkjet Printing to Tissue Engineering. *Biotechnol. J.* **2006**, *1*, 910–917. [CrossRef]

192. Yan, J.; Huang, Y.; Chrisey, D.B. Laser-Assisted Printing of Alginate Long Tubes and Annular Constructs. *Biofabrication* **2013**, *5*, 015002. [CrossRef] [PubMed]

193. Beyer, S.T.; Bsoul, A.; Ahmadi, A.; Walus, K. 3D Alginate Constructs for Tissue Engineering Printed Using a Coaxial Flow Focusing Microfluidic Device. In Proceedings of the 2013 Transducers & Eurosensors XXVII: The 17th International Conference on Solid-State Sensors, Actuators and Microsystems (TRANSDUCERS & EUROSENSORS XXVII), Barcelona, Spain, 6–20 June 2013; pp. 1206–1209. [CrossRef]

194. Gaetani, R.; Feyen, D.A.M.; Verhage, V.; Slaats, R.; Messina, E.; Christman, K.L.; Giacomello, A.; Doevendans, P.A.F.M.; Sluijter, J.P.G. Epicardial Application of Cardiac Progenitor Cells in a 3D-Printed Gelatin/Hyaluronic Acid Patch Preserves Cardiac Function after Myocardial Infarction. *Biomaterials* **2015**, *61*, 339–348. [CrossRef] [PubMed]

195. Maiullari, F.; Costantini, M.; Milan, M.; Pace, V.; Chirivì, M.; Maiullari, S.; Rainer, A.; Baci, D.; Marei, H.E.-S.; Seliktar, D.; et al. A Multi-Cellular 3D Bioprinting Approach for Vascularized Heart Tissue Engineering Based on HUVECs and iPSC-Derived Cardiomyocytes. *Sci. Rep.* **2018**, *8*, 13532. [CrossRef]

196. Lee, H.; Cho, D.-W. One-Step Fabrication of an Organ-on-a-Chip with Spatial Heterogeneity Using a 3D Bioprinting Technology. *Lab Chip* **2016**, *16*, 2618–2625. [CrossRef] [PubMed]

197. Bhardwaj, N.; Kundu, S.C. Electrospinning: A Fascinating Fiber Fabrication Technique. *Biotechnol. Adv.* **2010**, *28*, 325–347. [CrossRef]

198. Huang, S.; Yang, Y.; Yang, Q.; Zhao, Q.; Ye, X. Engineered Circulatory Scaffolds for Building Cardiac Tissue. *J. Thorac. Dis.* **2018**, *10*, S2312–S2328. [CrossRef] [PubMed]

199. Castilho, M.; Hochleitner, G.; Wilson, W.; van Rietbergen, B.; Dalton, P.D.; Groll, J.; Malda, J.; Ito, K. Mechanical Behavior of a Soft Hydrogel Reinforced with Three-Dimensional Printed Microfibre Scaffolds. *Sci. Rep.* **2018**, *8*, 1245. [CrossRef]

200. Yildirim, Y.; Naito, H.; Didie, M.; Karikkineth, B.C.; Biermann, D.; Eschenhagen, T.; Zimmermann, W. Development of a Biological Ventricular Assist Device: Preliminary Data from a Small Animal Model. *Circulation* **2007**, *116*, I16–I23. [CrossRef]

201. Lee, E.J.; Kim, D.E.; Azeloglu, E.U.; Costa, K.D. Engineered Cardiac Organoid Chambers: Toward a Functional Biological Model Ventricle. *Tissue Eng. Part A* **2008**, *14*, 215–225. [CrossRef]

202. Li, R.A.; Keung, W.; Cashman, T.J.; Backeris, P.C.; Johnson, B.V.; Bardot, E.S.; Wong, A.O.T.; Chan, P.K.W.; Chan, C.W.Y.; Costa, K.D. Bioengineering an Electro-Mechanically Functional Miniature Ventricular Heart Chamber from Human Pluripotent Stem Cells. *Biomaterials* **2018**, *163*, 116–127. [CrossRef]

203. MacQueen, L.A.; Sheehy, S.P.; Chantre, C.O.; Zimmerman, J.F.; Pasqualini, F.S.; Liu, X.; Goss, J.A.; Campbell, P.H.; Gonzalez, G.M.; Park, S.-J.; et al. A Tissue-Engineered Scale Model of the Heart Ventricle. *Nat. Biomed. Eng.* **2018**, *2*, 930–941. [CrossRef] [PubMed]

204. Lee, A.; Hudson, A.R.; Shiwarski, D.J.; Tashman, J.W.; Hinton, T.J.; Yerneni, S.; Bliley, J.M.; Campbell, P.G.; Feinberg, A.W. 3D Bioprinting of Collagen to Rebuild Components of the Human Heart. *Science* **2019**, *365*, 482–487. [CrossRef] [PubMed]

205. Kupfer, M.E.; Lin, W.-H.; Ravikumar, V.; Qiu, K.; Wang, L.; Gao, L.; Bhuiyan, D.; Lenz, M.; Ai, J.; Mahutga, R.R.; et al. In Situ Expansion, Differentiation and Electromechanical Coupling of Human Cardiac Muscle in a 3D Bioprinted, Chambered Organoid. *Circ. Res.* **2020**. [CrossRef] [PubMed]

206. Garoffolo, G.; Pesce, M. Mechanotransduction in the Cardiovascular System: From Developmental Origins to Homeostasis and Pathology. *Cells* **2019**, *8*, 1607. [CrossRef]

207. Fink, C.; Ergün, S.; Kralisch, D.; Remmers, U.; Weil, J.; Eschenhagen, T. Chronic Stretch of Engineered Heart Tissue Induces Hypertrophy and Functional Improvement. *FASEB J.* **2000**, *14*, 669–679. [CrossRef]

208. Lasher, R.A.; Pahnke, A.Q.; Johnson, J.M.; Sachse, F.B.; Hitchcock, R.W. Electrical Stimulation Directs Engineered Cardiac Tissue to an Age-Matched Native Phenotype. *J. Tissue Eng.* **2012**, *3*. [CrossRef]

209. Ruwhof, C. Mechanical Stress-Induced Cardiac Hypertrophy: Mechanisms and Signal Transduction Pathways. *Cardiovasc. Res.* **2000**, *47*, 23–37. [CrossRef]

210. Severs, N.J. The cardiac Muscle Cell. *Bioessays* **2000**, *22*, 188–199. [CrossRef]

211. Lie-Venema, H.; van den Akker, N.M.S.; Bax, N.A.M.; Winter, E.M.; Maas, S.; Kekarainen, T.; Hoeben, R.C.; DeRuiter, M.C.; Poelmann, R.E.; Gittenberger-de Groot, A.C. Origin, Fate, and Function of Epicardium-Derived Cells (EPDCs) in Normal and Abnormal Cardiac Development. *Sci. World J.* **2007**, *7*, 1777–1798. [CrossRef]

212. Goette, A.; Lendeckel, U. Electrophysiological Effects of Angiotensin II. Part I: Signal Transduction and Basic Electrophysiological Mechanisms. *Europace* **2008**, *10*, 238–241. [CrossRef]
213. Zhang, P.; Su, J.; Mende, U. Cross Talk between Cardiac Myocytes and Fibroblasts: From Multiscale Investigative Approaches to Mechanisms and Functional Consequences. *Am. J. Physiol. Heart Circ. Physiol.* **2012**, *303*, H1385–H1396. [CrossRef] [PubMed]
214. Brutsaert, D.L. Cardiac Endothelial-Myocardial Signaling: Its Role in Cardiac Growth, Contractile Performance, and Rhythmicity. *Physiol. Rev.* **2003**, *83*, 59–115. [CrossRef] [PubMed]
215. Iyer, R.K.; Odedra, D.; Chiu, L.L.Y.; Vunjak-Novakovic, G.; Radisic, M. Vascular Endothelial Growth Factor Secretion by Nonmyocytes Modulates Connexin-43 Levels in Cardiac Organoids. *Tissue Eng. Part A* **2012**, *18*, 1771–1783. [CrossRef] [PubMed]
216. Garzoni, L.R.; Rossi, M.I.D.; de Barros, A.P.D.N.; Guarani, V.; Keramidas, M.; Balottin, L.B.L.; Adesse, D.; Takiya, C.M.; Manso, P.P.; Otazú, I.B.; et al. Dissecting Coronary Angiogenesis: 3D Co-Culture of Cardiomyocytes with Endothelial or Mesenchymal Cells. *Exp. Cell Res.* **2009**, *315*, 3406–3418. [CrossRef] [PubMed]
217. Liau, B.; Zhang, D.; Bursac, N. Functional Cardiac Tissue Engineering. *Regen. Med.* **2012**, *7*, 187–206. [CrossRef] [PubMed]
218. Xue, T.; Cho, H.C.; Akar, F.G.; Tsang, S.-Y.; Jones, S.P.; Marbán, E.; Tomaselli, G.F.; Li, R.A. Functional Integration of Electrically Active Cardiac Derivatives from Genetically Engineered Human Embryonic Stem Cells with Quiescent Recipient Ventricular Cardiomyocytes: Insights into the Development of Cell-Based Pacemakers. *Circulation* **2005**, *111*, 11–20. [CrossRef] [PubMed]
219. Ponard, J.G.C.; Kondratyev, A.A.; Kucera, J.P. Mechanisms of Intrinsic Beating Variability in Cardiac Cell Cultures and Model Pacemaker Networks. *Biophys. J.* **2007**, *92*, 3734–3752. [CrossRef]
220. Radisic, M.; Fast, V.G.; Sharifov, O.F.; Iyer, R.K.; Park, H.; Vunjak-Novakovic, G. Optical Mapping of Impulse Propagation in Engineered Cardiac Tissue. *Tissue Eng. Part A* **2009**, *15*, 851–860.
221. Schaffer, P.; Ahammer, H.; Müller, W.; Koidl, B.; Windisch, H. Di-4-ANEPPS Causes Photodynamic Damage to Isolated Cardiomyocytes. *Pflug. Arch.* **1994**, *426*, 548–551. [CrossRef]
222. Hamill, O.P.; Marty, A.; Neher, E.; Sakmann, B.; Sigworth, F.J. Improved Patch-Clamp Techniques for high-Resolution Current Recording from Cells and Cell-Free Membrane Patches. *Pflug. Arch.* **1981**, *391*, 85–100. [CrossRef]
223. Del Corsso, C.; Srinivas, M.; Urban-Maldonado, M.; Moreno, A.P.; Fort, A.G.; Fishman, G.I.; Spray, D.C. Transfection of Mammalian Cells with Connexins and Measurement of Voltage Sensitivity of Their Gap Junctions. *Nat. Protoc.* **2006**, *1*, 1799–1809. [CrossRef] [PubMed]
224. Brown, K.T. *Advanced Micropipette Techniques for Cell Physiology*; Wiley: Hoboken, NJ, USA, 1986; ISBN 9780471909521.
225. Kehat, I.; Gepstein, A.; Spira, A.; Itskovitz-Eldor, J.; Gepstein, L. High-Resolution Electrophysiological Assessment of Human Embryonic Stem Cell-Derived Cardiomyocytes. *Circ. Res.* **2002**, *91*, 659–661. [CrossRef] [PubMed]
226. Reppel, M.; Pillekamp, F.; Brockmeier, K.; Matzkies, M.; Bekcioglu, A.; Lipke, T.; Nguemo, F.; Bonnemeier, H.; Hescheler, J. The Electrocardiogram of Human Embryonic Stem Cell-Derived Cardiomyocytes. *J. Electrocardiol.* **2005**, *38*, 166–170. [CrossRef] [PubMed]
227. Tung, L.; Zhang, Y. Optical Imaging of Arrhythmias in Tissue Culture. *J. Electrocardiol.* **2006**, *39*, S2–S6. [CrossRef]
228. Efimov, I.R.; Nikolski, V.P.; Salama, G. Optical Imaging of the Heart. *Circ. Res.* **2004**, *95*, 21–33. [CrossRef]
229. Lieu, D.K.; Turnbull, I.C.; Costa, K.D.; Li, R.A. Engineered Human Pluripotent Stem Cell-Derived Cardiac Cells and Tissues for Electrophysiological Studies. *Drug Discov. Today Dis. Models* **2012**, *9*, e209–e217. [CrossRef]
230. Van Der Velden, J.; Klein, L.J.; Zaremba, R.; Boontje, N.M.; Huybregts, M.A.; Stooker, W.; Eijsman, L.; de Jong, J.W.; Visser, C.A.; Visser, F.C.; et al. Effects of Calcium, Inorganic Phosphate, and pH on Isometric Force in Single Skinned Cardiomyocytes from Donor and Failing Human Hearts. *Circulation* **2001**, *104*, 1140–1146. [CrossRef]
231. Karakikes, I.; Ameen, M.; Termglinchan, V.; Wu, J.C. Human Induced Pluripotent Stem Cell-Derived Cardiomyocytes: Insights into Molecular, Cellular, and Functional Phenotypes. *Circ. Res.* **2015**, *117*, 80–88. [CrossRef]
232. Hunter, P.J.; Pullan, A.J.; Smaill, B.H. Modeling Total Heart Function. *Annu. Rev. Biomed. Eng.* **2003**, *5*, 147–177. [CrossRef]

International Journal of
Molecular Sciences

Review

Every Beat You Take—The Wilms' Tumor Suppressor WT1 and the Heart

Nicole Wagner * and Kay-Dietrich Wagner *

CNRS, INSERM, iBV, Université Côte d'Azur, 06107 Nice, France
* Correspondence: kwagner@unice.fr (K.-D.W.); nwagner@unice.fr (N.W.); Tel.: +33-493-377665

Abstract: Nearly three decades ago, the Wilms' tumor suppressor Wt1 was identified as a crucial regulator of heart development. Wt1 is a zinc finger transcription factor with multiple biological functions, implicated in the development of several organ systems, among them cardiovascular structures. This review summarizes the results from many research groups which allowed to establish a relevant function for Wt1 in cardiac development and disease. During development, Wt1 is involved in fundamental processes as the formation of the epicardium, epicardial epithelial-mesenchymal transition, coronary vessel development, valve formation, organization of the cardiac autonomous nervous system, and formation of the cardiac ventricles. Wt1 is further implicated in cardiac disease and repair in adult life. We summarize here the current knowledge about expression and function of Wt1 in heart development and disease and point out controversies to further stimulate additional research in the areas of cardiac development and pathophysiology. As re-activation of developmental programs is considered as paradigm for regeneration in response to injury, understanding of these processes and the molecules involved therein is essential for the development of therapeutic strategies, which we discuss on the example of WT1.

Keywords: Wilms' tumor suppressor 1 (Wt1); heart; cardiac development; coronary vessel formation; transcriptional regulation; cardiac malformation; epicardium; epicardial derived cells (EPDCs); epithelial mesenchymal transition (EMT); cardiac cell fate; regeneration

Citation: Wagner, N.; Wagner, K.-D. Every Beat You Take—The Wilms' Tumor Suppressor WT1 and the Heart. *Int. J. Mol. Sci.* **2021**, *22*, 7675. https://doi.org/10.3390/ijms22147675

Academic Editor: Christophe Chevillard

Received: 13 June 2021
Accepted: 16 July 2021
Published: 18 July 2021

1. Introduction

The Wilms' tumor 1 (WT1) gene was originally identified based on its mutational inactivation in Wilms' tumor (nephroblastoma) [1–3]. This first discovery of WT1 as the responsible gene in an autosomal-recessive condition classified it as a tumor-suppressor gene. Mutations of WT1 were associated with the development of kidney tumors and urogenital defects. However, later it became clear that mutations of WT1 only occur in a low frequency in nephroblastoma [4] and that most nephroblastomas [5] express high levels of WT1. Based on the overexpression of WT1 in leukemia and most solid cancers (reviewed in [6,7] and its cancer-promoting functions in the tumor stroma [8], WT1 is nowadays considered as an oncogene and attractive candidate for cancer therapy.

WT1 encodes a zinc finger transcription factor and RNA-binding protein [9–12]. As a transcriptional regulator, it can either activate or repress various target genes. Thus, WT1 influences cellular differentiation, growth, apoptosis, and metabolism. WT1 exists in multiple isoforms. Alternative splicing of exon 5 and exon 9 gives rise to major isoforms. Splicing of exon 9 generates the KTS isoforms, which either include or exclude three amino acids lysin, threonine, and serin (KTS) between zinc fingers 3 and 4 of the protein [13]. Although the majority of WT1 proteins are in the nucleus, some are present in the cytoplasm, located on actively translating polysomes. WT1 isoforms shuttle between the nucleus and cytoplasm [14]. The complexity of WT1 is further enhanced by post-translational modifications and a plethora of binding partners. WT1 directs the development of several organs and tissues, among them the heart.

The heart develops mostly from embryonic mesodermal germ layer cells and to some extent from ectoderm-derived cardiac neuronal crest (cushions of the outflow tract). The cardiogenic mesoderm differentiates into proepicardial, endocardial, and myocardial cells. The epicardium is formed from a subset of the proepicardial cells. Proepicardial cells also contribute subepicardial cells, interstitial fibroblast, pericytes, and a subset of the endothelial cells of the coronary vessels. The inner lining of the heart tube is formed by endocardial cells. The vertebrate heart forms as two concentric epithelial cylinders of myocardium and endocardium separated by an extended basement membrane matrix commonly referred to as cardiac jelly. The primitive heart tube is formed at embryonic day 8.5 (E8.5) in the mouse [15]. The primitive tube elongates and undergoes rightward looping. Further remodeling of the heart involves formation and expansion of the chambers, and formation of valves and septa, resulting in a heart with two atria and two ventricles [16]. The heart is the first organ to develop and is already functional at an early stage of fetal development, in line with its essential role for the distribution of oxygen and nutrients and removal of waste products and carbon dioxide. Several excellent reviews have already described cardiac development in detail [17–22]. Thus, we focus here only on the role of Wt1. Wt1 expression was first observed in a transitory cluster of cells—the proepicardium and the coelomic epithelium at E9.5. Wt1-expressing proepicardial cells contact the dorsal wall of the heart from which the proepicardial cell migrate over the myocardium of the heart tube to form the epicardial layer by E12.5 [23,24]. This view has been challenged recently by the detection of a common progenitor cell population of epicardium and myocardium using single-cell RNA sequencing [25]. How these common progenitors might migrate during cardiac development is currently an open question.

A proportion of epicardial cells undergoes epithelial-to-mesenchymal transition (EMT), which induces the formation of epicardial-derived cells (EPDCs), a population of multipotent mesenchymal cardiac progenitor cells, which might differentiate into cardiomyocytes, fibroblasts, smooth muscle, and endothelial cells [26–28], which is discussed in detail later. First indications for the indispensable role of Wt1 in heart homeostasis came from the observations made in Wt1 knockout embryonic mice which died at mid-gestation due to cardiac malformations [29].

Here, we review the history of investigations characterizing the role of WT1 (i) in cardiac development, (ii) in cardiac disease and regeneration, and (iii) in different cardiac cell types and transcriptional regulatory mechanisms. We indicate emerging notions and point out problems and promises in the field of development of therapeutic strategies for cardiac repair.

2. WT1 in Heart Development

Nearly thirty years ago, Armstrong and colleagues, using in situ mRNA hybridization, observed Wt1 expression in the differentiating heart mesothelium of the mouse embryo at embryonic day 9 [23]. In the same year, the group of Jaenisch introduced a mutation into the murine *Wt1* gene by gene targeting in embryonic stem cells. The embryos homozygous for this mutation died between days 13 and 15 of gestation. Besides the lack of kidney and gonad formation in *Wt1* mutant mice, the authors observed a severe heart hypoplasia with thinned right ventricular walls, a rounded apex, and a reduction of size of the left ventricles, signs of congestive heart failure, suggesting that cardiac malfunction was the cause of early embryonic death [29]. As Wt1 has been described before only to be expressed in the epicardium, but has not yet been observed in the myocardium, it remained unclear whether these features of cardiac malformation were due to primary defects in the myocardial tissue or a consequence of disturbed development in other tissues. A more detailed view on Wt1 expression during murine heart development was achieved using a lacZ reporter gene inserted into a YAC (yeast artificial chromosome) construct which demonstrated Wt1 expression in the early proepicardium, the epicardium, and subepicardial mesenchymal cells (SEMCs) throughout development. In Wt1-deficient animals, the epicardium did not form correctly, which results in disruption in the formation of the coronary vasculature, leading to pericardial bleeding and midgestational death of the embryo. Complementation of *Wt1* null embryos with a human *WT1* transgene rescued both embryonic heart defects and midgestational death, confirming that indeed heart failure causes the death of Wt1-deficient embryos [24].

Wt1-expressing cell types during heart development in different species are summarized in Table 1 and further described below. Of note, expression of Wt1 is limited to a subset of the identified cells. Functional differences between Wt1-expressing cells and the Wt1-negative counterparts remain mostly unknown at present.

Table 1. Wt1-expressing cell types during heart development.

Cell Type	Species	Reference
proepicardial cells	mouse, bird, zebrafish	[23,24,30,31]
epicardial cells	mouse, bird, zebrafish, human	[24,28,30–33]
endocardial cells	bird, human	[28,34]
subepicardial mesenchymal cells (SEMC), epicardial-derived cells (EPDCs)	mouse, bird	[24,28,31,32]
valvular interstitial cells	bird	[28]
smooth muscle cells	bird	[28]
endothelial cells	bird, mouse, human	[28,31,32,34,35]
fibroblasts	bird, mouse	[27,36]
cardiomyocytes	mouse, human	[26,33,37]

Studies in birds confirmed the expression of Wt1 in epicardium- and epicardial-derived cells (EPDCs) during embryonic development [31]. Using normal avian and quail-to-chick chimeric embryos, the origin and fate of Wt1-expressing EPDCs were later described and the effects of epicardial ablation on cardiac development investigated [28]. Wt1-expressing EPDCs were found to populate the subepicardial space and to invade the ventricular myocardium. Upon differentiation in smooth muscle and endothelial cells, Wt1 expression decreased in EPDCs. Undifferentiated EPDCs continued to express Wt1 and invaded the ventricular myocardium and the atrio-ventricular (AV) valves. Disruption of normal epicardial development either by proepicardial ablation or block reduced the number of invasive Wt1-positive EPDCs, and provoked anomalies in the coronary vessels, the ventricular myocardium, and the AV cushions. In addition to Wt1, EPDCs express retinaldehyde-dehydrogenase (Raldh) 2 [38,39]. It had been demonstrated that in humans WT1 transcriptionally regulates the retinoic acid receptor alpha (*RAR-α*) gene [40]. Transcriptional target genes of WT1 with relevance in the heart are summarized in Table 2 and discussed below.

The phenotype of the WT1-deficient mice further resembled that of retinoic acid (RA)-depleted mice. Depletion of RA from the diet is known to severely disturb heart development, causing hypoplasia of the ventricles [41]. The authors suggested therefore that Wt1 maintains the EPDCs in an undifferentiated, RA-producing state to contribute to ventricular myocardium compaction in the development of the myocardial wall [28]. Availability of retinoic acid during cardiac development is mediated by Raldh2. It has been shown that Wt1 transcriptionally activates Raldh2 [42]. Pericardium and sinus horn formation are coupled and are based on the expansion and exact temporal release of pleuropericardial membranes (PPM) from the underlying subcoelomic mesenchyme. Wt1-deficient mouse embryos displayed a failure to form myocardialized sinus horns and a loss of Raldh2 expression in the subcoelomic mesenchyme, pointing to a crucial role of Wt1 and downstream Raldh2/RA signaling in sinus horn development [43]. Furthermore, Wt1-mutant mice were shown to display unilateral partial PPM absence in the dorsomedial region. Failure of PPM release affects the closure of the remaining communication area between pericardial and pleural cavities, the bilateral pericardioperitoneal canals (PPCs), which is disturbed in Wt1-deficient embryos, leading to pleuropericardial communication and lateralization of the cardinal veins [44]. The group of Muñoz -Chapuli suggests that the proepicardium is an evolutionary derivative of the primordium of an ancient external pronephric glomerulus, initially based on the epicardial development in lampreys (Petromyzon), the most primitive living lineage of vertebrates [45]. Employing chick proepicardium, they propound that Wt1 could repress the nephrogenic potential of the proepicardium, while at the same time promote nephrogenesis in the intermediate mesoderm. This paradoxical function could be explained by the dual role of Wt1, which promotes mesenchymal to epithelial transition (MET) in the kidney and EMT in the epicardium [46], through a mechanism known as chromatin flip–flop [47]. Promotion of EMT in the developing epicardium and MET in the growing kidney is not only reflected by morphological cellular changes, but also differential expression of podocyte markers. In their study, the authors focused on podocalyxin, known to be transcriptionally regulated by Wt1, and to be activated by Wt1 in kidney podocytes [48], which they found in contrast to be upregulated in Wt1-deficient epicardium [46]. To further strengthen this theory it appears interesting to investigate the expression of other Wt1 transcriptional targets in kidney MET and epicardial EMT, such as nephrin [49], nestin [50], and podocin [51].

In addition, the relation between Wt1-expressing epicardial derivatives and the development of compact ventricular myocardium has been investigated. The differences in myocardial architecture specifically between the right ventricle (RV) and the left ventricle (LV) in association to epicardial formation and distribution of Wt1-expressing cells were studied. The authors demonstrated that the RV is less densely and later covered by the epicardium than the LV. They also observed that compact myocardial layer formation occurred in parallel with the presence of Wt1-expressing cells and was more pronounced in the LV than in the RV, and within the RV more accentuated in the postero-lateral wall than in the anterior wall, which might explain the lateralized differences in ventricular morphology of the heart [52]. The same group was able to identify a function of the epicardium in cardiac autonomic nervous system modulation, essential for proper cardiac activity by altering heart rate, conduction velocity, and force of contraction. They revealed expression of neuronal markers in the epicardium during early cardiac development, notably of tubulin beta-3 chain (Tubb3), which was colocalized with Wt1 in epicardium and the nervous system, neural cell adhesion molecule (Ncam), and the β2 adrenergic receptor (β2AR). Adrenaline (epinephrine), a catecholamine, is known to modulate heart rate, velocity of conduction, and contraction strength in the heart through its binding to β2AR. Inhibition of the outgrowth of the epicardium abolished the response to adrenaline administration, indicating that the epicardium is necessary for a normal response of the heart to adrenaline during early cardiac development [53]. This report further confirmed a role of Wt1 in neural function, as suggested by several studies [23,54–59].

In zebrafish, two orthologues of *wt1* have been described: *wt1a* [60,61] and *wt1b* [62]. Both of them were found to be expressed in adult zebrafish hearts, but exhibited a differential expression level in other organs, as well as a differing temporal patterning during development, suggesting distinctive functions during zebrafish development [62]. During zebrafish cardiac development, Wt1 is required for the proper development of the proepicardial organ and epicardial lineage [30]. A later study proposed that Wt1-interacting protein (Wtip), a protein identified as a Wt1-interacting partner by a yeast two-hybrid screen [63], signals in conjunction with WT1 for proepicardial organ specification and cardiac left/right asymmetry in the zebrafish heart [64]. Two main cardiac cell types were suggested to be involved in zebrafish heart regeneration using ex vivo cultures: epicardial cells, displaying a larger, prismatic morphology and Wt1/Gata4 (Gata-binding protein 4) expression, and endocardial small, rounded cells, positive for Nfat2 (nuclear factor of activated T-cells 2) and Gata4 [65].

Table 2. WT1-target genes related to cardiac development and disease.

Gene	Reference
Insulin like growth factor 1 receptor (IGF-1-R)	[66]
Epidermal growth factor receptor (EGFR)	[67]
Retinoic acid receptor alpha (RAR- α)	[40]
Retinaldehyde-dehydrogenase (Raldh) 2	[42]
Insulin receptor (IR)	[68]
Paired box gene 2 (Pax2)	[69]
Platelet-derived growth factor A (PDGFA)	[70,71]
Early growth response protein 1 (EGR-1)	[72]
Insulin like growth factor 2 (IGF-2)	[73]
Transforming growth factor beta (TGF-β)	[74]
Colony-stimulating factor-1 (CSF-1)	[75]
Syndecan 1	[76]
Midkine	[77]
Vitamin D receptor (Vdr)	[78,79]
Podocalyxin	[48]
Nephrin (Nphs1)	[49]
Podocin (Nphs2)	[51]
Tyrosinkinase receptor (Trk)B	[32]
Nestin	[50]
Erythropoietin (EPO)	[80]
α4 Integrin	[81]
Vascular endothelial growth factor (VEGF)	[82,83]
Vascular endothelial growth factor receptor (Vegfr) 2	[84,85]
ETS proto-oncogene (ETS)-1	[84]
Snail (Snai1)	[86]
Slug (Snai2)	[87]
E-Cadherin	[86,88]
VE-Cadherin	[89]
Coronin1B	[90]
Cxcl10 (C-X-C Motif Chemokine Ligand 10)	[91]
Ccl5 (C-C Motif Chemokine Ligand 5)	[91]
Interferon regulatory factor (Irf)7	[91]
c-Kit (tyrosine-protein kinase KIT)	[8]
Pecam-1 (platelet and endothelial cell adhesion molecule 1)	[8]
Telomere repeat binding factor (Trf) 2	[92]
Bone morphogenetic protein (Bmp) 4	[93]

3. WT1 in the Adult Heart and Cardiac Pathologies

Already in 1994, Wt1 transcripts were detected by Northern blot in adult rat heart tissues [94]. Whether modifications in Wt1 expression occur under pathophysiological conditions and which cell types express the protein remained open questions. Our group was the first to demonstrate that Wt1 is a useful early marker of myocardial infarction [95], a finding later confirmed by others [96–98]. We focused on the de novo Wt1 expression in the coronary vasculature of the ischemic myocardium. As Wt1 is essential for normal growth of the heart during development, we originally reasoned that it might also play a role in adult cardiac hypertrophy. To test this hypothesis, we analyzed the expression of Wt1 in normal hearts and in the hypertrophied left ventricles of spontaneously hypertensive rats (SHRs), with activation of the renin–angiotensin system by transgenic (over) expression of human renin and angiotensinogen genes, and with postinfarct remodeling of the heart after ligation of the left coronary artery (LAD). Interestingly, we detected an over two-fold increase of cardiac Wt1 mRNA expression after LAD ligation, but no differences for the two hypertrophy models compared to controls. Further experiments using LAD ligation demonstrated a rapid increase of cardiac Wt1 levels already 24 h after LAD ligation, which remained elevated for nine weeks following the ischemic injury. Strikingly, in addition to its expression in the epicardium, we observed Wt1 localized to the coronary vessels in proximity to the infarcted tissue. Coronary vessels of non-infarcted animals did not express Wt1. Wt1 was expressed in endothelial as well as in vascular smooth muscle cells in the border zone of infarcted tissues. We confirmed this finding also in human cardiac ischemic tissues (unpublished results). Interestingly, WT1 expression could also be detected in healthy adult human myocardium by others [99]. Colocalization of Wt1 with proliferating cell nuclear antigen (PCNA) and vascular endothelial growth factor (VEGF) suggests a role of Wt1 in the proliferative response of the coronary vasculature to cardiac hypoxia [95]. In a following study, we were the first to demonstrate that Wt1 expression is indeed triggered by hypoxia, which involves transcriptional activation of the Wt1 promoter by the hypoxia inducible factor 1 (HIF-1) [100]. Later studies confirmed our finding that Wt1 is a hypoxia-regulated gene [83,101]. Interestingly, it had been demonstrated that ischemia in vivo (through myocardial infarction in mice) or in vitro (hypoxia exposition of epicardial human explants) induced an embryonic reprogramming of the epicardial compartment, involving migration of epicardial-derived stem cell marker c-Kit expressing Wt1-positive cells which contributed to re-vascularization and cardiac remodeling [102]. As we identified c-Kit as a transcriptional target of Wt1 in the context of vascular formation [8], it seems conceivable that mobilization of c-Kit precursor cells represents one mechanism of Wt1-mediated cardiac neovascularization after ischemia. We further identified the telomere repeat-binding factor (Trf) 2 to be regulated by Wt1 [92]. Down-regulation of Trf2 has been demonstrated to provoke cardiomyocyte telomere erosion and apoptosis, linking telomere dysfunction to heart failure [103].

Thymosin β4 (Tβ4), a 43-amino-acid G-actin-sequestering peptide which is expressed in the embryonic heart and implicated in coronary vessel development in mice [104], has been shown to activate cardiac regeneration through stimulation of the expression of embryonic developmental genes in the adult epicardium, leading to de novo coronary vessel formation after myocardial infarction. However, a significant increase could only be reported for Vegf, Vegfr2, and TGFβ levels, whereas Wt1 levels were not significantly altered 24 h after MI compared to vehicle-treated animals [105]. A later study additionally revealed that adult Wt1$^+$ GFP$^+$ EPDCs cells obtained through Tβ4 priming and myocardial infarction are a heterogeneous population expressing cardiac progenitor and mesenchymal stem markers that can restore an embryonic gene program, but do not revert entirely to adopt an embryonic phenotype [106].

First suspicions for a role of Wt1 in human cardiac pathologies originated in 2004, with a case report from an adult XY karyotype patient with a N-terminal *WT1* missense mutation presenting a very unusual phenotype: ambiguous genitalia, but normal testosterone levels, absence of kidney disease, and an associated congenital heart defect [107]. Later, a role for WT1 in some cases of congenital diaphragmatic hernia associated with the Meacham syndrome phenotype had been suggested [108]. Meacham syndrome is a rare sporadically occurring multiple malformation syndrome characterized by male pseudo-hermaphroditism with abnormal internal female genitalia, complex congenital heart defects, including hypoplastic left hearts, and diaphragmatic abnormalities [109]. In a number of Meacham syndrome patients, heterozygous missense mutations in the C–terminal zinc finger domains of WT1 could be identified, suggesting that at least some cases displaying phenotypes of Meacham syndrome are caused by mutations at the *WT1* locus [108]. We reported the case of a 4-month-old girl, who presented with end-stage renal disease, nephroblastomatosis, thrombopenia, anemia, pericarditis, and cardiac hypertrophy accompanied by severe hypertension. Sequence analysis identified a heterozygous nonsense mutation in exon 9 of *WT1*, which leads to a truncation of the WT1 protein at the beginning of zinc finger 3 [110]. WT1 is a transcriptional regulator of erythropoietin, which might explain the persistent anemia in this patient [80]. Evolution over time showed severe and resistant high blood pressure, despite multi-drug therapy and bilateral nephrectomy, which did not result in the normalization of the blood pressure values. Acute episodes of high blood pressure were associated with cardiogenic shock and anemia. The little patient showed a severe concentric myocardial hypertrophy, with moderate signs of heart failure and intermittent pericarditis [110]. Still awaiting kidney transplantation, the child died due to myocardial infarction at the age of five years. Later, another case of cardiac pathology in a patient with a *WT1* mutation was reported: A 46, XY phenotypic male patient with isolated nephrotic syndrome, end-stage renal disease, and hypertension, presented at the age of 6.3 years. A mutation in exon 8 of the *WT1* gene was identified. After starting hemodialysis, manifestations of hypertension and renal failure improved, but he died at 6.8 years of age as a result of heart and respiratory failure [111]. Monozygotic twins with congenital nephrotic syndrome caused by a *WT1* mutation have been reported to have died due to sepsis and extensive thrombosis of central venous system and sepsis and sudden heart failure at ages 23 weeks/13.5 months, respectively [112]. WT1 misexpression has been reported in autopsy findings from two human fetuses, displaying congenital pulmonary airway malformation, bilateral renal agenesis, and congenital heart defects [113]. Shortly after, re-evaluation of autopsy data from fourteen additional fetuses with combined renal agenesis and cardiac anomalies revealed abnormalities of Wt1 expression, mostly in liver mesenchymal cells. As WT1 is widely expressed in mesothelium, it had been suggested that the defects could be caused by abnormal function of mesenchyme derived from mesothelial cells [114]. WT1 is further expressed in cardiac angiosarcomas, which is the most common malignant neoplasm of the heart in adults. As other primary cardiac malignancies such as synovial sarcoma, leiomyosarcoma, and unclassified sarcomas are frequently negative for WT1, this finding might be helpful for differential diagnosis. It further confirms the implication of WT1 in vascular formation [115].

Interestingly, it has been shown recently using patient biopsies that the thickening of the epicardium and migration of Wt1-positive EPDCs contributes to atrial fibro-fatty infiltration, a source of atrial fibrillation. Employing Wt1 genetic lineage mouse lines, the authors showed that adult EPDCs maintain an adipogenic potential in the epicardial layer and can shift to a fibrotic phenotype in response to distinct stimuli, identifying the epicardium as a central regulator of the balance between fat and fibrosis accumulation [116]. Additionally, the expression of TGFβ1 and FGFs (fibroblast growth factors) by EPDCs has been suggested to contribute to the pathogenesis of myocardial fibrosis, apoptosis, arrhythmias, and cardiac dysfunction in a mouse model of arrhythmogenic cardiomyopathy (ACM) [117].

4. WT1 in the Heart–Focus on Different Cell Types and Regulatory Mechanisms

Table 3 summarizes WT1-expressing cell types in the adult heart. Reported functions and regulatory mechanisms are discussed below.

Table 3. Wt1-expressing cell types in adult heart.

Cell Type	Species	Reference
epicardial cells	rat, mouse, human	[95,102]
endothelial cells	rat (MI, ischemia), mouse (MI, ischemia)	[95,118]
vascular smooth muscle cells	rat (MI, ischemia)	[95]
cardiomyocytes	mouse (priming with Tβ4, followed by MI, ischemia)	[119]
	mouse	[37]
fibroblasts	mouse (MI, ischemia)	[118]
adipocytes	mouse (MI, ischemia)	[120]
macrophages	zebrafish (cardiac injury)	[121]

Although a relationship between Wt1 and myocardial blood vessel development had already been suggested [24,28], it remained unclear whether Wt1 is indeed necessary for normal vascularization of the heart. Coronary vessel formation is organized through a series of tightly regulated events. The epicardial cells undergo an epithelial-to-mesenchymal transition [122–124] to become subepicardial mesenchymal cells. The subepicardial mesenchymal cells then migrate into the myocardium, where they differentiate into endothelial cells, smooth muscle cells, and perivascular fibroblasts of the coronary vessels [124,125]. Further steps in coronary vessel formation include stabilization of the newly formed vessels and remodeling to connect the vessels to the main coronary arteries, originating from the aorta (reviewed in [126]). In contrast to this classical view, mostly lineage tracing experiments suggested an important contribution of sinus venosus-derived endothelial cells [127,128] or the endocardium [129] for cardiac vessel endothelial cells, which has been questioned later [130] (reviewed in [131,132]).

In addition to the epicardium, we clearly observed nuclear Wt1 protein expression in the coronary vessels of mouse embryos at E12.5, E15.5, and E18.5. Notably, we detected endogenous Wt1 protein but not reporter gene activity. Wt1-deficient embryos (E12.5) failed to form subepicardial coronary vessels. To identify candidate target genes of Wt1 in the process of coronary vessel formation, we performed a transcriptome analysis of differentially expressed genes from hearts of wild-type and *Wt1*-deficient mice. One of the genes found to be differentially expressed was *Ntrk2*, the gene encoding for the tyrosine kinase type B receptor (TrkB) [32]. TrkB is a tyrosine kinase receptor with high affinity for brain-derived neurotrophic factor (BDNF) and neurotrophin 4/5 (NT4/5) (reviewed in [133]). A role for BDNF signaling in coronary blood vessel formation had emerged based on the observation that BDNF-deficient mice displayed abnormal myocardial vessel formation due to endothelial cell apoptosis [134]. TrkB and Wt1 co-localized in the epicardium and the coronary vessels of mouse embryonic hearts at E12.5. TrkB expression was absent from Wt1-deficient embryonic hearts. TrkB-deficient mouse embryos revealed a reduction of coronary vessel formation along with enhanced apoptosis. In addition to these lines of evidence which suggested that *Ntrk2*, the gene encoding the TrkB neurotrophin receptor, represents a target of Wt1 in the process of myocardial vascularization, molecular approaches employing transient transfections, Dnase1 footprint analyses, and electrophoretic mobility shift assays helped to identify a binding site for Wt1 in the *Ntrk2* promoter. This binding site was necessary for transgenic expression of a lacZ reporter in the developing myocardial vasculature and other known sites of Wt1 expression as the gonads and the coelomic epithelium. Activation of TrkB expression by Wt1 appears therefore to be a critical step for the proper development of the coronary vessels [32]. Another protein that is expressed by newly forming vessels is the intermediate filament protein nestin [135]. We could demonstrate the regulation of nestin by Wt1; nestin further colocalized with Wt1 in the epicardium and the forming coronary vessels, and was undetectable in Wt1 knockout hearts [50]. Nestin has been found to be highly expressed in proangiogenic capillaries after myocardial infarction and has been proposed to play a role in remodeling the cytoskeleton of cells in the human postinfarcted myocardium [136]. Moreover, the transmembrane cell adhesion molecule α4 integrin has been identified to be a transcriptional target of Wt1 in cardiac development [81]. α4-integrin-deficient mouse embryos display epicardial and coronary vessel formation defects, similar to those observed in Wt1 knockout embryos [137]. The transcriptional activation of the *α4 integrin* gene by Wt1 could therefore be an additional regulatory mechanism for the formation of the epicardium. We further identified the major podocyte protein nephrin as a transcriptional target of Wt1 [49]. Nephrin is not only required for kidney podocyte function [138], but also is crucial for cardiac vessel formation during development. We found nephrin to be expressed during human and mouse cardiac development. Nephrin-deficient mice displayed epicardial defects, a disturbed coronary vessel formation, and an increased apoptosis predominantly in the developing epicardium. Direct interaction of nephrin with the low affinity neurotrophin receptor p75NTR and subsidiary upregulation of p75NTR are critically involved in the cardiac phenotype of nephrin-deficient embryos [139]. Cardiac abnormalities have been reported in FinMajor nephrin mutation patients, which presented mainly with mild cardiac hypertrophy [140–142]. Another protein critically involved in cardiac vessel formation, which is transcriptionally regulated by Wt1, is the transcription factor Ets-1 [84]. Like Wt1, Ets-1 deficiency results in a failure of epicardial differentiation, a disturbed coronary vessel formation, and myocardial defects [143]. Recently, extracardiac septum transversum/proepicardium (ST/PE)-derived endothelial cells have been shown to be required for proper coronary vascular morphogenesis. Conditional deletion of Wt1 from both, the ST/PE and the endothelium disrupted embryonic coronary transmural patterning, leading to embryonic death. The ST/PE contributes a significant fraction of cells ($\approx$20%) to the coronary endothelium during embryogenesis required for proper coronary vascular development [144].

Using Wt1^{GFPCre} and inducible Wt1$^{\text{CreERT2}}$ mouse lines, the group of William Pu suggested in 2008 that Wt1-expressing epicardial cells contribute to the cardiomyocyte lineage during normal heart development. Wt1 epicardial cells located on the heart at E10.5–E11.5 differentiated in vivo into fully functional cardiomyocytes, as evidenced ex vivo by spontaneous contractility and calcium oscillations with kinetics, amplitude, and frequency characteristic of cardiomyocytes [26]. Moreover, in 2008, Cai and colleagues reported the identification of a cardiac myocyte lineage that derives from the proepicardial organ. These T-box transcription factor (Tbx) 18-expressing progenitor cells migrated onto the outer cardiac surface to form the epicardium, and then contributed to myocytes in the ventricular septum and the atrial and ventricular walls [145]. Shortly after, Christoffels et al. showed that Tbx18 is expressed in left ventricular and interventricular septum cardiomyocytes independent of a epicardial contribution [146]. Both groups used independent Tbx18 Cre knock-in lines, which differed considerably in sensitivity and specificity. The results of the Tbx18 reporter system used by Christoffels et al. correspond to endogenous Tbx18 expression data reported earlier [147]. Studying in detail the migration and differentiation of epicardium-derived cells, the group of Poelman already observed in 1998 that EPDCs migrated to the subendocardium, myocardium, and atrioventricular cushions. The functional role of these novel EPDCs remained however unclear [148]. Employing chick proepicardial explant cultures, it had further been demonstrated that proepicardial cells were able to differentiate into cardiac muscle cells in vitro, reflecting the pluripotency of the pericardial mesoderm [149]. In 2011, a reactivation of *Wt1* expression after myocardial infarction resulting in cardiomyocyte restitution has been proposed. Using Wt1^{CreEGFP} and inducible Wt1$^{\text{CreERT2}}$ lines, the authors observed only a mild increase in Wt1 expression upon cardiac injury and no initiation of cardiac vessel formation, which is in contrast to our previous findings [95]. Only by using thymosin β4 (Tβ4) priming which had before been reported to initiate expression of embryonic developmental genes in the epicardium [105], a significant reactivation of Wt1 expression after myocardial infarction was achieved resulting in the appearance of some Wt1-positive cardiomyocytes in the border zone of the infarcted area. These findings suggest a contribution of Wt1-positive EPDCs to the myocardium after myocardial infarction in the artificial setting of thymosin β4 priming before infarction [119]. However, co-authors from this study showed one year later, using the epicardium genetic lineage tracing line Wt1$^{\text{CreERT2/+}}$ and double reporter line Rosa26$^{\text{mTmG/+}}$, that epicardial cells do not differentiate into cardiomyocytes following myocardial infarction and Tβ4 treatment. Their study clearly raises cautions regarding a potential clinical use of Tβ4 with the goal to increase cardiac repair [150]. An additional Wt1 Cre using a BAC clone has been described, which again gave different results. Without using Tβ4, the authors observed a significant increase of Wt1 expression and proliferation in the epicardium shortly after myocardial infarction, leading to the formation of a Wt1-lineage-positive subepicardial mesenchyme until two weeks post-infarction. These mesenchymal cells were shown to contribute to the fibroblast population, myofibroblasts, and coronary endothelium in the infarct zone, a few of them later also differentiated into cardiomyocytes [118]. C. Rudat and A. Kispert undertook a substantial effort to identify Wt1-expressing cardiac cell types and to clarify the contribution of Wt1-expressing progenitor cells to differentiated cardiac cells. Using in situ hybridization and immunofluorescence, they revealed expression of Wt1 mRNA and protein not before E9.5 in the (pro)epicardium and endothelial cells throughout development. They further determined that neither Wt1^{CreEGFP} nor Wt1$^{\text{CreERT2}}$ lineage tracing systems are reliable epicardial fate-defining approaches due to ectopic recombination and poor recombination efficiency. Endogenous expression of Wt1 in the endothelium and eventually the myocardium in the developing heart eliminates Wt1-based Cre lines to trace the epicardial contribution to myocardial, and endothelial cells in the murine heart. The proposed epicardial origin of myocardial and endothelial cells in the heart using Wt1-based Cre/loxP lines appears therefore not justified [35]. Accordingly, Wt1 had already been excluded from epicardial cell fate mapping approaches in zebrafish due to its non-epicardial expression in the fish hearts [151]. A population of adult cardiac resident colony forming unit fibroblasts (cCFUs), most likely originating from the proepicardium/epicardium has been identified, giving rise mainly to cellular components of the coronary vasculature. A differentiation into cardiomyocytes in vivo could, however, not be supported [152]. Our group found Wt1 to be expressed in cardiomyocytes during development (first time point studied E10.5) and throughout lifespan. The number of Wt1-positive cardiomyocytes as well as the individual cellular intranuclear Wt1 expression decreased during development and was very low in adulthood, but we propose that low levels of Wt1 expression are sufficient to maintain a cardiac progenitor subset from terminal differentiation. Myocardial infarction strongly up-regulated Wt1-expressing cardiomyocytes, as well as individual nuclear Wt1 expression in cardiomyocytes. Interestingly, in contrast to the expression pattern in epicardial cells, Wt1 was expressed in a speckled manner in cardiomyocytes [37], suggesting the presence of the Wt1 + KTS variant [153]. We detected Wt1+ cardiomyocytes already 48 h after myocardial infarction. Given the distance to the epicardium and the differing expression pattern of Wt1, speckled in the nucleus in cardiomyocytes versus diffuse nuclear expression in epicardial cells, it is unlikely that these cardiomyocytes are epicardium-derived [37]. It has been shown that de novo cardiomyocytes arise in adjacent areas of a myocardial infarction [154,155]. This could support cardiac tissue regeneration by Wt1 reactivation in response to stimuli as hypoxia/inflammation, inducing progenitor cell proliferation and cell survival. We observed that upon cardiac differentiation of mouse embryonic stem cells (mESCs), Wt1 expression increased. Overexpression of Wt1 in mESCs reduced phenotypic cardiomyocyte differentiation in vitro keeping the cells in a more progenitor-like stage [37]. Recently, a common progenitor pool of the epicardium and myocardium has been identified by single cell transcriptomic analyses. Most of the clusters expressed Wt1, which explains expression in some cardiomyocytes and epicardium later in life [25], suggesting that these few cells with low level Wt1 expression are sufficient to maintain a cardiac progenitor subset from terminal differentiation, which becomes reactivated in cardiac repair.

It is necessary to underline, that the initial proposition of the importance of the epicardium in the formation of fibroblasts was based on experimental studies using avian model systems [36,156] until differentiation of Wt1-expressing EPDCs in cardiac fibroblasts in the mouse has been demonstrated in 2012. By employing a mWt1/IRES/GFP-Cre (Wt1^{Cre}) mouse line, Wessels and coworkers proposed that Wt1$^+$EPDCs contribute to the majority of cardiac fibroblasts [27]. It has further been shown that epicardial knockout using an inducible *Wt1*^{CreERT2} mouse line of serum response factor (Srf) or myocardin related transcription factors (Mrtfs), which function as Srf co-activators, disrupts EPDC migration in development, leading to subepicardial hemorrhage probably due to a reduction in EPDC-derived pericytes which stabilize the coronary vessels [157]. Pericytes are mural cells and are found residing within the basement membrane in microvessels, they are able to differentiate in adipocytes, vascular smooth muscle cells (VSMCs), and myofibroblasts, and consequently modulate the vascular network [158]. However, a clear indication that Wt1 marks pericytes, is still missing until now. This is also because there is no single molecular marker that unequivocally identifies pericytes and distinguishes them from vascular smooth muscle or other mesenchymal cells. In adult mice following myocardial infarction, epicardial specific deletion of Srf or Mrtfs resulted in an improved functional outcome after MI and decreased left ventricular fibrosis [159]. In line with this, Zhang and coworkers suggested that the embryonic epicardium and derived mesenchymal cells were the major source of fibroblasts in endocardial fibroelastosis (EFE), a pathological condition characterized by diffuse profound thickening of the endocardium with abnormal deposition of collagen and elastin predominantly in the left ventricle, often associated with hypoplastic left heart syndrome (HLHS) [160]. The developmental origins and activation of cardiac fibroblasts in response to injury have been reviewed recently [161].

Regarding a possible implication of epicardial-derived Wt1-expressing progenitor cells for cardiac repair, the opinions are diverging. Some studies suggest an important role after myocardial infarction [118,119,162–164], while others did not confirm these results [150]. These controversial results derive from the different experimental approaches, staining procedures, and limitations of the Wt1-Cre mouse models used [35,165]. Re-activated epicardium is heterogenous and different from developmental epicardial cells [106], only a few cells in adult epicardium express Wt1 and are reliable targeted by the Wt1Cre lines [35,166].

Epicardial epithelial-mesenchymal transition (EMT) generates the formation of epicardial-derived cells (EPDCs), a population of multipotent mesenchymal cardiac progenitor cells, that may differentiate into various cardiac cell types. The concrete role of Wt1 in this process remained to be elucidated. Martinez-Estrada and colleagues demonstrated that an epicardial specific knockout of Wt1 resulted in a reduction of mesenchymal progenitor cells and their resultant differentiated cell types in the heart. Using immortalized tamoxifen-inducible WT1-knockout epicardial cells, the authors identified direct transcriptional regulation of Snail (Snai1), a master regulator of EMT (reviewed in [167]) and E-Cadherin by Wt1 as the underlying molecular mechanism of EMT. Wt1 activated the Snail promoter while epithelial E-Cadherin appears to be repressed through a double mechanism: first, through direct inhibition by Wt1, second, through repression by Snail, which itself is activated by Wt1 [86]. However, in a study investigating the cardiovascular defects in platelet-derived growth factor receptor (Pdgfr) α-deficient mice, the authors observed a significant increase of Wt1 expression both in Pdgfrα-deficient hearts as well as in small-hairpin RNA PDGFRα-transduced human adult epicardial cells which was not accompanied by an altered expression of E-Cadherin [168]. Accordingly, the effect of WT1 overexpression on E-cadherin expression seems not directly correlated to the effect of WT1 knockdown as described by Martinez-Estrada. Further discrepancies to the findings of Martinez-Estrada emerged from a following study investigating EMT in human epicardial cells. Human adult epicardial cells lost their epithelial character and gained α smooth actin (α SMA) expression when stimulated by transforming growth factor receptor β (TGFβ). This TGFβ-induced EMT was accompanied by a decrease of WT1 and E-Cadherin expression, and an increase of Snail and PDGFRα expression. Similar results were obtained using WT1 knockdown in the human epicardial cells. The contradiction between these data and the findings from Martinez-Estrada and colleagues might be due to the differences in the EMT process of human adult epicardial cells and mouse embryonic epicardial cells [169]. However, Casanova and colleagues established an epicardial-specific knockout model for Snail, which neither demonstrated any cardiac abnormalities nor abnormalities in epicardial EMT, indicating that Snail is simply not required for cardiac epicardial EMT [170]. Using primary murine epicardial cells, Takeichi and colleagues suggested that epicardial EMT is regulated through the bi-directional regulation of Slug (Snai2) by Wt1 and Tbx18 (T-box18). Tbx18 upregulated, Wt1 downregulated Slug by binding to its promoter and affecting its activity [87]. However, no epicardial defects have been reported in Tbx18 knockout or transgenic overexpression models [171], which questions the relevance for Tbx18 in epicardial EMT.

The group of W. Pu did not observe any alterations of E-Cadherin expression in Wt1-deficient epicardium compared to control epicardium. They evidenced in contrast a reduction of Lef1 and Ctnnb1 (β-catenin), components of the Wnt/β signaling pathway, as well as decreased Wnt5 and Raldh2 expression in the epicardium of Wt1 knockout animals, suggesting that Wt1 regulates EMT through β-catenin and retinoic acid signaling [172]. In a very nice study, the group of Kispert took advantage of the expression of Wt1 in the mesothelial lining of the heart and analyzed the mechanisms of mesothelial mobilization in cardiac development by inducible genetic lineage tracing. They observed that epicardial mobilization occurs from E12.5 to E14.5 at relatively constant rates. To exclude that proepicardial and eventually myocardial activity of the Cre driver lines influenced extra-epicardial lineages which in turn could affect the epicardium, they activated the Wt1$^{\text{CreERT2}}$ driver line in a way that enabled to achieve epicardial recombination at E10.5, when the mesothelial lining has just completed its formation. Using this approach, they demonstrated a functional requirement of platelet-derived growth factor receptor (Pdgfr)α, fibroblast growth factor receptor (Fgfr) 1/2, Hedgehog (Hh), and Smoothened (Smo), but not for Notch, canonical Wnt, and Tgf-β/bone morphogenetic protein (Bmp) signaling in this process [173]. Expression analysis of five epicardial markers, Wt1, Transcription factor (Tcf) 21, Tbx18, Semaphorin (Sema) 3d, and Scleraxis BHLH Transcription Factor (Scx) in development revealed overlapping expression in the proepicardial organ and the epicardium until E13.5, suggesting that epicardium-derived cell fate is specified after EMT [174]. A microarray-based expression analysis of transcriptional changes associated with *Wt1* deletion in *Cre*$^+$ embryonic epicardial cells and subsequent molecular approaches identified the chemokines Ccl5, Cxcl10, and the interferon regulatory factor Irf7 as transcriptional targets of Wt1 in the heart. Cxcl 10 was shown to inhibit epicardial cell migration and Ccl5 impaired cardiomyocyte proliferation, which could partially explain the reduced number of EPDCs and the thinned myocardium of Wt1-deficient animals. The regulation of Irf7 by Wt1 was proposed to be implied in cardiac repair mechanisms [91]. Transcriptional regulation of bone morphogenetic protein (Bmp) 4 by Wt1 is required for the transition of epicardial cells from a cuboidal morphology to a squamous, flattened cell shape [93]. A very recent study identified the proteoglycan agrin and its receptor dystroglycan, components of the extracellular matrix (ECM) to be required for proper epicardial EMT. Agrin deficiency impaired EMT and disturbed development of the epicardium, also reflected by a down-regulation of Wt1 [175].

Mesothelium, including epicardium, had been shown to contribute to visceral fat [176]. In line with this finding, it had been demonstrated that epicardial progenitors contribute to adipocytes during development, a process termed epicardium to fat transition (EFT). In adult, this programming is quiescent, but can be reactivated by cardiac injury [120]. An elegant study focused on how mechanical stimuli in cardiomyocytes can influence cardiomyocyte cell fates to transdifferentiate into adipose tissue. They demonstrated that Wt1, although not sufficient to provoke the myocyte-to-adipocyte switch, is essential for the conversion process induced by peroxisome proliferator receptor (PPAR)γ. Further, a physical co-interaction of WT1 and PPARγ in the nucleus of myocyte–adipocyte converting cells has been measured, suggesting a cooperative role of the two transcription factors in regulating the adipogenic program [177]. Tang and colleagues reported an up- regulation of Wt1 expression in the pericardial adipose tissue following myocardial infarction. They further characterized Wt1-expressing pericardial adipose-derived stem cells (pADSCs) for cardiac repair. In vitro experiments demonstrated the capability of Wt1+ pADSCs to differentiate into vessel-like structures and contracting cardiomyocytes. In vivo transplantation of Wt1+ pADSCs into infarcted hearts resulted in significant cardiac benefits through Hgf (hepatocyte growth factor) mediated proangiogenic and antiapoptotic effects [178].

Only recently, a function of Wt1 for immune cell regulation in the heart emerged. Wt1 expression in the epicardium seems to be required for the establishment of macrophages originating from fetal yolk sac which reside below and within the epicardium, which later become cardiac resident macrophages [179], implicated in cardiac repair and homeostasis [180,181]. Whether Wt1 acts as a transcriptional regulator of the epicardial program required for fetal yolk sac macrophage recruitment to the heart or if Wt1 impacts recruitment indirectly through its prominent role in the formation of the epicardial structure remains to be further elucidated [179]. *Wt1*$^+$ stromal cells in peritoneal, pleural, and pericardial organs, through retinoic acid metabolism, assure homeostasis of resident large cavity macrophages which are involved in tissue repair [182]. In zebrafish, a subpopulation of Wt1b-expressing macrophages could be identified, which contributed to organ repair, especially in the context of cardiac injury [121]. Given our observation that Wt1 is expressed in immune cells in the context of cancer [8], it will be interesting to further identify an implication of Wt1 in immune responses in the context of cardiovascular diseases.

Not much is known how Wt1 is regulated in the heart. Apart from hypoxia and direct transcriptional activation by HIF-1 [100], which are likely to be involved in cardiac development and repair, Hippo signaling components have been proposed to regulate Wt1 expression, epicardial EMT and epicardial cell proliferation and differentiation into coronary endothelial cells [183]. Vieira and colleagues identified an epigenetic mechanism implicating chromatin remodeling of the *Wt1* locus as a critical event in epicardial activity in the developing and adult heart after cardiac injury. They suggested that Wt1 is dynamically controlled by SWItch/sucrose nonfermentable (SWI/SNF) chromatin-remodeling complexes containing Brahma-related gene 1 (BRG1) and Tβ4 [184].

Identification of WT1-modulating factors is of great interest regarding a potential role for cardiac repair in vivo, which is limited due to the lack of techniques to isolate, expand, differentiate, and transfer Wt1+ progenitor cells. However, Wt1 upregulation to enhance cardiac repair might promote tumor growth in patients at risk and should be cautiously monitored. Further research is necessary to delineate the intricacies of modulating WT1 for an optimized therapeutic benefit.

5. Conclusions

Wt1 has now been firmly established as a key player in cardiovascular development. It regulates critical steps in heart development, such as formation of the epicardium from the proepicardium, epithelial to mesenchymal transition of epicardial cells, establishment of the coronary vessels, cardiac autonomic nervous system modulation, and the compaction of the ventricles. A very important clinical task remains to establish a profound cardiac evaluation of patients presenting WT1 mutations; *WT1* has long regarded as a gene implicated solely in kidney disorders. We are convinced that WT1 is implicated apart from cardiac diseases also in neurological, ophthalmological, as well as hematological disorders. A function for Wt1 in cardiac repair and regeneration has been proposed based on the finding that it is strongly re-expressed in epicardial, endothelial, and myocardial cells after myocardial infarction. However, to translate these expression patterns in potential therapeutic approaches, more knowledge about the different cardiac cell types and the consequences of the upregulation of WT1 is required. Finally, caution must be taken regarding the tumor-promoting role of this protein in patients susceptible to cancer.

Author Contributions: Conceptualization, N.W. and K.-D.W.; writing—original draft preparation, N.W. and K.-D.W.; writing—review and editing, N.W. and K.-D.W.; funding acquisition, N.W. and K.-D.W. All authors have read and agreed to the published version of the manuscript.

Funding: The authors of this review were funded by Fondation pour la Recherche Medicale, grant number FRM DPC20170139474 (K.-D.W.), Fondation ARC pour la recherche sur le cancer", grant number n°PJA 20161204650 (N.W.), Gemluc (N.W.), and Plan Cancer INSERM (K.-D.W.)

Institutional Review Board Statement: Not applicable.

Informed Consent Statement: Not applicable.

Acknowledgments: The authors thank J.-F. Michiels for teaching us to see the hidden.

Conflicts of Interest: The authors declare no conflict of interest.

Abbreviations

ACM	Arrhythmogenic cardiomyopathy
αSMA	α smooth actin
AV	Atrio-ventricular
β2AR	β2 adrenergic receptor

BDNF	Brain-derived neurotrophic factor
Bmp	Bone morphogenetic protein
Brg1	Brahma-related gene 1
Ccl5	C-C motif chemokine ligand 5
C-Kit	Tyrosine-protein kinase KIT
CSF-1	Colony-stimulating factor-1
Ctnnb1	β-catenin
Cxcl10	C-X-C motif chemokine ligand 10
E	Embryonic day
EFE	Endocardial fibroelastosis
EFT	Epicardium to fat transition
EMT	Epithelial mesenchymal transition
EPDCs	Epicardial-derived cells
EGFR	Epidermal growth factor receptor
Egr 1	Early growth response protein 1
EPO	Erythropoietin
Ets	ETS proto-oncogene
Fgf	Fibroblast growth factor
Fgfr	Fibroblast growth factor receptor
Gata	Gata-binding protein
HH	Hedgehog
Hif-1	Hypoxia inducible factor 1
HLHS	Hypoplastic left heart syndrome
IGF-1-R	Insulin like growth factor 1 receptor
IGF 2	Insulin like growth factor 2
Irf	Interferon regulatory factor
IR	Insulin receptor
LAD	Ligation of the left coronary artery
LV	Left ventricle
MESCs	Mouse embryonic stem cells
MET	Mesenchymal epithelial transition
MI	Myocardial infarction
Mrtfs	Myocardin related transcription factors
Ncam	Neural cell adhesion molecule
Nfat	Nuclear factor of activated T cells
Nphs1	Nephrin
Nphs2	Podocin
NT	Neurotrophin
P	Postnatal day
pADSCs	Pericardial adipose-derived stem cells
Pax 2	Paired box gene 2
Pdgfa	Platelet-derived growth factor A
Pdgfr	Platelet-derived growth factor receptor
Pecam-1	Platelet and endothelial cell adhesion molecule 1
Ppar	Peroxisome proliferator receptor
PPC	Pericardioperitoneal canal
PPM	Pleuropericardial membrane
Raldh	Retinaldehyde-dehydrogenase
RA	Retinoic acid
RAR	Retinoic acid receptor
RV	Right ventricle
Scx	Scleraxis BHLH transcription factor
Sema	Semaphorin
SEMCs	Subepicardial mesenchymal cells
SHR	Spontaneously hypertensive rat
Smo	Smoothened
Snai1	Snail
Snai2	Slug

Srf	Serum response factor
ST/PE	Septum transversum/proepicardium
SWI/SNF	SWItch/sucrose nonfermentable
Tbx	T-box transcription factor
Tcf	Transcription factor
Tgf-β	Transforming growth factor beta
Tβ4	Thymosin β4
Trf	Telomere repeat binding factor
Trk	Tyrosinkinase receptor
Tubb3	Tubulin beta-3 chain
Vdr	Vitamin D receptor
Vegf	Vascular endothelial growth factor
Vegfr	Vascular endothelial growth factor receptor
Wt1	Wilms' tumor suppressor 1
Wtip	Wt1 interacting protein
YAC	Yeast artificial chromosome

References

1. Haber, D.A.; Buckler, A.J.; Glaser, T.; Call, K.M.; Pelletier, J.; Sohn, R.L.; Douglass, E.C.; Housman, D.E. An internal deletion within an 11p13 zinc finger gene contributes to the development of Wilms' tumor. *Cell* **1990**, *61*, 1257–1269. [CrossRef]
2. Call, K.M.; Glaser, T.; Ito, C.Y.; Buckler, A.J.; Pelletier, J.; Haber, D.A.; Rose, E.A.; Kral, A.; Yeger, H.; Lewis, W.H. Isolation and characterization of a zinc finger polypeptide gene at the human chromosome 11 Wilms' tumor locus. *Cell* **1990**, *60*, 509–520. [CrossRef]
3. Gessler, M.; Poustka, A.; Cavenee, W.; Neve, R.L.; Orkin, S.H.; Bruns, G.A. Homozygous deletion in Wilms tumours of a zinc-finger gene identified by chromosome jumping. *Nature* **1990**, *343*, 774–778. [CrossRef] [PubMed]
4. Little, M.; Wells, C. A clinical overview of WT1 gene mutations. *Hum. Mutat.* **1997**, *9*, 209–225. [CrossRef]
5. Rivera, M.N.; Haber, D.A. Wilms' tumour: Connecting tumorigenesis and organ development in the kidney. *Nat. Rev. Cancer* **2005**, *5*, 699–712. [CrossRef]
6. Yang, L.; Han, Y.; Suarez Saiz, F.; Saurez Saiz, F.; Minden, M.D. A tumor suppressor and oncogene: The WT1 story. *Leukemia* **2007**, *21*, 868–876. [CrossRef]
7. Sugiyama, H. Wilms' tumor gene WT1: Its oncogenic function and clinical application. *Int. J. Hematol.* **2001**, *73*, 177–187. [CrossRef] [PubMed]
8. Wagner, K.D.; Cherfils-Vicini, J.; Hosen, N.; Hohenstein, P.; Gilson, E.; Hastie, N.D.; Michiels, J.F.; Wagner, N. The Wilms' tumour suppressor Wt1 is a major regulator of tumour angiogenesis and progression. *Nat. Commun.* **2014**, *5*, 5852. [CrossRef]
9. Larsson, S.H.; Charlieu, J.P.; Miyagawa, K.; Engelkamp, D.; Rassoulzadegan, M.; Ross, A.; Cuzin, F.; van Heyningen, V.; Hastie, N.D. Subnuclear localization of WT1 in splicing or transcription factor domains is regulated by alternative splicing. *Cell* **1995**, *81*, 391–401. [CrossRef]
10. Bardeesy, N.; Pelletier, J. Overlapping RNA and DNA binding domains of the wt1 tumor suppressor gene product. *Nucleic Acids Res.* **1998**, *26*, 1784–1792. [CrossRef] [PubMed]
11. Laity, J.H.; Dyson, H.J.; Wright, P.E. Molecular basis for modulation of biological function by alternate splicing of the Wilms' tumor suppressor protein. *Proc. Natl. Acad. Sci. USA* **2000**, *97*, 11932–11935. [CrossRef] [PubMed]
12. Ladomery, M.; Sommerville, J.; Woolner, S.; Slight, J.; Hastie, N. Expression in Xenopus oocytes shows that WT1 binds transcripts in vivo, with a central role for zinc finger one. *J. Cell Sci.* **2003**, *116*, 1539–1549. [CrossRef] [PubMed]
13. Haber, D.A.; Sohn, R.L.; Buckler, A.J.; Pelletier, J.; Call, K.M.; Housman, D.E. Alternative splicing and genomic structure of the Wilms tumor gene WT1. *Proc. Natl. Acad. Sci. USA* **1991**, *88*, 9618–9622. [CrossRef] [PubMed]
14. Niksic, M.; Slight, J.; Sanford, J.R.; Caceres, J.F.; Hastie, N.D. The Wilms' tumour protein (WT1) shuttles between nucleus and cytoplasm and is present in functional polysomes. *Hum. Mol. Genet.* **2004**, *13*, 463–471. [CrossRef] [PubMed]
15. Männer, J.; Wessel, A.; Yelbuz, T.M. How does the tubular embryonic heart work? Looking for the physical mechanism generating unidirectional blood flow in the valveless embryonic heart tube. *Dev. Dyn.* **2010**, *239*, 1035–1046. [CrossRef] [PubMed]
16. Moorman, A.F.; Christoffels, V.M. Cardiac chamber formation: Development, genes, and evolution. *Physiol. Rev.* **2003**, *83*, 1223–1267. [CrossRef]
17. Miquerol, L.; Kelly, R.G. Organogenesis of the vertebrate heart. *Wiley Interdiscip. Rev. Dev. Biol.* **2013**, *2*, 17–29. [CrossRef] [PubMed]
18. Kelly, R.G.; Buckingham, M.E.; Moorman, A.F. Heart fields and cardiac morphogenesis. *Cold Spring Harb. Perspect. Med.* **2014**, *4*, a015750. [CrossRef]
19. Sylva, M.; van den Hoff, M.J.; Moorman, A.F. Development of the human heart. *Am. J. Med. Genet. A* **2014**, *164A*, 1347–1371. [CrossRef]

20. Wolters, R.; Deepe, R.; Drummond, J.; Harvey, A.B.; Hiriart, E.; Lockhart, M.M.; van den Hoff, M.J.B.; Norris, R.A.; Wessels, A. Role of the Epicardium in the Development of the Atrioventricular Valves and Its Relevance to the Pathogenesis of Myxomatous Valve Disease. *J. Cardiovasc. Dev. Dis.* **2021**, *8*, 54. [CrossRef]

21. Buijtendijk, M.F.J.; Barnett, P.; van den Hoff, M.J.B. Development of the human heart. *Am. J. Med. Genet. C Semin. Med. Genet.* **2020**, *184*, 7–22. [CrossRef]

22. Meilhac, S.M.; Buckingham, M.E. The deployment of cell lineages that form the mammalian heart. *Nat. Rev. Cardiol.* **2018**, *15*, 705–724. [CrossRef]

23. Armstrong, J.F.; Pritchard-Jones, K.; Bickmore, W.A.; Hastie, N.D.; Bard, J.B. The expression of the Wilms' tumour gene, WT1, in the developing mammalian embryo. *Mech. Dev.* **1993**, *40*, 85–97. [CrossRef]

24. Moore, A.W.; McInnes, L.; Kreidberg, J.; Hastie, N.D.; Schedl, A. YAC complementation shows a requirement for Wt1 in the development of epicardium, adrenal gland and throughout nephrogenesis. *Development* **1999**, *126*, 1845–1857. [CrossRef]

25. Tyser, R.C.V.; Ibarra-Soria, X.; McDole, K.; Arcot Jayaram, S.; Godwin, J.; van den Brand, T.A.H.; Miranda, A.M.A.; Scialdone, A.; Keller, P.J.; Marioni, J.C.; et al. Characterization of a common progenitor pool of the epicardium and myocardium. *Science* **2021**, *371*, eabb2986. [CrossRef]

26. Zhou, B.; Ma, Q.; Rajagopal, S.; Wu, S.M.; Domian, I.; Rivera-Feliciano, J.; Jiang, D.; von Gise, A.; Ikeda, S.; Chien, K.R.; et al. Epicardial progenitors contribute to the cardiomyocyte lineage in the developing heart. *Nature* **2008**, *454*, 109–113. [CrossRef]

27. Wessels, A.; van den Hoff, M.J.; Adamo, R.F.; Phelps, A.L.; Lockhart, M.M.; Sauls, K.; Briggs, L.E.; Norris, R.A.; van Wijk, B.; Perez-Pomares, J.M.; et al. Epicardially derived fibroblasts preferentially contribute to the parietal leaflets of the atrioventricular valves in the murine heart. *Dev. Biol.* **2012**, *366*, 111–124. [CrossRef]

28. Pérez-Pomares, J.M.; Phelps, A.; Sedmerova, M.; Carmona, R.; González-Iriarte, M.; Muñoz-Chápuli, R.; Wessels, A. Experimental studies on the spatiotemporal expression of WT1 and RALDH2 in the embryonic avian heart: A model for the regulation of myocardial and valvuloseptal development by epicardially derived cells (EPDCs). *Dev. Biol.* **2002**, *247*, 307–326. [CrossRef] [PubMed]

29. Kreidberg, J.A.; Sariola, H.; Loring, J.M.; Maeda, M.; Pelletier, J.; Housman, D.; Jaenisch, R. WT-1 is required for early kidney development. *Cell* **1993**, *74*, 679–691. [CrossRef]

30. Serluca, F.C. Development of the proepicardial organ in the zebrafish. *Dev. Biol.* **2008**, *315*, 18–27. [CrossRef]

31. Carmona, R.; González-Iriarte, M.; Pérez-Pomares, J.M.; Muñoz-Chápuli, R. Localization of the Wilm's tumour protein WT1 in avian embryos. *Cell Tissue Res.* **2001**, *303*, 173–186. [CrossRef]

32. Wagner, N.; Wagner, K.D.; Theres, H.; Englert, C.; Schedl, A.; Scholz, H. Coronary vessel development requires activation of the TrkB neurotrophin receptor by the Wilms' tumor transcription factor Wt1. *Genes Dev.* **2005**, *19*, 2631–2642. [CrossRef]

33. Ambu, R.; Vinci, L.; Gerosa, C.; Fanni, D.; Obinu, E.; Faa, A.; Fanos, V. WT1 expression in the human fetus during development. *Eur. J. Histochem.* **2015**, *59*, 2499. [CrossRef]

34. Duim, S.N.; Smits, A.M.; Kruithof, B.P.; Goumans, M.J. The roadmap of WT1 protein expression in the human fetal heart. *J. Mol. Cell Cardiol.* **2016**, *90*, 139–145. [CrossRef]

35. Rudat, C.; Kispert, A. Wt1 and epicardial fate mapping. *Circ. Res.* **2012**, *111*, 165–169. [CrossRef]

36. Männer, J. Does the subepicardial mesenchyme contribute myocardioblasts to the myocardium of the chick embryo heart? A quail-chick chimera study tracing the fate of the epicardial primordium. *Anat. Rec.* **1999**, *255*, 212–226. [CrossRef]

37. Wagner, N.; Ninkov, M.; Vukolic, A.; Cubukcuoglu Deniz, G.; Rassoulzadegan, M.; Michiels, J.F.; Wagner, K.D. Implications of the Wilms' Tumor Suppressor Wt1 in Cardiomyocyte Differentiation. *Int. J. Mol. Sci.* **2021**, *22*, 4346. [CrossRef]

38. Xavier-Neto, J.; Shapiro, M.D.; Houghton, L.; Rosenthal, N. Sequential programs of retinoic acid synthesis in the myocardial and epicardial layers of the developing avian heart. *Dev. Biol.* **2000**, *219*, 129–141. [CrossRef]

39. Moss, J.B.; Xavier-Neto, J.; Shapiro, M.D.; Nayeem, S.M.; McCaffery, P.; Dräger, U.C.; Rosenthal, N. Dynamic patterns of retinoic acid synthesis and response in the developing mammalian heart. *Dev. Biol.* **1998**, *199*, 55–71. [CrossRef]

40. Goodyer, P.; Dehbi, M.; Torban, E.; Bruening, W.; Pelletier, J. Repression of the retinoic acid receptor-alpha gene by the Wilms' tumor suppressor gene product, wt1. *Oncogene* **1995**, *10*, 1125–1129.

41. WILSON, J.G.; ROTH, C.B.; WARKANY, J. An analysis of the syndrome of malformations induced by maternal vitamin A deficiency. Effects of restoration of vitamin A at various times during gestation. *Am. J. Anat.* **1953**, *92*, 189–217. [CrossRef]

42. Guadix, J.A.; Ruiz-Villalba, A.; Lettice, L.; Velecela, V.; Muñoz-Chápuli, R.; Hastie, N.D.; Pérez-Pomares, J.M.; Martínez-Estrada, O.M. Wt1 controls retinoic acid signalling in embryonic epicardium through transcriptional activation of Raldh2. *Development* **2011**, *138*, 1093–1097. [CrossRef]

43. Norden, J.; Grieskamp, T.; Lausch, E.; van Wijk, B.; van den Hoff, M.J.; Englert, C.; Petry, M.; Mommersteeg, M.T.; Christoffels, V.M.; Niederreither, K.; et al. Wt1 and retinoic acid signaling in the subcoelomic mesenchyme control the development of the pleuropericardial membranes and the sinus horns. *Circ. Res.* **2010**, *106*, 1212–1220. [CrossRef]

44. Norden, J.; Grieskamp, T.; Christoffels, V.M.; Moorman, A.F.; Kispert, A. Partial absence of pleuropericardial membranes in Tbx18- and Wt1-deficient mice. *PLoS ONE* **2012**, *7*, e45100. [CrossRef] [PubMed]

45. Pombal, M.A.; Carmona, R.; Megías, M.; Ruiz, A.; Pérez-Pomares, J.M.; Muñoz-Chápuli, R. Epicardial development in lamprey supports an evolutionary origin of the vertebrate epicardium from an ancestral pronephric external glomerulus. *Evol. Dev.* **2008**, *10*, 210–216. [CrossRef] [PubMed]

46. Cano, E.; Carmona, R.; Velecela, V.; Martínez-Estrada, O.; Muñoz-Chápuli, R. The proepicardium keeps a potential for glomerular marker expression which supports its evolutionary origin from the pronephros. *Evol. Dev.* **2015**, *17*, 224–230. [CrossRef]

47. Essafi, A.; Webb, A.; Berry, R.L.; Slight, J.; Burn, S.F.; Spraggon, L.; Velecela, V.; Martinez-Estrada, O.M.; Wiltshire, J.H.; Roberts, S.G.; et al. A wt1-controlled chromatin switching mechanism underpins tissue-specific wnt4 activation and repression. *Dev. Cell* **2011**, *21*, 559–574. [CrossRef] [PubMed]

48. Palmer, R.E.; Kotsianti, A.; Cadman, B.; Boyd, T.; Gerald, W.; Haber, D.A. WT1 regulates the expression of the major glomerular podocyte membrane protein Podocalyxin. *Curr. Biol.* **2001**, *11*, 1805–1809. [CrossRef]

49. Wagner, N.; Wagner, K.D.; Xing, Y.; Scholz, H.; Schedl, A. The major podocyte protein nephrin is transcriptionally activated by the Wilms' tumor suppressor WT1. *J. Am. Soc. Nephrol.* **2004**, *15*, 3044–3051. [CrossRef]

50. Wagner, N.; Wagner, K.D.; Scholz, H.; Kirschner, K.M.; Schedl, A. Intermediate filament protein nestin is expressed in developing kidney and heart and might be regulated by the Wilms' tumor suppressor Wt1. *Am. J. Physiol. Regul. Integr. Comp. Physiol.* **2006**, *291*, R779–R787. [CrossRef] [PubMed]

51. Dong, L.; Pietsch, S.; Tan, Z.; Perner, B.; Sierig, R.; Kruspe, D.; Groth, M.; Witzgall, R.; Gröne, H.J.; Platzer, M.; et al. Integration of Cistromic and Transcriptomic Analyses Identifies Nphs2, Mafb, and Magi2 as Wilms' Tumor 1 Target Genes in Podocyte Differentiation and Maintenance. *J. Am. Soc. Nephrol.* **2015**, *26*, 2118–2128. [CrossRef]

52. Vicente-Steijn, R.; Scherptong, R.W.; Kruithof, B.P.; Duim, S.N.; Goumans, M.J.; Wisse, L.J.; Zhou, B.; Pu, W.T.; Poelmann, R.E.; Schalij, M.J.; et al. Regional differences in WT-1 and Tcf21 expression during ventricular development: Implications for myocardial compaction. *PLoS ONE* **2015**, *10*, e0136025. [CrossRef] [PubMed]

53. Kelder, T.P.; Duim, S.N.; Vicente-Steijn, R.; Végh, A.M.; Kruithof, B.P.; Smits, A.M.; van Bavel, T.C.; Bax, N.A.; Schalij, M.J.; Gittenberger-de Groot, A.C.; et al. The epicardium as modulator of the cardiac autonomic response during early development. *J. Mol. Cell Cardiol.* **2015**, *89*, 251–259. [CrossRef] [PubMed]

54. Wagner, K.D.; Wagner, N.; Vidal, V.P.; Schley, G.; Wilhelm, D.; Schedl, A.; Englert, C.; Scholz, H. The Wilms' tumor gene Wt1 is required for normal development of the retina. *EMBO J.* **2002**, *21*, 1398–1405. [CrossRef]

55. Wagner, N.; Wagner, K.D.; Hammes, A.; Kirschner, K.M.; Vidal, V.P.; Schedl, A.; Scholz, H. A splice variant of the Wilms' tumour suppressor Wt1 is required for normal development of the olfactory system. *Development* **2005**, *132*, 1327–1336. [CrossRef]

56. Wagner, K.D.; Wagner, N.; Schley, G.; Theres, H.; Scholz, H. The Wilms' tumor suppressor Wt1 encodes a transcriptional activator of the class IV POU-domain factor Pou4f2 (Brn-3b). *Gene* **2003**, *305*, 217–223. [CrossRef]

57. Gao, Y.; Toska, E.; Denmon, D.; Roberts, S.G.; Medler, K.F. WT1 regulates the development of the posterior taste field. *Development* **2014**, *141*, 2271–2278. [CrossRef]

58. Schnerwitzki, D.; Perry, S.; Ivanova, A.; Caixeta, F.V.; Cramer, P.; Günther, S.; Weber, K.; Tafreshiha, A.; Becker, L.; Vargas Panesso, I.L.; et al. Neuron-specific inactivation of. *Life Sci. Alliance* **2018**, *1*, e201800106. [CrossRef] [PubMed]

59. Schnerwitzki, D.; Hayn, C.; Perner, B.; Englert, C. Wt1 Positive dB4 Neurons in the Hindbrain Are Crucial for Respiration. *Front. NeuroSci.* **2020**, *14*, 529487. [CrossRef]

60. Serluca, F.C.; Fishman, M.C. Pre-pattern in the pronephric kidney field of zebrafish. *Development* **2001**, *128*, 2233–2241. [CrossRef] [PubMed]

61. Drummond, I.A., Majumdar, A.; Hentschel, H.; Elger, M.; Solnica-Krezel, L.; Schier, A.F.; Neuhauss, S.C.; Stemple, D.L.; Zwartkruis, F.; Rangini, Z.; et al. Early development of the zebrafish pronephros and analysis of mutations affecting pronephric function. *Development* **1998**, *125*, 4655–4667. [CrossRef]

62. Bollig, F.; Mehringer, R.; Perner, B.; Hartung, C.; Schäfer, M.; Schartl, M.; Volff, J.N.; Winkler, C.; Englert, C. Identification and comparative expression analysis of a second wt1 gene in zebrafish. *Dev. Dyn.* **2006**, *235*, 554–561. [CrossRef] [PubMed]

63. Srichai, M.B.; Konieczkowski, M.; Padiyar, A.; Konieczkowski, D.J.; Mukherjee, A.; Hayden, P.S.; Kamat, S.; El-Meanawy, M.A.; Khan, S.; Mundel, P.; et al. A WT1 co-regulator controls podocyte phenotype by shuttling between adhesion structures and nucleus. *J. Biol. Chem.* **2004**, *279*, 14398–14408. [CrossRef] [PubMed]

64. Powell, R.; Bubenshchikova, E.; Fukuyo, Y.; Hsu, C.; Lakiza, O.; Nomura, H.; Renfrew, E.; Garrity, D.; Obara, T. Wtip is required for proepicardial organ specification and cardiac left/right asymmetry in zebrafish. *Mol. Med. Rep.* **2016**, *14*, 2665–2678. [CrossRef]

65. Romano, N.; Ceci, M. The face of epicardial and endocardial derived cells in zebrafish. *Exp. Cell Res.* **2018**, *369*, 166–175. [CrossRef] [PubMed]

66. Werner, H.; Rauscher, F.J.; Sukhatme, V.P.; Drummond, I.A.; Roberts, C.T.; LeRoith, D. Transcriptional repression of the insulin-like growth factor I receptor (IGF-I-R) gene by the tumor suppressor WT1 involves binding to sequences both upstream and downstream of the IGF-I-R gene transcription start site. *J. Biol. Chem.* **1994**, *269*, 12577–12582. [CrossRef]

67. Englert, C.; Hou, X.; Maheswaran, S.; Bennett, P.; Ngwu, C.; Re, G.G.; Garvin, A.J.; Rosner, M.R.; Haber, D.A. WT1 suppresses synthesis of the epidermal growth factor receptor and induces apoptosis. *EMBO J.* **1995**, *14*, 4662–4675. [CrossRef]

68. Webster, N.J.; Kong, Y.; Sharma, P.; Haas, M.; Sukumar, S.; Seely, B.L. Differential effects of Wilms tumor WT1 splice variants on the insulin receptor promoter. *BioChem. Mol. Med.* **1997**, *62*, 139–150. [CrossRef]

69. Ryan, G.; Steele-Perkins, V.; Morris, J.F.; Rauscher, F.J.; Dressler, G.R. Repression of Pax-2 by WT1 during normal kidney development. *Development* **1995**, *121*, 867–875. [CrossRef]

70. Wang, Z.Y.; Qiu, Q.Q.; Deuel, T.F. The Wilms' tumor gene product WT1 activates or suppresses transcription through separate functional domains. *J. Biol. Chem.* **1993**, *268*, 9172–9175. [CrossRef]

71. Wang, Z.Y.; Madden, S.L.; Deuel, T.F.; Rauscher, F.J. The Wilms' tumor gene product, WT1, represses transcription of the platelet-derived growth factor A-chain gene. *J. Biol. Chem.* **1992**, *267*, 21999–22002. [CrossRef]

72. Madden, S.L.; Cook, D.M.; Morris, J.F.; Gashler, A.; Sukhatme, V.P.; Rauscher, F.J. Transcriptional repression mediated by the WT1 Wilms tumor gene product. *Science* **1991**, *253*, 1550–1553. [CrossRef] [PubMed]

73. Ward, A.; Pooler, J.A.; Miyagawa, K.; Duarte, A.; Hastie, N.D.; Caricasole, A. Repression of promoters for the mouse insulin-like growth factor II-encoding gene (Igf-2) by products of the Wilms' tumour suppressor gene wt1. *Gene* **1995**, *167*, 239–243. [CrossRef]

74. Dey, B.R.; Sukhatme, V.P.; Roberts, A.B.; Sporn, M.B.; Rauscher, F.J.; Kim, S.J. Repression of the transforming growth factor-beta 1 gene by the Wilms' tumor suppressor WT1 gene product. *Mol. Endocrinol.* **1994**, *8*, 595–602. [CrossRef]

75. Harrington, M.A.; Konicek, B.; Song, A.; Xia, X.L.; Fredericks, W.J.; Rauscher, F.J. Inhibition of colony-stimulating factor-1 promoter activity by the product of the Wilms' tumor locus. *J. Biol. Chem.* **1993**, *268*, 21271–21275. [CrossRef]

76. Cook, D.M.; Hinkes, M.T.; Bernfield, M.; Rauscher, F.J. Transcriptional activation of the syndecan-1 promoter by the Wilms' tumor protein WT1. *Oncogene* **1996**, *13*, 1789–1799.

77. Adachi, Y.; Matsubara, S.; Pedraza, C.; Ozawa, M.; Tsutsui, J.; Takamatsu, H.; Noguchi, H.; Akiyama, T.; Muramatsu, T. Midkine as a novel target gene for the Wilms' tumor suppressor gene (WT1). *Oncogene* **1996**, *13*, 2197–2203.

78. Wagner, K.D.; Wagner, N.; Sukhatme, V.P.; Scholz, H. Activation of vitamin D receptor by the Wilms' tumor gene product mediates apoptosis of renal cells. *J. Am. Soc. Nephrol.* **2001**, *12*, 1188–1196. [CrossRef] [PubMed]

79. Maurer, U.; Jehan, F.; Englert, C.; Hubinger, G.; Weidmann, E.; DeLuca, H.F.; Bergmann, L. The Wilms' tumor gene product (WT1) modulates the response to 1,25-dihydroxyvitamin D3 by induction of the vitamin D receptor. *J. Biol. Chem.* **2001**, *276*, 3727–3732. [CrossRef]

80. Dame, C.; Kirschner, K.M.; Bartz, K.V.; Wallach, T.; Hussels, C.S.; Scholz, H. Wilms tumor suppressor, Wt1, is a transcriptional activator of the erythropoietin gene. *Blood* **2006**, *107*, 4282–4290. [CrossRef]

81. Kirschner, K.M.; Wagner, N.; Wagner, K.D.; Wellmann, S.; Scholz, H. The Wilms tumor suppressor Wt1 promotes cell adhesion through transcriptional activation of the alpha4integrin gene. *J. Biol. Chem.* **2006**, *281*, 31930–31939. [CrossRef] [PubMed]

82. Hanson, J.; Gorman, J.; Reese, J.; Fraizer, G. Regulation of vascular endothelial growth factor, VEGF, gene promoter by the tumor suppressor, WT1. *Front. BioSci.* **2007**, *12*, 2279–2290. [CrossRef]

83. McCarty, G.; Awad, O.; Loeb, D.M. WT1 protein directly regulates expression of vascular endothelial growth factor and is a mediator of tumor response to hypoxia. *J. Biol. Chem.* **2011**, *286*, 43634–43643. [CrossRef]

84. Wagner, N.; Michiels, J.F.; Schedl, A.; Wagner, K.D. The Wilms' tumour suppressor WT1 is involved in endothelial cell proliferation and migration: Expression in tumour vessels in vivo. *Oncogene* **2008**, *27*, 3662–3672. [CrossRef]

85. Kirschner, K.M.; Sciesielski, L.K.; Krueger, K.; Scholz, H. Wilms tumor protein-dependent transcription of VEGF receptor 2 and hypoxia regulate expression of the testis-promoting gene. *J. Biol. Chem.* **2017**, *292*, 20281–20291. [CrossRef]

86. Martínez-Estrada, O.M.; Lettice, L.A.; Essafi, A.; Guadix, J.A.; Slight, J.; Velecela, V.; Hall, E.; Reichmann, J.; Devenney, P.S.; Hohenstein, P.; et al. Wt1 is required for cardiovascular progenitor cell formation through transcriptional control of Snail and E-cadherin. *Nat. Genet.* **2010**, *42*, 89–93. [CrossRef] [PubMed]

87. Takeichi, M.; Nimura, K.; Mori, M.; Nakagami, H.; Kaneda, Y. The transcription factors Tbx18 and Wt1 control the epicardial epithelial-mesenchymal transition through bi-directional regulation of Slug in murine primary epicardial cells. *PLoS ONE* **2013**, *8*, e57829. [CrossRef]

88. Hosono, S.; Gross, I.; English, M.A.; Hajra, K.M.; Fearon, E.R.; Licht, J.D. E-cadherin is a WT1 target gene. *J. Biol. Chem.* **2000**, *275*, 10943–10953. [CrossRef] [PubMed]

89. Kirschner, K.M.; Sciesielski, L.K.; Scholz, H. Wilms' tumour protein Wt1 stimulates transcription of the gene encoding vascular endothelial cadherin. *Pflugers Arch.* **2010**, *460*, 1051–1061. [CrossRef]

90. Hsu, W.H.; Yu, Y.R.; Hsu, S.H.; Yu, W.C.; Chu, Y.H.; Chen, Y.J.; Chen, C.M.; You, L.R. The Wilms' tumor suppressor Wt1 regulates Coronin 1B expression in the epicardium. *Exp. Cell Res.* **2013**, *319*, 1365–1381. [CrossRef]

91. Velecela, V.; Lettice, L.A.; Chau, Y.Y.; Slight, J.; Berry, R.L.; Thornburn, A.; Gunst, Q.D.; van den Hoff, M.; Reina, M.; Martínez, F.O.; et al. WT1 regulates the expression of inhibitory chemokines during heart development. *Hum. Mol. Genet.* **2013**, *22*, 5083–5095. [CrossRef]

92. Maï, M.E.; Wagner, K.D.; Michiels, J.F.; Gilson, E.; Wagner, N. TRF2 acts as a transcriptional regulator in tumor angiogenesis. *Mol. Cell Oncol.* **2015**, *2*, e988508. [CrossRef]

93. Velecela, V.; Torres-Cano, A.; García-Melero, A.; Ramiro-Pareta, M.; Müller-Sánchez, C.; Segarra-Mondejar, M.; Chau, Y.Y.; Campos-Bonilla, B.; Reina, M.; Soriano, F.X.; et al. Epicardial cell shape and maturation are regulated by Wt1 via transcriptional control of. *Development* **2019**, *146*. [CrossRef] [PubMed]

94. Walker, C.; Rutten, F.; Yuan, X.; Pass, H.; Mew, D.M.; Everitt, J. Wilms' tumor suppressor gene expression in rat and human mesothelioma. *Cancer Res.* **1994**, *54*, 3101–3106.

95. Wagner, K.D.; Wagner, N.; Bondke, A.; Nafz, B.; Flemming, B.; Theres, H.; Scholz, H. The Wilms' tumor suppressor Wt1 is expressed in the coronary vasculature after myocardial infarction. *FASEB J.* **2002**, *16*, 1117–1119. [CrossRef] [PubMed]

96. Camici, G.G.; Stallmach, T.; Hermann, M.; Hassink, R.; Doevendans, P.; Grenacher, B.; Hirschy, A.; Vogel, J.; Lüscher, T.F.; Ruschitzka, F.; et al. Constitutively overexpressed erythropoietin reduces infarct size in a mouse model of permanent coronary artery ligation. *Methods Enzymol.* **2007**, *435*, 147–155. [CrossRef] [PubMed]

97. Duim, S.N.; Kurakula, K.; Goumans, M.J.; Kruithof, B.P. Cardiac endothelial cells express Wilms' tumor-1: Wt1 expression in the developing, adult and infarcted heart. *J. Mol. Cell Cardiol.* **2015**, *81*, 127–135. [CrossRef]

98. Braitsch, C.M.; Kanisicak, O.; van Berlo, J.H.; Molkentin, J.D.; Yutzey, K.E. Differential expression of embryonic epicardial progenitor markers and localization of cardiac fibrosis in adult ischemic injury and hypertensive heart disease. *J. Mol. Cell Cardiol.* **2013**, *65*, 108–119. [CrossRef] [PubMed]

99. Orlandi, A.; Ciucci, A.; Ferlosio, A.; Genta, R.; Spagnoli, L.G.; Gabbiani, G. Cardiac myxoma cells exhibit embryonic endocardial stem cell features. *J. Pathol.* **2006**, *209*, 231–239. [CrossRef]

100. Wagner, K.D.; Wagner, N.; Wellmann, S.; Schley, G.; Bondke, A.; Theres, H.; Scholz, H. Oxygen-regulated expression of the Wilms' tumor suppressor Wt1 involves hypoxia-inducible factor-1 (HIF-1). *FASEB J.* **2003**, *17*, 1364–1366. [CrossRef] [PubMed]

101. McCarty, G.; Loeb, D.M. Hypoxia-sensitive epigenetic regulation of an antisense-oriented lncRNA controls WT1 expression in myeloid leukemia cells. *PLoS ONE* **2015**, *10*, e0119837. [CrossRef]

102. Limana, F.; Bertolami, C.; Mangoni, A.; Di Carlo, A.; Avitabile, D.; Mocini, D.; Iannelli, P.; De Mori, R.; Marchetti, C.; Pozzoli, O.; et al. Myocardial infarction induces embryonic reprogramming of epicardial c-kit(+) cells: Role of the pericardial fluid. *J. Mol. Cell Cardiol.* **2010**, *48*, 609–618. [CrossRef]

103. Oh, H.; Wang, S.C.; Prahash, A.; Sano, M.; Moravec, C.S.; Taffet, G.E.; Michael, L.H.; Youker, K.A.; Entman, M.L.; Schneider, M.D. Telomere attrition and Chk2 activation in human heart failure. *Proc. Natl. Acad. Sci. USA* **2003**, *100*, 5378–5383. [CrossRef] [PubMed]

104. Smart, N.; Risebro, C.A.; Melville, A.A.; Moses, K.; Schwartz, R.J.; Chien, K.R.; Riley, P.R. Thymosin beta4 induces adult epicardial progenitor mobilization and neovascularization. *Nature* **2007**, *445*, 177–182. [CrossRef]

105. Bock-Marquette, I.; Shrivastava, S.; Pipes, G.C.; Thatcher, J.E.; Blystone, A.; Shelton, J.M.; Galindo, C.L.; Melegh, B.; Srivastava, D.; Olson, E.N.; et al. Thymosin beta4 mediated PKC activation is essential to initiate the embryonic coronary developmental program and epicardial progenitor cell activation in adult mice in vivo. *J. Mol. Cell Cardiol.* **2009**, *46*, 728–738. [CrossRef] [PubMed]

106. Bollini, S.; Vieira, J.M.; Howard, S.; Dubè, K.N.; Balmer, G.M.; Smart, N.; Riley, P.R. Re-activated adult epicardial progenitor cells are a heterogeneous population molecularly distinct from their embryonic counterparts. *Stem Cells Dev.* **2014**, *23*, 1719–1730. [CrossRef] [PubMed]

107. Köhler, B.; Pienkowski, C.; Audran, F.; Delsol, M.; Tauber, M.; Paris, F.; Sultan, C.; Lumbroso, S. An N-terminal WT1 mutation (P181S) in an XY patient with ambiguous genitalia, normal testosterone production, absence of kidney disease and associated heart defect: Enlarging the phenotypic spectrum of WT1 defects. *Eur. J. Endocrinol.* **2004**, *150*, 825–830. [CrossRef]

108. Suri, M.; Kelehan, P.; O'neill, D.; Vadcyar, S.; Grant, J.; Ahmed, S.F.; Tolmie, J.; McCann, E.; Lam, W.; Smith, S.; et al. WT1 mutations in Meacham syndrome suggest a coelomic mesothelial origin of the cardiac and diaphragmatic malformations. *Am. J. Med. Genet. A* **2007**, *143A*, 2312–2320. [CrossRef]

109. Meacham, L.R.; Winn, K.J.; Culler, F.L.; Parks, J.S. Double vagina, cardiac, pulmonary, and other genital malformations with 46,XY karyotype. *Am. J. Med. Genet.* **1991**, *41*, 478–481. [CrossRef]

110. Wagner, N.; Wagner, K.D.; Afanetti, M.; Nevo, F.; Antignac, C.; Michiels, J.F.; Schedl, A.; Berard, E. A novel Wilms' tumor 1 gene mutation in a child with severe renal dysfunction and persistent renal blastema. *Pediatr. Nephrol.* **2008**, *23*, 1445–1453. [CrossRef]

111. Yang, Y.; Feng, D.; Huang, J.; Nie, X.; Yu, Z. A child with isolated nephrotic syndrome and WT1 mutation presenting as a 46, XY phenotypic male. *Eur. J. Pediatr.* **2013**, *172*, 127–129. [CrossRef]

112. Fencl, F.; Malina, M.; Stará, V.; Zieg, J.; Mixová, D.; Seeman, T.; Bláhová, K. Discordant expression of a new WT1 gene mutation in a family with monozygotic twins presenting with congenital nephrotic syndrome. *Eur. J. Pediatr.* **2012**, *171*, 121–124. [CrossRef] [PubMed]

113. Loo, C.K.; Algar, E.M.; Payton, D.J.; Perry-Keene, J.; Pereira, T.N.; Ramm, G.A. Possible role of WT1 in a human fetus with evolving bronchial atresia, pulmonary malformation and renal agenesis. *Pediatr. Dev. Pathol.* **2012**, *15*, 39–44. [CrossRef]

114. Loo, C.K.; Pereira, T.N.; Ramm, G.A. Abnormal WT1 expression in human fetuses with bilateral renal agenesis and cardiac malformations. *Birth Defects Res. A Clin. Mol. Teratol.* **2012**, *94*, 116–122. [CrossRef]

115. Ge, Y.; Ro, J.Y.; Kim, D.; Kim, C.H.; Reardon, M.J.; Blackmon, S.; Zhai, J.; Coffey, D.; Benjamin, R.S.; Ayala, A.G. Clinicopathologic and immunohistochemical characteristics of adult primary cardiac angiosarcomas: Analysis of 10 cases. *Ann. Diagn. Pathol.* **2011**, *15*, 262–267. [CrossRef]

116. Suffee, N.; Moore-Morris, T.; Jagla, B.; Mougenot, N.; Dilanian, G.; Berthet, M.; Proukhnitzky, J.; Le Prince, P.; Tregouet, D.A.; Pucéat, M.; et al. Reactivation of the Epicardium at the Origin of Myocardial Fibro-Fatty Infiltration During the Atrial Cardiomyopathy. *Circ. Res.* **2020**, *126*, 1330–1342. [CrossRef]

117. Yuan, P.; Cheedipudi, S.M.; Rouhi, L.; Fan, S.; Simon, L.; Zhao, Z.; Hong, K.; Gurha, P.; Marian, A.J. Single Cell RNA-Sequencing Uncovers Paracrine Functions of the Epicardial-Derived Cells in Arrhythmogenic Cardiomyopathy. *Circulation* **2021**. [CrossRef]

118. van Wijk, B.; Gunst, Q.D.; Moorman, A.F.; van den Hoff, M.J. Cardiac regeneration from activated epicardium. *PLoS ONE* **2012**, *7*, e44692. [CrossRef] [PubMed]

119. Smart, N.; Bollini, S.; Dubé, K.N.; Vieira, J.M.; Zhou, B.; Davidson, S.; Yellon, D.; Riegler, J.; Price, A.N.; Lythgoe, M.F.; et al. De novo cardiomyocytes from within the activated adult heart after injury. *Nature* **2011**, *474*, 640–644. [CrossRef] [PubMed]

120. Liu, Q.; Huang, X.; Oh, J.H.; Lin, R.Z.; Duan, S.; Yu, Y.; Yang, R.; Qiu, J.; Melero-Martin, J.M.; Pu, W.T.; et al. Epicardium-to-fat transition in injured heart. *Cell Res.* **2014**, *24*, 1367–1369. [CrossRef] [PubMed]

121. Sanz-Morejón, A.; García-Redondo, A.B.; Reuter, H.; Marques, I.J.; Bates, T.; Galardi-Castilla, M.; Große, A.; Manig, S.; Langa, X.; Ernst, A.; et al. Wilms Tumor 1b Expression Defines a Pro-regenerative Macrophage Subtype and Is Required for Organ Regeneration in the Zebrafish. *Cell Rep.* **2019**, *28*, 1296–1306.e1296. [CrossRef]

122. Mikawa, T.; Gourdie, R.G. Pericardial mesoderm generates a population of coronary smooth muscle cells migrating into the heart along with ingrowth of the epicardial organ. *Dev. Biol.* **1996**, *174*, 221–232. [CrossRef] [PubMed]

123. Pérez-Pomares, J.M.; Macías, D.; García-Garrido, L.; Muñoz-Chápuli, R. The origin of the subepicardial mesenchyme in the avian embryo: An immunohistochemical and quail-chick chimera study. *Dev. Biol.* **1998**, *200*, 57–68. [CrossRef]

124. Vrancken Peeters, M.P.; Gittenberger-de Groot, A.C.; Mentink, M.M.; Poelmann, R.E. Smooth muscle cells and fibroblasts of the coronary arteries derive from epithelial-mesenchymal transformation of the epicardium. *Anat. Embryol.* **1999**, *199*, 367–378. [CrossRef] [PubMed]

125. Dettman, R.W.; Denetclaw, W.; Ordahl, C.P.; Bristow, J. Common epicardial origin of coronary vascular smooth muscle, perivascular fibroblasts, and intermyocardial fibroblasts in the avian heart. *Dev. Biol.* **1998**, *193*, 169–181. [CrossRef]

126. Reese, D.E.; Mikawa, T.; Bader, D.M. Development of the coronary vessel system. *Circ. Res.* **2002**, *91*, 761–768. [CrossRef]

127. Riley, P.R.; Smart, N. Vascularizing the heart. *Cardiovasc. Res.* **2011**, *91*, 260–268. [CrossRef] [PubMed]

128. Red-Horse, K.; Ueno, H.; Weissman, I.L.; Krasnow, M.A. Coronary arteries form by developmental reprogramming of venous cells. *Nature* **2010**, *464*, 549–553. [CrossRef]

129. Wu, B.; Zhang, Z.; Lui, W.; Chen, X.; Wang, Y.; Chamberlain, A.A.; Moreno-Rodriguez, R.A.; Markwald, R.R.; O'Rourke, B.P.; Sharp, D.J.; et al. Endocardial cells form the coronary arteries by angiogenesis through myocardial-endocardial VEGF signaling. *Cell* **2012**, *151*, 1083–1096. [CrossRef]

130. Zhang, H.; Pu, W.; Li, G.; Huang, X.; He, L.; Tian, X.; Liu, Q.; Zhang, L.; Wu, S.M.; Sucov, H.M.; et al. Endocardium Minimally Contributes to Coronary Endothelium in the Embryonic Ventricular Free Walls. *Circ. Res.* **2016**, *118*, 1880–1893. [CrossRef]

131. Sharma, B.; Chang, A.; Red-Horse, K. Coronary Artery Development: Progenitor Cells and Differentiation Pathways. *Annu. Rev. Physiol.* **2017**, *79*, 1–19. [CrossRef]

132. Lupu, I.E.; De Val, S.; Smart, N. Coronary vessel formation in development and disease: Mechanisms and insights for therapy. *Nat. Rev. Cardiol.* **2020**, *17*, 790–806. [CrossRef]

133. Dechant, G. Molecular interactions between neurotrophin receptors. *Cell Tissue Res.* **2001**, *305*, 229–238. [CrossRef] [PubMed]

134. Donovan, M.J.; Lin, M.I.; Wiegn, P.; Ringstedt, T.; Kraemer, R.; Hahn, R.; Wang, S.; Ibañez, C.F.; Rafii, S.; Hempstead, B.L. Brain derived neurotrophic factor is an endothelial cell survival factor required for intramyocardial vessel stabilization. *Development* **2000**, *127*, 4531–4540. [CrossRef] [PubMed]

135. Mokrý, J.; Cízková, D.; Filip, S.; Ehrmann, J.; Osterreicher, J.; Kolár, Z.; English, D. Nestin expression by newly formed human blood vessels. *Stem Cells Dev.* **2004**, *13*, 658–664. [CrossRef]

136. Mokry, J.; Pudil, R.; Ehrmann, J.; Cizkova, D.; Osterreicher, J.; Filip, S.; Kolar, Z. Re-expression of nestin in the myocardium of postinfarcted patients. *Virchows Arch.* **2008**, *453*, 33–41. [CrossRef] [PubMed]

137. Yang, J.T.; Rayburn, H.; Hynes, R.O. Cell adhesion events mediated by alpha 4 integrins are essential in placental and cardiac development. *Development* **1995**, *121*, 549–560. [CrossRef]

138. Kestilä, M.; Lenkkeri, U.; Männikkö, M.; Lamerdin, J.; McCready, P.; Putaala, H.; Ruotsalainen, V.; Morita, T.; Nissinen, M.; Herva, R.; et al. Positionally cloned gene for a novel glomerular protein—Nephrin—Is mutated in congenital nephrotic syndrome. *Mol. Cell* **1998**, *1*, 575–582. [CrossRef]

139. Wagner, N.; Morrison, H.; Pagnotta, S.; Michiels, J.F.; Schwab, Y.; Tryggvason, K.; Schedl, A.; Wagner, K.D. The podocyte protein nephrin is required for cardiac vessel formation. *Hum. Mol. Genet.* **2011**, *20*, 2182–2194. [CrossRef] [PubMed]

140. Patrakka, J.; Kestilä, M.; Wartiovaara, J.; Ruotsalainen, V.; Tissari, P.; Lenkkeri, U.; Männikkö, M.; Visapää, I.; Holmberg, C.; Rapola, J.; et al. Congenital nephrotic syndrome (NPHS1): Features resulting from different mutations in Finnish patients. *Kidney Int.* **2000**, *58*, 972–980. [CrossRef]

141. Bernardor, J.; Faudeux, C.; Chaussenot, A.; Antignac, C.; Goldenberg, A.; Gubler, M.C.; Wagner, N.; Bérard, E. Nephrotic syndrome and mitochondrial disorders: Questions. *Pediatr. Nephrol.* **2019**, *34*, 1373–1374. [CrossRef] [PubMed]

142. Bernardor, J.; Faudeux, C.; Chaussenot, A.; Antignac, C.; Goldenberg, A.; Gubler, M.C.; Wagner, N.; Bérard, E. Nephrotic syndrome and mitochondrial disorders: Answers. *Pediatr. Nephrol.* **2019**, *34*, 1375–1377. [CrossRef] [PubMed]

143. Lie-Venema, H.; Gittenberger-de Groot, A.C.; van Empel, L.J.; Boot, M.J.; Kerkdijk, H.; de Kant, E.; DeRuiter, M.C. Ets-1 and Ets-2 transcription factors are essential for normal coronary and myocardial development in chicken embryos. *Circ. Res.* **2003**, *92*, 749–756. [CrossRef]

144. Cano, E.; Carmona, R.; Ruiz-Villalba, A.; Rojas, A.; Chau, Y.Y.; Wagner, K.D.; Wagner, N.; Hastie, N.D.; Muñoz-Chápuli, R.; Pérez-Pomares, J.M. Extracardiac septum transversum/proepicardial endothelial cells pattern embryonic coronary arterio-venous connections. *Proc. Natl. Acad. Sci. USA* **2016**, *113*, 656–661. [CrossRef] [PubMed]

145. Cai, C.L.; Martin, J.C.; Sun, Y.; Cui, L.; Wang, L.; Ouyang, K.; Yang, L.; Bu, L.; Liang, X.; Zhang, X.; et al. A myocardial lineage derives from Tbx18 epicardial cells. *Nature* **2008**, *454*, 104–108. [CrossRef]

146. Christoffels, V.M.; Grieskamp, T.; Norden, J.; Mommersteeg, M.T.; Rudat, C.; Kispert, A. Tbx18 and the fate of epicardial progenitors. *Nature* **2009**, *458*, E8–E9. [CrossRef] [PubMed]

147. Franco, D.; Meilhac, S.M.; Christoffels, V.M.; Kispert, A.; Buckingham, M.; Kelly, R.G. Left and right ventricular contributions to the formation of the interventricular septum in the mouse heart. *Dev. Biol.* **2006**, *294*, 366–375. [CrossRef]

148. Gittenberger-de Groot, A.C.; Vrancken Peeters, M.P.; Mentink, M.M.; Gourdie, R.G.; Poelmann, R.E. Epicardium-derived cells contribute a novel population to the myocardial wall and the atrioventricular cushions. *Circ. Res.* **1998**, *82*, 1043–1052. [CrossRef] [PubMed]
149. Kruithof, B.P.; van Wijk, B.; Somi, S.; Kruithof-de Julio, M.; Pérez Pomares, J.M.; Weesie, F.; Wessels, A.; Moorman, A.F.; van den Hoff, M.J. BMP and FGF regulate the differentiation of multipotential pericardial mesoderm into the myocardial or epicardial lineage. *Dev. Biol.* **2006**, *295*, 507–522. [CrossRef]
150. Zhou, B.; Honor, L.B.; Ma, Q.; Oh, J.H.; Lin, R.Z.; Melero-Martin, J.M.; von Gise, A.; Zhou, P.; Hu, T.; He, L.; et al. Thymosin beta 4 treatment after myocardial infarction does not reprogram epicardial cells into cardiomyocytes. *J. Mol. Cell Cardiol.* **2012**, *52*, 43–47. [CrossRef] [PubMed]
151. Kikuchi, K.; Gupta, V.; Wang, J.; Holdway, J.E.; Wills, A.A.; Fang, Y.; Poss, K.D. tcf21+ epicardial cells adopt non-myocardial fates during zebrafish heart development and regeneration. *Development* **2011**, *138*, 2895–2902. [CrossRef] [PubMed]
152. Chong, J.J.; Chandrakanthan, V.; Xaymardan, M.; Asli, N.S.; Li, J.; Ahmed, I.; Heffernan, C.; Menon, M.K.; Scarlett, C.J.; Rashidianfar, A.; et al. Adult cardiac-resident MSC-like stem cells with a proepicardial origin. *Cell Stem Cell* **2011**, *9*, 527–540. [CrossRef]
153. Englert, C.; Vidal, M.; Maheswaran, S.; Ge, Y.; Ezzell, R.M.; Isselbacher, K.J.; Haber, D.A. Truncated WT1 mutants alter the subnuclear localization of the wild-type protein. *Proc. Natl. Acad. Sci. USA* **1995**, *92*, 11960–11964. [CrossRef] [PubMed]
154. Senyo, S.E.; Steinhauser, M.L.; Pizzimenti, C.L.; Yang, V.K.; Cai, L.; Wang, M.; Wu, T.D.; Guerquin-Kern, J.L.; Lechene, C.P.; Lee, R.T. Mammalian heart renewal by pre-existing cardiomyocytes. *Nature* **2013**, *493*, 433–436. [CrossRef]
155. Hsieh, P.C.; Segers, V.F.; Davis, M.E.; MacGillivray, C.; Gannon, J.; Molkentin, J.D.; Robbins, J.; Lee, R.T. Evidence from a genetic fate-mapping study that stem cells refresh adult mammalian cardiomyocytes after injury. *Nat. Med.* **2007**, *13*, 970–974. [CrossRef]
156. Gittenberger-de Groot, A.C.; Vrancken Peeters, M.P.; Bergwerff, M.; Mentink, M.M.; Poelmann, R.E. Epicardial outgrowth inhibition leads to compensatory mesothelial outflow tract collar and abnormal cardiac septation and coronary formation. *Circ. Res.* **2000**, *87*, 969–971. [CrossRef] [PubMed]
157. Trembley, M.A.; Velasquez, L.S.; de Mesy Bentley, K.L.; Small, E.M. Myocardin-related transcription factors control the motility of epicardium-derived cells and the maturation of coronary vessels. *Development* **2015**, *142*, 21–30. [CrossRef] [PubMed]
158. Armulik, A.; Genové, G.; Betsholtz, C. Pericytes: Developmental, physiological, and pathological perspectives, problems, and promises. *Dev. Cell* **2011**, *21*, 193–215. [CrossRef]
159. Quijada, P.; Misra, A.; Velasquez, L.S.; Burke, R.M.; Lighthouse, J.K.; Mickelsen, D.M.; Dirkx, R.A.; Small, E.M. Pre-existing fibroblasts of epicardial origin are the primary source of pathological fibrosis in cardiac ischemia and aging. *J. Mol. Cell Cardiol.* **2019**, *129*, 92–104. [CrossRef] [PubMed]
160. Zhang, H.; Huang, X.; Liu, K.; Tang, J.; He, L.; Pu, W.; Liu, Q.; Li, Y.; Tian, X.; Wang, Y.; et al. Fibroblasts in an endocardial fibroelastosis disease model mainly originate from mesenchymal derivatives of epicardium. *Cell Res.* **2017**, *27*, 1157–1177. [CrossRef]
161. Tallquist, M.D.; Molkentin, J.D. Redefining the identity of cardiac fibroblasts. *Nat. Rev. Cardiol.* **2017**, *14*, 484–491. [CrossRef]
162. Marín-Juez, R.; El-Sammak, H.; Helker, C.S.M.; Kamezaki, A.; Mullapuli, S.T.; Bibli, S.I.; Foglia, M.J.; Fleming, I.; Poss, K.D.; Stainier, D.Y.R. Coronary Revascularization During Heart Regeneration Is Regulated by Epicardial and Endocardial Cues and Forms a Scaffold for Cardiomyocyte Repopulation. *Dev. Cell* **2019**, *51*, 503–515.e504. [CrossRef] [PubMed]
163. Liu, Y.H.; Lai, L.P.; Huang, S.Y.; Lin, Y.S.; Wu, S.C.; Chou, C.J.; Lin, J.L. Developmental origin of postnatal cardiomyogenic progenitor cells. *Future Sci. OA* **2016**, *2*, FSO120. [CrossRef] [PubMed]
164. Huang, G.N.; Thatcher, J.E.; McAnally, J.; Kong, Y.; Qi, X.; Tan, W.; DiMaio, J.M.; Amatruda, J.F.; Gerard, R.D.; Hill, J.A.; et al. C/EBP transcription factors mediate epicardial activation during heart development and injury. *Science* **2012**, *338*, 1599–1603. [CrossRef]
165. Redpath, A.N.; Smart, N. Recapturing embryonic potential in the adult epicardium: Prospects for cardiac repair. *Stem Cells Transl. Med.* **2021**, *10*, 511–521. [CrossRef] [PubMed]
166. Zhou, B.; Honor, L.B.; He, H.; Ma, Q.; Oh, J.H.; Butterfield, C.; Lin, R.Z.; Melero-Martin, J.M.; Dolmatova, E.; Duffy, H.S.; et al. Adult mouse epicardium modulates myocardial injury by secreting paracrine factors. *J. Clin. Investig.* **2011**, *121*, 1894–1904. [CrossRef] [PubMed]
167. Nieto, M.A. The snail superfamily of zinc-finger transcription factors. *Nat. Rev. Mol. Cell Biol.* **2002**, *3*, 155–166. [CrossRef]
168. Bax, N.A.; Bleyl, S.B.; Gallini, R.; Wisse, L.J.; Hunter, J.; Van Oorschot, A.A.; Mahtab, E.A.; Lie-Venema, H.; Goumans, M.J.; Betsholtz, C.; et al. Cardiac malformations in Pdgfralpha mutant embryos are associated with increased expression of WT1 and Nkx2.5 in the second heart field. *Dev. Dyn.* **2010**, *239*, 2307–2317. [CrossRef]
169. Bax, N.A.; van Oorschot, A.A.; Maas, S.; Braun, J.; van Tuyn, J.; de Vries, A.A.; Groot, A.C.; Goumans, M.J. In vitro epithelial-to-mesenchymal transformation in human adult epicardial cells is regulated by TGFβ-signaling and WT1. *Basic Res. Cardiol.* **2011**, *106*, 829–847. [CrossRef]
170. Casanova, J.C.; Travisano, S.; de la Pompa, J.L. Epithelial-to-mesenchymal transition in epicardium is independent of Snail1. *Genesis* **2013**, *51*, 32–40. [CrossRef]
171. Greulich, F.; Farin, H.F.; Schuster-Gossler, K.; Kispert, A. Tbx18 function in epicardial development. *Cardiovasc. Res.* **2012**, *96*, 476–483. [CrossRef]

172. von Gise, A.; Zhou, B.; Honor, L.B.; Ma, Q.; Petryk, A.; Pu, W.T. WT1 regulates epicardial epithelial to mesenchymal transition through β-catenin and retinoic acid signaling pathways. *Dev. Biol.* **2011**, *356*, 421–431. [CrossRef] [PubMed]

173. Lüdtke, T.H.; Rudat, C.; Kurz, J.; Häfner, R.; Greulich, F.; Wojahn, I.; Aydoğdu, N.; Mamo, T.M.; Kleppa, M.J.; Trowe, M.O.; et al. Mesothelial mobilization in the developing lung and heart differs in timing, quantity, and pathway dependency. *Am. J. Physiol. Lung Cell Mol. Physiol.* **2019**, *316*, L767–L783. [CrossRef]

174. Menendez, J.A.; Lupu, R. Fatty acid synthase and the lipogenic phenotype in cancer pathogenesis. *Nat. Rev. Cancer* **2007**, *7*, 763–777. [CrossRef] [PubMed]

175. Sun, X.; Malandraki-Miller, S.; Kennedy, T.; Bassat, E.; Klaourakis, K.; Zhao, J.; Gamen, E.; Vieira, J.M.; Tzahor, E.; Riley, P.R. The extracellular matrix protein agrin is essential for epicardial epithelial-to-mesenchymal transition during heart development. *Development* **2021**, *148*, dev197525. [CrossRef]

176. Chau, Y.Y.; Bandiera, R.; Serrels, A.; Martínez-Estrada, O.M.; Qing, W.; Lee, M.; Slight, J.; Thornburn, A.; Berry, R.; McHaffie, S.; et al. Visceral and subcutaneous fat have different origins and evidence supports a mesothelial source. *Nat. Cell Biol.* **2014**, *16*, 367–375. [CrossRef] [PubMed]

177. Dorn, T.; Kornherr, J.; Parrotta, E.I.; Zawada, D.; Ayetey, H.; Santamaria, G.; Iop, L.; Mastantuono, E.; Sinnecker, D.; Goedel, A.; et al. Interplay of cell-cell contacts and RhoA/MRTF-A signaling regulates cardiomyocyte identity. *EMBO J.* **2018**, *37*, e98133. [CrossRef] [PubMed]

178. Tang, J.; Wang, X.; Tan, K.; Zhu, H.; Zhang, Y.; Ouyang, W.; Liu, X.; Ding, Z. Injury-induced fetal reprogramming imparts multipotency and reparative properties to pericardial adipose stem cells. *Stem Cell Res. Ther.* **2018**, *9*, 218. [CrossRef]

179. Stevens, S.M.; von Gise, A.; VanDusen, N.; Zhou, B.; Pu, W.T. Epicardium is required for cardiac seeding by yolk sac macrophages, precursors of resident macrophages of the adult heart. *Dev. Biol.* **2016**, *413*, 153–159. [CrossRef]

180. Epelman, S.; Lavine, K.J.; Beaudin, A.E.; Sojka, D.K.; Carrero, J.A.; Calderon, B.; Brija, T.; Gautier, E.L.; Ivanov, S.; Satpathy, A.T.; et al. Embryonic and adult-derived resident cardiac macrophages are maintained through distinct mechanisms at steady state and during inflammation. *Immunity* **2014**, *40*, 91–104. [CrossRef]

181. Lavine, K.J.; Epelman, S.; Uchida, K.; Weber, K.J.; Nichols, C.G.; Schilling, J.D.; Ornitz, D.M.; Randolph, G.J.; Mann, D.L. Distinct macrophage lineages contribute to disparate patterns of cardiac recovery and remodeling in the neonatal and adult heart. *Proc. Natl. Acad. Sci. USA* **2014**, *111*, 16029–16034. [CrossRef] [PubMed]

182. Buechler, M.B.; Kim, K.W.; Onufer, E.J.; Williams, J.W.; Little, C.C.; Dominguez, C.X.; Li, Q.; Sandoval, W.; Cooper, J.E.; Harris, C.A.; et al. A Stromal Niche Defined by Expression of the Transcription Factor WT1 Mediates Programming and Homeostasis of Cavity-Resident Macrophages. *Immunity* **2019**, *51*, 119–130.e115. [CrossRef] [PubMed]

183. Singh, A.; Ramesh, S.; Cibi, D.M.; Yun, L.S.; Li, J.; Li, L.; Manderfield, L.J.; Olson, E.N.; Epstein, J.A.; Singh, M.K. Hippo Signaling Mediators Yap and Taz Are Required in the Epicardium for Coronary Vasculature Development. *Cell Rep.* **2016**, *15*, 1384–1393. [CrossRef] [PubMed]

184. Vieira, J.M.; Howard, S.; Villa Del Campo, C.; Bollini, S.; Dubé, K.N.; Masters, M.; Barnette, D.N.; Rohling, M.; Sun, X.; Hankins, L.E.; et al. BRG1-SWI/SNF-dependent regulation of the Wt1 transcriptional landscape mediates epicardial activity during heart development and disease. *Nat. Commun.* **2017**, *8*, 16034. [CrossRef] [PubMed]

MDPI
St. Alban-Anlage 66
4052 Basel
Switzerland
Tel. +41 61 683 77 34
Fax +41 61 302 89 18
www.mdpi.com

International Journal of Molecular Sciences Editorial Office
E-mail: ijms@mdpi.com
www.mdpi.com/journal/ijms

9 783036 539980